Computation, Cognition, and Pylyshyn

Computation, Cognition, and Pylyshyn

edited by Don Dedrick and Lana Trick

A Bradford Book
The MIT Press
Cambridge, Massachusetts
London, England

For information about special quantity discounts, please email special_sales@mitpress.mit.edu

This book was set in Stone Sans and Stone Serif by SNP Best-set Typesetter Ltd., Hong Kong.

Printed and bound in the United States of America.

Library of Congress Cataloging-in-Publication Data

Computation, cognition, and Pylyshyn / edited by Don Dedrick and Lana Trick.
   p. cm.
Includes bibliographical references and index.
ISBN 978-0-262-01284-3 (hardcover : alk. paper)—ISBN 978-0-262-51242-8 (pbk. : alk. paper)
1. Cognition. 2. Cognitive science. 3. Pylyshyn, Zenon W., 1937–. I. Dedrick, Don. II. Trick, Lana, 1957–.
BF311.C593 2009
153—dc22

                                                                    2008042147

10 9 8 7 6 5 4 3 2 1

Contents

Preface

Lana Trick was a Ph.D. student of Zenon's at the University of Western Ontario. Don Dedrick read Pylyshyn's work as a philosophy graduate student at the University of Toronto. Coming together in Guelph in 2004 (the twentieth anniversary of the publication of *Computation and Cognition*), the two of us imagined a conference that not only honored Pylyshyn's important work—work that still offers the best model for what can be called "classical cognitive science"—but work that would engage in dialogue with Pylyshyn as well. From the start we thought the idea of a *festschrift* for Pylyshyn to be less than perfect. Not because he was undeserving of delightful respect (he is), but because his views were, and are, far from fossilized. They were, and are, still influential as a real model of the way the mind might work. As Jerry Fodor argues, in the introduction to this book, Pylyshyn's recent work may well solve one of the fundamental problems and puzzles in cognitive science: how our minds are, after all, connected to the external world.

Supported by the Social Sciences and Humanities Research Council of Canada (SSHRC), we organized a conference unofficially called "Zencon" at the University of Guelph, which is west of Toronto in Ontario. It took place April 29 to May 1, 2005. We brought together important cognitive scientists from a range of disciplines, and the conference, as with the book, was divided in a somewhat arbitrary way into contributions dealing either with vision or with the foundations of cognitive science. Some participants, such as Susan Carey and Brian Cantwell Smith, were not able to contribute to this volume. But consider those who have contributed: the philosophers Austen Clark, John Bickel, and Andrew Brook (invited to contribute, after the fact), all engage critically with Pylyshyn's work. Brian Keane, a philosopher converted (or converged) to cognitive science, mixes the conceptual with the empirical (we invited him too, after the conference). The neuroscientist Mel Goodale, a colleague-in-arms with

Pylyshyn at the University of Western Ontario, and Stevan Harnad, the one-time, long-time editor of the journal *Behavioral and Brain Sciences*, write about cognition and action: the former from the perspective of vision, the latter from that of computation. Zenon's erstwhile student, Mike Dawson, has himself written important books on the foundations of cognitive science, and has an essay in this book about what connectionism might be used for. There is also an essay on the foundations of linguistics, by Charles Reiss. Reiss has been so impressed by Pylyshyn's work that he gave his son this middle name: Zenon. Claudia Uller, once a postdoctoral fellow at Rutgers Center for Cognitive Science, writes about number as it is understood by animals, human and otherwise, and Brian Scholl, another former student of Pylyshyn's, continues on with important work in the multiple-object tracking tradition, as does Lana Trick, while Richard Wright exploits Pylyshyn's influential ideas about cognitive penetrability in a paper dealing with visual search. (Trick and Wright, like Brook and Keane, did not present at Zencon.)

The conference, a great success, included posters by graduate students, a keynote address by Zenon, and a great deal of excellent argument and conversation. We hope the reader will find this book, which is derived from those sessions, to be a valuable resource for thinking about computation, cognition, and Pylyshyn.

Introduction: So What's So Good about Pylyshyn?

Jerry Fodor

Good question. I shall attempt to explain.

There are, I think, four foundational questions for which a viable cognitive science must provide answers (foundational in the sense that they arise in every department of cognitive science; from [as it might be] perception, to problem solving, to cognitive development, and so forth through the whole catalog). They are these:

i.   What is the nature of mental processes?
ii. What kinds of things are mental representations?
iii. How do mental representations have content?
iv. How do mental representations attach to the world?

"Classical" cognitive science, to which Zenon has given his unswerving, career-long allegiance, got started with Turing's suggestion for solving (i): Mental processes are computations. In particular, they are computations rather than associations. Much joy; great relief. For the associationist tradition, which had been the heart of British empiricism ever since Aristotle (I know, I know) had gradually but unmistakably revealed itself to be bankrupt. And, until the "computer analogy" appeared on the horizon, there seemed to be nothing to replace it with: No wonder so many psychologists gave up on the mind and turned behaviorist.

I'm not, myself, convinced that the computational theory of mind (CTM) will do the whole job over the long haul (for why I'm not, see Fodor 1983, 2000). But I'm quite certain (inter alia for reasons we set forth in Fodor and Pylyshyn 1988) that associationism is dead. The fruitless attempts of connectionists to resuscitate the corpse have made that clear.

One of the nice things about the computational answer to (i) is that it quite radically constrains the possible answers to (ii). In principle just about anything can enter into associations: All that's required for Ys to be associated to Xs is that Xs are reliable causes of Ys, and there are, of course, many plausible candidates for reliable causal connections. So one finds in the tradition all sorts of views about what sorts of things associative relations hold among. Not just "Ideas" but, for example, neural firings, unconditioned stimuli and unconditioned responses, conditioned stimuli and conditioned responses, stimuli and behavioral dispositions, nodes in networks, words in sentences, reflexes and their releasers, names and faces, percepts and motor gestures, and so on, without end. It's one of the disappointments of associationism that, if it were true, the theory of mental processes would tell us so very little about the nature of mental states. Not so the computational theory. If mental processes are computations, then mental states have to be the kinds of representational states to which computations can apply. Not everything qualifies.

That is a long and very interesting story, but the short version is that computations are, by definition, operations defined on structured objects; in particular, on objects that have constituent structure. It follows that there is an intrinsic relation between the computational theory of mind and the theory that mental representations are sentence-like; in effect, that there is a "language of thought." I think it's because he is acutely aware of this connection that Zenon has very deep suspicions about the thesis that mental representations are picture-like. But however the data on mental images finally turn out, and Zenon has argued pretty convincingly that they are thus far inconclusive (see, e.g., Pylyshyn 2003), mental images can't be more than a sideshow in the main story about mental representation.[1] Not, at least, if mental processes are to be computations. Everybody has known for ages that images don't have the right sort of structure to be the bearers of truth-values (for example, they offer no structural analogues to predication). But it turns out that they also don't have the right sort of structure to be the domains of mental processes; not, at least, if Turning was on the right track. Sentences, however, do.

So, then, mental processes are something like computations and mental representations are something like sentences. Those are the outlines of the answers that classical cognitive science offered as replacements for associationism; we've spent the last fifty years or so working on the details. Some progress is discernible. But what about questions (iii) and (iv)? Here, so it seems to me, classical cognitive science has found itself in something of a pickle; a pickle that's so deep (if I may mix a metaphor) that most of its practitioners haven't so much as noticed that they are in it. What's so good about Pylyshyn—in particular, what's so good about Pylyshyn's recent work—is that maybe, just possibly maybe, it shows us the way out of the pickle we're in.

There are standard answers to (iii) and (iv) to be found in the cognitive science literature (including, by the way, the connectionist literature); practically everybody takes it for granted that these standard answers are more or less true. Our pickle, however, is that they aren't. This is a very long story, but I'll try to say enough to suggest the outlines. Then I'll sketch what I take to be Pylyshyn's Way Out. I won't, however, even try to convince you that Pylyshyn's is the right way out. The arguments for that are mainly of the last-log-afloat variety.

So, then, how do mental representations have content? Classical cognitive science hoped to explain content by forging an alliance with inferential role semantics (IRS). IRS says that the content of mental representations is determined by (or is identical to, or supervenes on, or whatever; take your pick) their inferential roles. Nobody knows exactly what that means because nobody knows exactly what inferential roles are; but at least in the philosophical literature, the paradigms are inferential relations among sentences (mutatis mutandis, among beliefs) that turn on the "logical" vocabulary; "and," for example. These inferential relations are controlled by rules that determine which such inferences are valid (or, more generally, "good" or "warranted"). So, the story about the English word "and" is that it means what it does because English speakers are disposed to construct and accept such arguments as "if 'P and Q' is true, then 'P' is true"; "if 'P and Q' is true, then 'Q' is true" and "if 'P' is true and 'Q' is true, then 'P and Q' is true."

Considering that it leaks at every seam, it is simply remarkable how many people in cognitive science believe some version of this inferential role story about content; and how intractable their belief in it has been. This is a twice told tale, and I won't bother you with much of it here. Suffice it to remark on what is hardly ever noticed: IRS actually comports very badly with CTM. The problem is looming circularity. The computational story says that mental processes are inferences, and inferences (as opposed to the mere scrambling of syntactic objects) are the sorts of things that preserve relevant aspects of semantic values. The inference from "P and Q" to "P" is supposed to preserve part of the content of "P and Q"; the inference from "is a dog" to "is an animal" is supposed to preserve part of the content of "dog"; the inference from "John arrived on Tuesday" to "John arrived" is supposed to preserve part of the content of "John arrived on Tuesday"; and so forth. In short, to think of mental processes as computations is, inter alia, to presuppose some or other notion of the content of a mental representation. This is, I think, unavoidable if inference is to play the role in a theory of mental processes that CTM commends; that is, to explain how it is that mental processes often lead to the truth of our beliefs and the success or our actions.

But if you are going to presuppose the notion of content when you say what an inference is, you must not also presuppose the notion of inference when you say what content is; as, of course, inferential role semantics proposes to do: IRS consists of the claim that the content of a representation is a function of its inferential role. That would surely be blatantly circular, and circularity is ipso facto pickle-making. (Perhaps I should explain why philosophers, of all people, should have failed to notice this problem. It's because, almost without exception, philosophers who care about what content is don't much care about what mental processes are and vice versa. For reasons that strike me as deeply obscure, they think that the first is a "conceptual" issue, but the second isn't. God only knows what they think this distinction amounts to. I certainly don't.)

It's also worth remarking that, if you accept IRS as the answer to (iii), you are to likely to find yourself in a difficult position—a pickle, in fact—when you turn to (iv). The reason is that inferences are relations among "intentional" objects (objects that have contents, like beliefs, propositions, mental representations, or whatever). In particular they aren't relations between intentional objects and things in the world. An inference may take you from one belief about chairs to another belief about chairs; but it can't take you from beliefs about chairs to chairs. There are no inferences that have chairs as their premises or their conclusions. The upshot is among the major anomalies of current versions of classical cognitive science: Lots of people who think that IRS is the answer to (iii) also think that there is no answer to (iv). That's to say, for example, that there is no relevant semantic relation between the concept LONDON and the city of that name. Likewise there is no relevant semantic relation between the thought London is in England and the fact that London is in England. In particular, it's not the case that the latter is what makes the former true; and it's not the case that London's being in England (together with "in" being transitive) is what makes the inference from "you are in London" to "you are in England" sound. (If you don't believe me that anyone could deny such truisms, Jackendoff [1997] provides a clear example. He is, alas, by no means the only one who does so.)

There is an alternative; probably it's the answer to (iv) that's most widely endorsed by psychologists (and by practitioners of AI in the classical tradition). It's that though it is inferential role that attaches one's mental representations to their content, it's some or other aspect of one's behaviors (or behavioral capacities) that attaches them to the world. Your concept DOG has the content that it does because of many facts like its being inferentially connected to your concept ANIMAL. (Which such facts? IRS doesn't say; apparently as a matter of principle.) But it's attached to the world—in particular, to dogs—by the fact that to have the concept is ipso facto to be capable of certain behaviors; primarily, it's to have the capacity to distinguish dogs from other things. So, every psychologist believes unblinkingly and, I fear, unshakably that testing for discriminated responses to dogs is the way to test for possession of the concept DOG. And anybody in AI will tell you that for a machine to have (/acquire) the concept DOG is for it to have (/acquire) the ability to discriminate between a dog and everything else.

But there are all sorts of reasons why that can't be the right answer to (iv). For one thing nobody (except, possibly, God) does have the capacity to discriminate dogs from everything else. Your discriminative capacities in respect of dogs depends, in large part, on what true beliefs you have about dogs; and having beliefs about dogs presupposes having the concept DOG. So the right order of explanation runs from concept possession to discriminative behavior, not the other way around. Only the self-evidence of this truth can account for its going so widely unnoticed.

The other reason why having a concept can't be having a discriminative capacity is that, unlike concepts, discriminative capacities don't compose. I am tired of saying this and you, no doubt, are tired of hearing it. If you want more, look it up. (Fodor and Lepore 1992 is a place to start.)

So that's the pickle we're in. Classical cognitive science has maybe found a way into questions (i) and (ii). But there's a hopeless mess about (iii) and (iv); it's a reliable rule of thumb that practically everything that cognitive science believes about (iii) and (iv) is false. It's long overdue for somebody to worry about that. Enter Pylyshyn's recent work, for which I am an unabashed enthusiast.

Let's put (iii) to one side. I think (and I suspect that Pylyshyn does too, though he doesn't always write that way) that there simply is no such thing as "conceptual content" in the sense that (iii) has in mind. I think that the only semantic property that concepts (/words) have is reference,[2] which is in effect to say that (iii) reduces to (iv). Accordingly, a semantics for a natural language is a specification of what its referring expressions refer to and of the conditions under which its sentences would be true. Ditto for mental representations, of course; so here's a sketch of what a cognitive science might look like on those assumptions.

There is a relation of reference that holds between structurally simple ("primitive") mental representations and things in the world. Presumably the paradigm of such a relation is in perceptual recognition, where whatever it is that a mental representation refers to causes the tokening of that representation in the perceiver's head. In the crudest possible formulation, this means something like: The mental representation DOG refers to dogs because encounters with dogs reliably cause the mind to entertain instances of that mental representation.

The reference of complex mental representations (BROWN DOG, for example) is determined compositionally by the referents of its primitive constituents together with its structure. BROWN refers to (the color) brown, DOG refers to dogs, and the referent BROWN DOG is the set of all and only the dogs that are brown.

If I'm right about (iv) being the real issue about the semantics of mental representation and (iii) being a red herring, and if Turing is right about mental processes being computations, that answers all the fundamental questions of cognitive science. A Very Good Day's Work, so it seems to me.

Except: I've thus far left out something important that's been lurking in the wings through the discussion, namely that the cognitive science we want shouldn't beg the questions that it's supposed to answer. In particular, it mustn't presuppose any semantic (or psychological) concepts when it says what reference is and what refers to what. For example, it won't do to say that the reference of a concept is whatever the concept applies to. That's arguably true enough, but it exploits notions like "concept" and "applies to," which a semantic theory is supposed to explain, not just take for granted.

Here, then, is our pickle in a nutshell. The whole project that I've been sketching collapses unless we can give some account of the reference of a concept that doesn't, overtly or otherwise, presuppose such concepts as REFERENCE or CONCEPT. We need, for example, an account of how you might refer to a cow even if you don't have the concept COW.[3] In short, we need to explain how there can be nonconceptual reference.[4] I think Pylyshyn's recent work shows how one might proceed in the direction of such an account, thereby extracting the pickle from the nutshell. If that's so, then it's the best idea about the semantics of mental representation than anybody has ever had.

Here, as I understand it, is the basic idea.

To begin with, the model for nonconceptual reference is the semantics of "bare" demonstratives ("this" or "that" as opposed to "this cow" or "that kangaroo"). Mental representations that function as bare demonstratives denote unconceptualized objects-in-the-world. So, you can demonstrate that without representing it as that cow, or that kangaroo or, indeed, as anything but that.[5] Lots of philosophers think that there can't be bare demonstratives. And, of course, lots of philosophers think that there can. (I don't suppose you'll find that surprising if you know lots of philosophers.) Their argument is that real demonstratives (unlike the hypothesized bare ones) pick out their referents as things that satisfy some conceptualization that the demonstrator has in mind. Very crudely put, what sticks "that" to a cow when somebody demonstrates a cow by saying "that" is his having a certain representation of that in mind (viz. COW) when he says it. Reference, according to this view, is always under some description or another; that's so even in cases where the vehicle of reference doesn't make the description explicit.

This suggests two subsidiary questions to add to our list:

v. Are there bare demonstrative representations in the language of thought?
vi. If there are, what sticks them to their referents?

Zenon's answer to (v) is that, in the case of bare demonstration, the relation of a referent to the symbol that refers to it is purely causal; no conceptualizations need apply. Under certain conditions (presumed to be psychophysically specifiable) things in the world "grab" mental indexes that then remain attached to them through various kinds of transformations (like, for example, brief occlusion of the perceptual object). The terminology is illuminating; conceptualization runs from outside in; it's a kind of action; one applies concepts to things in the world which, if all goes well, satisfy the concepts that one applies to them. "Grabbing" works the other way around; the world reaches in and latches onto a mental representation, which it holds onto under circumstances that, in principle, psychophysical experiments should be able to reveal. This should all sound familiar; in effect, it revives a very old epistemological tradition according to which the mind is active in perception but passive in sensation. If that general story is right, then Zenon's "FINSTS" (see, e.g., Pylyshyn 2003) are the paradigms of bare demonstratives,[6] and FINSTING something is the basic operation by which mind-to-world reference is established. FINSTS are where the intentional gets its grip on the physical; it's where cognitive psychology starts to get "naturalized."

Will anything of this sort actually work? Search me. But, for what it's worth, my prediction is that slowly, perhaps over the next decade or so, cognitive science will come to understand: first, that IRS can't be right about the content of mental representation; second, that reference is the crux of the problem about how the mental order could connect with the natural order; and third, that the cognitive science currently in situ doesn't have the foggiest idea what to do about all that. At that point, Zenon's story will appear as a star rising in the East; it gives us exactly what we very badly need, a place to start from. Any old port in a pickle, is what I always say.

Notes

1. I should add that, quite aside from the light Zenon's critique has thrown on the imagery issue per se, it has also occasioned the current interest in "architectural" properties of cognitive systems and in questions about the encapsulation of cognitive processes. Both are now standard topics in the cognitive science literature.

2. Actually I don't think that; there are special problems about "logical" words (see above) and, perhaps, about words that refer to mathematical objects (numbers and the like). All I need for present purposes is that reference is *among* the symbol–world relations to which a viable semantic theory for mental representations must be committed.

3. That may sound like a paradox, but it isn't. Or at least it isn't obvious that it is. I do take it to be self-evident that you can't refer to a cow *as a cow* unless you have the concept COW. But why couldn't somebody who doesn't have the concept COW nevertheless refer to a cow by saying (or thinking) *that* in a situation where, as a matter of fact, *that* is a cow. (E.g., you might say, or think, *that thing moos* in a situation where, unbeknownst to you, that thing is a cow—indeed, in a situation where you don't even have the concept COW.)

   This is really another of those issues about the right order of explanation. Are we to take having concepts as basic and explain reference in terms of it? Or are we to take reference as basic and use it to explain what it is to have a concept? I'm suggesting that the latter is the right way to proceed; but, of course, the proof is in the pudding.

4. The literature generally calls it *pre*conceptual reference because it presumably happens prior to conceptualization in such processes as, for example, perceptual recognition.

5. A less radical suggestion is that nonconceptual reference isn't *entirely* nonconceptual; rather it's mediated by (and only by) *the application of spatiotemporal concepts*. In one way, this is a very Kantian sort of view: spatiotemporal representation comes first in the order of perceptual processing. In another way, it's not: Kant

thought space and time are "modes of intuition," so locating a percept in space-time isn't, according to Kant, a species of conceptualization. There is no reason to dogmatize; in principle the issue is empirical rather than (as Kant would have said) "transcendental."

6. Take-home questions: are FINSTS the *only* kind of bare demonstratives in the language of thought? Should unmodified demonstratives in natural language be thought of as bare demonstratives or, along traditional lines, as implicit descriptions? The reader will *not* find the answers supplied at the end of the book.

References

Fodor, J. (1983). *Modularity of Mind*. Cambridge, Mass.: MIT Press.

Fodor, J. (2000). *The Mind Doesn't Work That Way*. Cambridge, Mass.: MIT Press.

Fodor, J., and E. Lepore (1992). *Holism: A Shopper's Guide*. Oxford: Blackwell.

Fodor, J., and Z. Pylyshyn (1988). Connectionism and cognitive architecture: A critical analysis. *Cognition* 28: 3–71.

Jackendoff, R. (1997). *Languages of the Mind*. Cambridge, Mass.: MIT Press.

Pylyshyn, Z. (2003). *Seeing and Visualizing: It's Not What You Think*. Cambridge, Mass.: MIT Press.

I Vision

1 Perception, Representation, and the World: The FINST That Binds

Zenon W. Pylyshyn

1 Some Historical Background

I recently discovered that work I was doing in the laboratory and in theoretical writings was implicitly taking a position on a set of questions that philosophers had been worrying about for much of the past thirty or more years. My clandestine involvement in philosophical issues began when a computer science colleague and I were trying to build a model of geometrical reasoning that would draw a diagram and notice things in the diagram as it drew it (Pylyshyn et al. 1978). One problem we found we had to face was that if the system discovered a right angle it had no way to tell whether this was the intersection of certain lines it had drawn earlier while constructing a certain figure, and if so which particular lines they were. Moreover, the model had no way of telling whether this particular right angle was identical to some bit of drawing it had encountered earlier and represented as, say, the base of a particular triangle. There was, in other words, no way to determined the identity of an element[1] at two different times if it was represented differently at those times. This led to some speculation about the need for what we called a "finger" that could be placed at a particular element of interest and that could be used to identify it as a particular token thing (the way you might identify a particular feature on paper by labeling it). In general we needed something like a finger that would stay attached to a particular element and could be used to maintain a correspondence between the individual element that was just noticed now and one that had been represented in some fashion at an earlier time. The idea of such fingers (which came to be called "FINgers of INSTantiation" or FINSTs) then suggested some empirical studies to see if humans had anything like this capability. Thus began a series of experimental investigations of FINSTs that occupied me and my students for much of the past twenty-seven years.

The idea of FINSTs as constituents of perceptual representations is a departure from the view of perceptual representation I had taken in *Computation and Cognition* some twenty-five years ago (Pylyshyn 1984) because it postulated a mental symbol that was not connected to the world by the semantic relation of satisfaction but by a causal or informational link. In the course of my work since that book I found myself thinking about why vision needed the sort of link provided by FINSTs to connect cognitive representations and the sensible world. My initial interest in FINSTs was a response to the fact that diagrams did not come into existence all of a sudden, but were constructed over time. It soon became clear that it does not matter how the figure came into existence, since the representation of the figure is itself built up over time. We clearly don't notice all there is to notice about a scene in an instant—we notice different things over a period of time as we move our eyes and our focal attention around. Consequently we may notice and represent the very same token element differently at different times. There is plenty of evidence that even when there are no eye movements we construct perceptual representations incrementally over time (Calis, Sterenborg, and Maarse 1984; Frohlich and Laux 1969; Kimchi 2000; Nakatani 1995; Nesmith and Rodwan 1967; Parks 1995; Reynolds 1978a; Sekuler and Palmer 1992; Tucker and Broota 1985), so we cannot escape the need to keep track of individual objects qua individuals over time.

Around the same time as we undertook these experiments (initially reported in Pylyshyn 1989; Pylyshyn and Storm 1988) another set of experiments was independently published by Daniel Kahneman (Kahneman, Treisman, and Gibbs 1992), who introduced the concept of an *object file*. An object file contains the conceptual representation of a (visual) object with which it is associated. Although this was not stressed in the Kahneman et al. report, object files are connected to individual visual objects and keep accumulating information about the individuals as they track them. Our view is similar to that of Kahneman et al. with two notable exceptions: (1) We were concerned primarily with the question of how an object file is associated with its appropriate object (answer: through the primitive index mechanism I call FINSTs); and (2) We assumed that the FINST index does not itself use the contents of object files in order to track the individual object token with which it is associated.

As with many ideas, it took a long time to appreciate that the basic idea was actually a proposal that introduced nonconceptual representation. Eventually it began to strike me that FINSTs had to be a very special sort of world–mind connection, different from what psychologists had been studying under the term "attention" and different from the semantic connection of "satisfaction" with which philosophers have had a long-standing but perplexing relationship. FINSTs differ from what psychologists call focal attention in several respects: (1) there is a small number of them; (2) they are generally data driven—that is, assigned by events taking place in the visual field; (3) they pick out individual things as opposed to regions; (4) they adhere to (stay connected to) the same individual thing (whatever exactly that turns out to be) as the thing moves around and changes any or all of its properties; and (5) their attachment is not mediated by a description of (i.e., an encoding of properties of) the thing in question. There are two theoretical reasons why these indexes function without an encoding of objects' properties. One is that there generally is no fixed (temporally unmarked) description that uniquely characterizes a particular token thing. Another is that one of the main purposes of FINSTs is to keep track of things qua individuals, independent of what properties they may in fact have. Although these assumptions largely reflect empirical facts about vision that have since been supported by experiments, they are inherent in the function that FINSTs were called upon to perform in our initial analysis (which I will explore using several different examples in this essay). The above five properties already mark FINSTs as being quite different from the sorts of mind–world (or representation–world) connections that psychologists (and AI people) had postulated in the past, because they not only serve to refer to token things but do so without representing the thing as falling under a concept or a description: The relation between the representation and the thing (or visual object) represented is not one in which the object satisfies some description. Rather, it is purely causal.

The FINST, according to this story, is an instrument of reference by which one can pick out and refer to things. The reference is nonconceptual because it does not refer to things that have certain properties or that fall under certain conceptual categories. Thus it is very similar to a demonstrative (such as "this" or "that"), the only exception being that in the case of words, the referent is conditioned by the intentions of the speaker as well as by other contextual factors, such as pointing or gazing at the referent. FINSTs may be thought of as demonstrative terms in the language of thought that allow a person to think about something in the world that was selected in perception (especially vision) because something drew attention to itself or, as I prefer to say, grabbed a FINST index. Once a FINST reference is established, it can be used to bind arguments of mental predicates, or conceptual information about the referent can be entered into the associated object file. Although the FINST idea may seem simple enough, it has surprising consequences. To give you a sense of how far-reaching this idea is, note that I have assumed that FINSTs provide a mechanism for referring to visual objects without appealing to their conceptual properties, which means that, in an important sense, the referrer does not know what he or she is referring to! To refer to something (say, that object in the corner of my room) without referring to it as a cat, or as some mass with a particular shape, or as a patch of tawny color, or (as Quine might put it) as a collection of undetached cat parts, is a strange phenomenon. Yet there must be a stage in the visual process where something like this happens, otherwise we could not construct our conceptual representations on a foundation of causal connections to the world, as we must to avoid circularity.

The issue of whether it makes sense to postulate a nonconceptual form of reference has been much debated in philosophy and elsewhere in cognitive science. Among those who support the idea of nonconceptual representations are certain AI practitioners (e.g., Brooks 1991) or philosophers (e.g., Clark 1999) who speak of embodied or situated cognition (and in fact some of these writers shun the use of the term "representation" entirely, although I believe that their view leads naturally to a form of nonconceptual representation). My position is closer to that of philosophers who speak of essential indexicals (e.g., Perry 1979) and logicians who argue for bare demonstratives (e.g., Lepore and Ludwig 2000), which are closely related to FINSTs. Many philosophers who write about the mind–world interface wish to ward off skeptical arguments by claiming that the most primitive reference must be accessible to conscious experience. John Campbell (2003) uses the phrase "conscious attention" to emphasize the essential conscious character of attention-based reference. Many writers also assume that the most basic form of reference must pick out locations or at least regions, believing that a mental grip on a region is the more acceptable form of contact between mind and world since it is possible to imagine regions being picked out by a "spotlight of attention." Still other philosophers deny that the mind–world link requires a nonconceptual representation at all (e.g., McDowell 1994). At this point I simply want to alert the reader to the fact that much philosophical baggage hangs on how we describe what goes on in the earliest stages of visual perception (where by earliest I mean logically, neurologically, and temporally, though not necessarily developmentally). I will return to these questions later but will begin by setting the stage for the view I have been defending in recent years.

2   Why Do We Need Nonconceptual Reference?

The most general view of what vision does is that it computes a representation of a scene that then becomes available to cognition so that we can think about it—we can draw inferences from it or decide what it is or what to do with it (and there may perhaps be a somewhat different version of this representation that becomes available for the immediate control of motor actions). This form of representation represents a visual scene "under a description," that is, it represents the visual objects as members of some category or as falling under a certain concept. This is a fundamental characteristic of cognitive or intentional theories that distinguishes them from physical theories (Pylyshyn 1984). We need this sort of representation because what determines our behavior is not the physical properties of the things around us, but how we interpret or classify them—or more generally what we take them to be. It is not the bright spots we see in the sky that determine which way we set out when we are lost, but the fact that we see them (or represent them) in a certain way or under a certain concept (e.g., as the pointer stars in the Big Dipper or as the North Star). It is because we represent them as members of a certain category that our perception is brought into contact with our knowledge of such things as astronomy and navigation. Moreover, what we represent need not even exist, as in the case of the Holy Grail, in order to determine our behavior. In other words, it is the fact that we perceive or conceptualize it in certain ways that allows us to think about it. This is common ground for virtually all contemporary theories of cognition.

Although I have emphasized the representation-governed nature of cognition, this is not the whole story, even if augmented with sensory transducers (as I assumed in Pylyshyn 1984). It turns out that the sort of description-building view of perception is missing a critical piece: how the descriptors connect with what they describe. Although it is not often recognized, we can, under certain conditions, also refer to or represent some things without representing them in terms of concepts. We can refer to some things preconceptually (the preferred term in philosophy appears to be nonconceptually). For example, in the presence of a visual stimulus, we can think thoughts that involve individual things by using a term such as "that" and thinking "that is a pen," where the term "that" (in mentalese) refers to something we have picked out in our field of view without reference to what conceptual category it falls under or what properties it has. A term such as "this" or "that" is called a *demonstrative*. Demonstratives in natural language work slightly differently than FINSTs because, as a tool for communication, they are tied to the intentions of the speaker and may even require pointing or some other directional gesture (such as direction of gaze), none of which concerns FINSTs.

Philosophers like John Perry (see, e.g., Perry 1979) have argued that demonstratives are ineliminable from language and thought. The reason for the ineliminability of demonstratives also applies in the case of visual representations. Not only can we represent visual scenes in which parts are not classified according to some category, but there are good reasons why at least some things must be referenced in this nonconceptual way. If we could only refer to things in terms of their category membership, how would the category be defined? It would presumably be defined in terms of other conceptual properties, and so on. In that case our concepts would always be rooted only in other concepts and would never be grounded in experience. Sooner or later the regress of specifying concepts in terms of other concepts has to bottom out. Traditionally, the "bottoming out" was assumed to occur at sensory properties, but this "sense data" view of concepts has never been able to account for the grounding of anything more than simple sensory concepts and has been largely abandoned.[2] The present proposal is that the grounding begins at the point where something is picked out directly by a mechanism that works like a demonstrative. What I propose is that FINST indexes do the picking out, and the things they pick out in the case of vision are what many people have been calling *visual objects* or *proto-objects*.

A second closely related problem with the view that representations consist solely of concepts or descriptions arises when we need to pick out a particular token individual. If our visual representations encoded a scene solely in terms of concepts or categories, then we would have no way to pick out or refer to particular individuals in a scene except through concepts or descriptions involving other concepts, and so on. In what follows I will suggest a number of ways in which such a recursion is inadequate, especially if our theory of vision is to be situated, in the sense of making bidirectional contact with the world—that is, contact in which individual visual objects in a scene causally invoke certain visual objects in a representation, and in which the visual objects in the representation can in turn be used to refer to particular individuals in the world. The need to pick out and refer to individual things is not something that arises only under arcane circumstances; it happens every time you look out and see the world. It arises for a number of very good reasons and is generally associated with what is referred to in psychology as *focal* or *selective attention*. This is not the place to analyze why focal attention is essential for organisms like us (but see Pylyshyn 2007), but it may be useful to at least list them since they are not always recognized or appreciated.

2.1 Some Reasons Why We Need a Mechanism for Selecting or Picking Out Token Things

The limited capacity of the mind to process information   Because information processing is limited, some selection is required. The proper way to characterize the dimension along which the mind is limited and consequently the basis for selection are important empirical questions on which there is now interesting convergent evidence (later I will consider the evidence pointing to objecthood as the unit of attention or the things over which attention selects).

Incremental construction of representations   In encoding or conceptualizing a scene it is necessary to keep track of individual tokens in order to build a consistent representation. This arises in part because a representation must be constructed incrementally over time as parts of the representation that are encoded (or noticed) at different times and must be put into correspondence.

Solving the binding problem   Information about the world is "packaged" or presented in certain ways; specifying how this happens leads to what Austen Clark (Clark 2000) calls the *binding problem* (after Treisman 1995, who introduced the term) or the *many properties problem* (after Jackson 1997). Very early in the visual information-processing stream we must distinguish between properties present in a scene and conjunctions of these properties present on individual objects (for example, we distinguish a scene containing a red square and a green circle from a scene containing a red circle and a green square). This occurs at an extremely primitive level in vision (Clark would say it occurs at the level of sentience, but I prefer to say it occurs in early vision or in the visual module), and the informational basis for this encoding must be present prior to the application of concepts like "circle" and "square" and even "red" and "green." It must be evident in the way the perceptual world is primitively parsed—otherwise that information would be fused and unrecoverable. I return to this topic below.

Detection of patterns defined in terms of parts   Visually discriminable patterns that are made up of parts cannot be represented unless we can specify which things partake in that pattern. The predicates Collinear(x,y,z),

Inside(x,y), Above(x,y), or even Location(x,y,z) cannot be evaluated unless the arguments x, y, and z are instantiated by objects in the scene (i.e., unless the variables are bound, in the computer science sense of that term, where this means bound to the values of their arguments rather than bound by a quantifier).

Tagging of individuals in a scene to mark them during visual processing
Many visual patterns can only be discriminated if a serial process operates over the visual objects, which requires that token visual objects be somehow "marked" so they may be referred to by what Ullman calls *visual routines*. Predicates such as "containing *n* items" or "is inside a closed contour" or "are on the same contour" all require the operation of a serial process over the scene, and this process requires that certain things in the scene be picked out and referenced (most psychologists refer to this picking out as *marking* or *tagging*, but that is a very misleading way of talking since nothing is done to the distal scene or to any representation of it—the visual system simply picks out and refers to certain token things).

In this essay I focus on the problem of establishing a correspondence between individual things in the world and their counterparts in a visual representation, since this is where the notion of a FINST index or FINST played its first theoretical role in our work. Before I describe how FINSTs are relevant to this connection, I offer a few remarks about what these things might be and also a few illustrations of how this sort of direct reference is missing from the usual representations that visual theories provide. Although I am concerned with the initial steps of the process that begins with nonconceptual connections between mind and world, the process eventually encodes a visual scene in terms of some conceptual structure. In that context we see FINSTs as a mechanism for connecting the mind with real physical objects in the world. But a FINST as a nonconceptual connection cannot, by its very nature, be guaranteed to pick out all and only individual physical objects, because "physical object" is a conceptual category. Something is an individual physical object (or any other sort of individual) if it meets certain conditions (see any dictionary for a largely inadequate attempt to lay out such conditions). In particular it has to meet what Clark (2000) has called *Strawsonian strictures*: It has to meet conditions of individuation and identity. To decide whether something is an individual physical object one must bring to bear criteria of identity (see the discussion of this point in Strawson 1963). What FINST indexes do is pick out a class of things that in our kind of world are very often coextensive with physical objects, yet which can be picked out without using criteria of identity. The visual system very often yields a fast and automatic parsing of the world that provides a starting point for conceptual categories—even categories like "cause," which can be nonconceptually recognized in certain circumstances (and the nonconceptual category can be distinguished from the conceptual one; see Schlottman and Shanks 1992). FINST indexes serve the function, in the overall operation of the visual system, of connecting minds with physical objects (even though they may fail to do so sometimes). This is why I often speak of FINST indexes as referring to visual objects or even just objects. They do, however, sometimes fail to select a physical object (e.g., if it is too small or too big, if the lighting is poor, or if it is an illusion, such as provided by holograms). What one does about such errors is a question that faces every theorist, since even with Strawsonian strictures there will inevitably be illusions and other sources of error and failures of reidentification. We simply recognize that there may be P-detectors even if they do not always detect all and only Ps.

Before moving on to an explication of the theory and the experiments I would like to provide some additional background by way of motivation for the principles of selection and nonconceptual indexing listed above. Theories of visual perception universally attempt to provide an effective (i.e., computable) mapping from dynamic 2-D patterns of proximal (retinal) stimulation to a representation of a 3-D scene. Both the world and its visual representation contain certain individuals. The world contains objects, or whatever your ontology takes to be the relevant individuals, while the representation contains symbols or symbol structures (or codes, nodes, geons, logogens, engrams, etc., as the theory specifies). The problem of keeping tokens of the representing elements in correspondence with tokens of individual things in the world turns out to be rather more difficult than one might have expected.

With the typical sort of conceptual representation, there is no way to pick out an individual in the world other than by finding those tokens in a scene that fall under a particular concept, or satisfy a particular description, or that possess the properties that are encoded in the representation. What I will try to show is that this cannot be what goes on in general; it can't be the case that the visual system can only pick out things in the scene by finding instances that satisfy its conceptual representation. There are phenomena that suggest that the visual system must be able to pick out individuals in a more direct manner, without using encoded properties or categories. If this claim is correct, then the visual system needs a mechanism for selecting and keeping track of individual visual objects that works more like demonstrative reference than description. And that, I suggest, is why we must have something like a FINST indexing mechanism that non-conceptually picks out a small number of individuals, keeps track of them, and provides a means by which the cognitive system can further examine them in order to encode their properties, move focal attention to them, or carry out a motor command in relation to them (e.g., point to them).

3 The Need for Individuating and Indexing: Empirical Motivations

There are two general problems[3] raised by the "description" view of visual representations, that is, the view that we pick out and refer to objects solely in terms of their categories or their encoded properties. One problem is that there is always an unlimited number of things in the world that can satisfy any particular category or description, so that if it is necessary to refer to a unique token individual among many similar ones in the visual field (especially when its location or properties are changing), a description will not do. The second problem is deeper. The visual system needs to be able to pick out a particular individual regardless of what properties the individual happens to have at any instant of time. It is often necessary to pick out something in the visual field as a particular enduring individual, rather than as whatever happens to have a certain set of properties or happens to occupy a particular location in space. An individual remains the same individual when it moves about or when it changes any (or even all) of its visible properties. Yet being the same individual is something that the visual system often needs to compute, as we shall see in the examples below.

I appreciate that being a particular individual encumbers the individuation process with the need for conditions of individuation, and real full-blooded individuals must meet this condition and in order to be conceptualized as that individual. But the visual system, in its encapsulated ignorance, appears to solve a subset or a scaled-down version of the individuation problem that is sufficient for its purposes, and which more often than not does correspond to real individuals (or real objects) in our kind of world or in our ecological niche. That is the beauty and the ingenuity of the visual module—it does things expeditiously that turn out to be the right things to do in this sort of world: a world populated mostly by objects that move in certain rigid ways, in which discontinuities in lightness and in depth have arbitrarily low probability because real scene edges occupy a vanishingly small part of the universe, in which precise but accidental alignments have a very low probability of occurring, in which the light tends to come from above and casts shadows downward, and so on. Vision is attuned to just the right properties, which it picks out without benefit of knowledge and expectations of what is likely to be in some particular scene at some particular time. It is blissfully ignorant but superlatively successful in our sort of world.

So I claim that a very important and neglected aspect of vision is the nonconceptual connection by which it picks out what I have been calling visual objects. In arguing for the insufficiency of conceptual (or descriptive) representations as the sole form of visual representation, I appeal to three empirical assumptions about early vision: (1) that individuation of object tokens is primitive and nonconceptual and precedes the detection of properties; (2) that detection of visual properties is the detection of properties-of-objects, as opposed to the detection either of properties tout court or properties-at-locations; and (3) that visual representations are generally constructed incrementally over time.

3.1 Assumption 1: Individuation of Object Tokens Is Primitive and Precedes the Detection of Properties

(a) Evaluating visual predicates The process of individuating visual object tokens is distinct from the process of recognizing and encoding the objects' types or their properties. Clearly, the visual system can distinguish two or more distinct token individuals regardless of the type to which each belongs, or to put it slightly differently, we can tell visually that there are several distinct individuals independent of the particular properties that each has; we can distinguish distinct objects (and count them) even if their visible properties are identical. What is usually diagnostic of (though not essential to) there being several token individuals is that they have different spatiotemporal properties (or locations). Without a mechanism for individuating objects independent of encoding their properties, it is hard to see how one could judge that the six visual objects in figure 1.1 are arranged linearly, especially if the visual objects in the figure were gradually changing their properties or if the figure as a whole was moving while maintaining the collinear arrangement. In general, featural properties of visual objects tend to be factored out when computing global patterns, regardless of the size and complexity of the global pattern (Navon 1977). Computing global patterns such as collinearity, or others discussed by Ullman (1984), requires that visual objects be registered as individuals while their local properties are ignored. Whatever the particular algorithm used to detect collinearity among visual objects, it is clear that specifying which points form a collinear pattern is a necessary part of the computation.

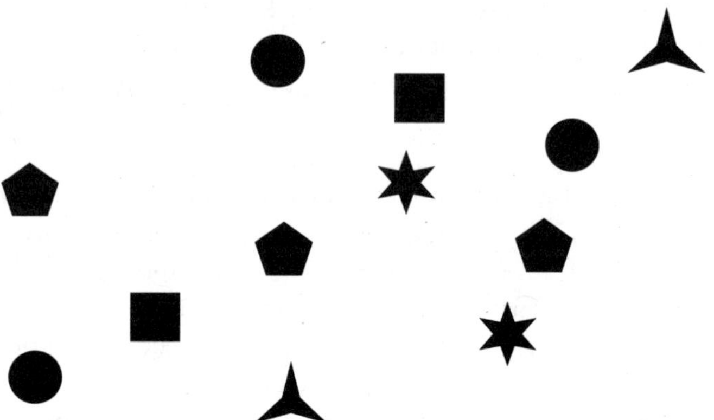

Figure 1.1
Find four or more items that are collinear. Judging collinearity requires selecting the
relevant individual objects and ignoring all their intrinsic (local) properties.

Here is another way to think of the process of computing relational
properties among a set of objects. In order to recognize a relational property, such as Collinear(X1, X2, . . . Xn) or Inside(X1, C1) or Part-of(F1, F2),
which apply over a number of particular individual objects, there must be
some way to specify which objects are the ones referred to in the relationship. For example, we cannot recognize the collinearity relation without
somehow picking out which objects are collinear. If there are many objects
in a scene only some of them may be collinear, so we must bind the objects
in question to argument positions in the relational predicate. Shimon
Ullman (see Ullman 1984), as well as many other investigators (e.g., Ballard
et al. 1997; Watson and Humphreys 1997; Yantis and Jones 1991), refers
to the objects in such examples as being "marked" or "tagged." The notion
of a tag is an intuitively appealing one since it suggests a way of labeling
objects to allow us to subsequently refer to them. Yet the operation of
tagging only makes sense if there is something on which a tag literally can
be placed. It does no good to tag an internal representation since the relation we wish to encode holds in the world and may not yet be encoded
in the representation. So we need a way of "tagging" that enables us to get
back to tagged objects in the world to update our representation of them.
But how do we tag parts of the world? It appears that what we need is
what labels give us in diagrams: a way to name or refer to individual parts
of a scene independent of their properties or their locations. This label-like
function that goes along with object individuation is an essential aspect
of the indexing mechanism that will be described in greater detail later.

(b) Visual individuation is different from visual discrimination There are numerous other sources of evidence suggesting that individuation is distinct from discrimination and recognition. For example, individuation has its own psychophysical discriminability function. James Intriligator's dissertation (described in Intriligator and Cavanagh 2001) showed that even at separations where objects can be visually resolved, they may nonetheless fail to be individuated or attentionally resolved, preventing the individual objects from being picked out from among the others. Without such individuation one could not count objects or carry out a sequence of commands that requires shifting attention from one to another. Given a 2-D array of points lying closer than their threshold of attentional resolution, one could not successfully follow such instructions as "move up one, right one, right one, down one, . . ." and so on. Such instructions were used by Intriligator and Cavanagh to measure attentional resolution. Figure 1.2 illustrates another difference between individuating and recognizing. It shows that one may be able to recognize the shape of objects and distinguish between a group of objects and a single (larger) object, and yet not be able to focus attention on an individual object within the group (in order to, say, pick out the third object from the left). Studies reported in He, Cavanagh, and Intriligator 1997 show that the process of individuating objects is separate from that of recognizing or encoding the properties of the objects.

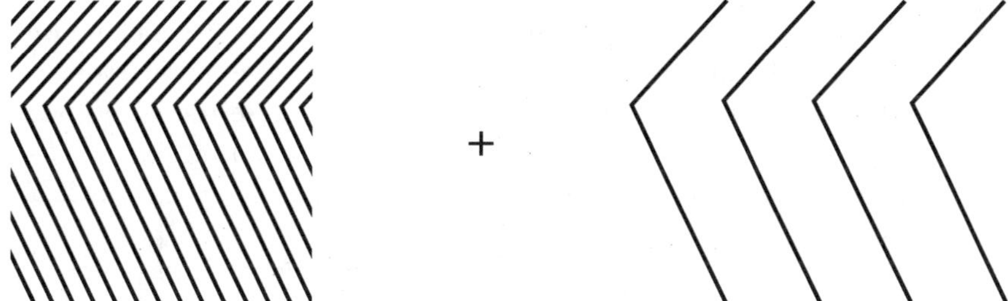

Figure 1.2
At a certain distance if you fixate on the cross you can easily tell which groups consist of similar-shaped lines, although you can only individuate lines in the group on the right. For example, while holding the page at arms length and fixating on the central cross you cannot count the lines or pick out the third line from the left, etc., in the panel on the left. (Based on Intriligator and Cavanagh 2001.)

(c) Rapid enumeration requires automatic individuation Studies of rapid enumeration (called *subitizing*), described by Lana Trick (Trick and Pylyshyn 1994), also show that individuating is distinct from (and prior to) computing the cardinality of a small set of objects. Trick and Pylyshyn showed that items arranged so that they cannot be preattentively individuated (or items that require focal attention in order to be individuated—as in the case of items lying on a particular curve or specified in terms of conjunctions of features) cannot be subitized, even when there are only a few of them (i.e., the signature break in the function relating reaction time to number of items is not observed in those cases). For example, in figure 1.3, when the squares are arranged concentrically (as on the left) they cannot be subitized, whereas the same squares arranged side by side can easily be subitized. According to our explanation of the subitizing phenomenon, small sets are enumerated faster than large sets when items are preattentively individuated because in that case each item attracts an index, so observers only need to count the number of active indexes without having to first search for the items. Thus we also predicted that precueing the location of preattentively individuated items would not affect the speed at which they were subitized, though it would affect counting larger numbers of items—a prediction borne out by our experiments.

(d) Subset selection The following experiment by Jacquie Burkell (Burkell and Pylyshyn 1997) illustrates and provides evidence in favor of the assumption that the visual system has a mechanism for picking out and accessing individuals prior to encoding their properties. Burkell showed that sudden-onset location cues (which we assume cause the assignment

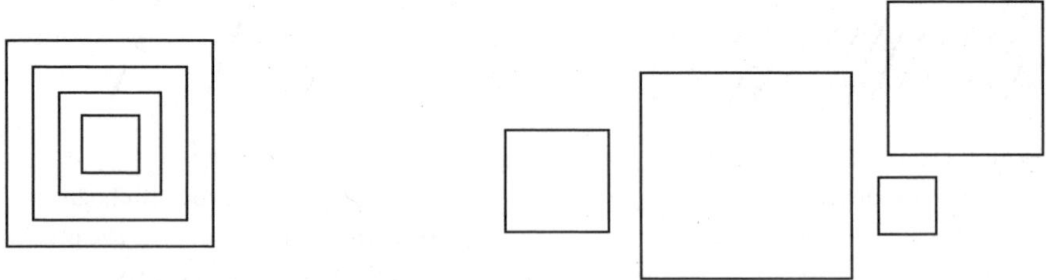

Figure 1.3
Squares arranged so they cannot be preattentively individuated (on the left) cannot be subitized, whereas the ones on the right are easily subitized. (Based on Trick and Pylyshyn 1994.)

of indexes) could be used to control search so that only the locations precued in this way are visited in the course of the search. This is what we would expect if the onset of such cues draws indexes and indexes can be used to determine where to direct focal attention.

In these studies (illustrated in figure 1.4) a number of placeholders (11 in the case illustrated), consisting of black Xs, appeared on the screen and remained there for one second. Then an additional 3 to 5 placeholders (which we refer to as the "late-onset cues") were displayed. After 100 ms one of the segments of each X disappeared and the remaining segment changed color, producing a display of right-oblique and left-oblique lines in either green or red. The subject had to search through only the cued subset for a line segment with a particular color and orientation (say, a left-oblique green line). Since the entire display had exemplars of all four combinations of color and orientation, search through the entire display is what is known as a *conjunction-search task* (which produces longer search times that increase as the number of items in the display increases). As expected, the target was detected more rapidly when it was one of the subset that had been precued by a late-onset cue, suggesting that subjects could directly access those items and ignore the rest.

There were, however, two additional findings that are even more relevant to the present discussion. The first depends on the fact that we manipulated the nature of the precued subset to be either a single-feature search task (i.e., in which the target differed from all other items in the search set by only one feature) or a conjunction-search task (in which only a combination of two features could identify the target because some of the nontargets differed from it in one feature and others differed from it in another feature). Although a search through the entire display would always constitute a conjunction-feature search, the subset that was precued by late onset cues could be either a simple or a conjunction-feature subset. So the critical question is: Is it the property of the entire display or the property of only the subset that determines the observed search behavior? We found clear evidence that only the property of the subset (whether it constituted a simple-search or a conjunction-search task) determined the relation between number of search items and reaction time. This provides strong evidence that only the cued subset is being selected as the search set. Notice that the distinction between a single-feature and a conjunction-feature search is a distinction that depends on the entire search set, so it must be the case that the entire precued subset is being treated as the search set: the subset effect could not be the result of the items in the subset being visited or otherwise processed one by one.

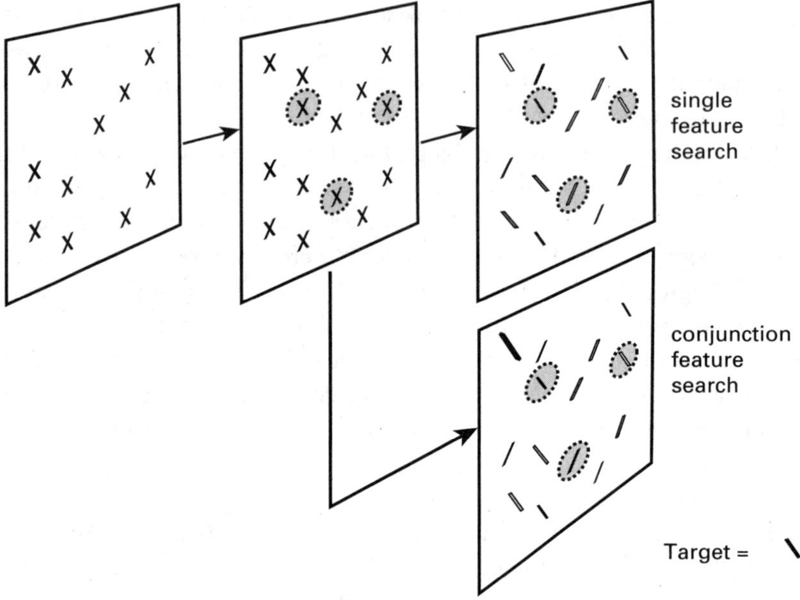

Figure 1.4
Sequence of events in the Burkell and Pylyshyn (1997) study. The observer sees a set of placeholder Xs, then three to five "late onset" placeholders appear briefly, signaling the items that will constitute the search items. Then all Xs change to search items (left or right oblique red or green line segments, shown here with circles around them for expository purposes) and the subject must try to find the specified target in one of two conditions. In the top display the target differs from all the nontargets by one feature, whereas in the bottom display, a combination of two features is required to distinguish the target.

The second item of particular relevance to the present discussion was the additional finding that when we systematically increased the distance between precued items there was no increase in search time per item, contrary to what one would expect if subset items were being spatially searched for. This is precisely what one would expect if the cued items are indexed and indexes are used to access the items directly, without having to scan the display. We also carried out the above experiment under rather technically difficult conditions in which subjects had to shift their gaze in the brief period between getting the late-onset cues and the start of the search process. We were able to show that indexes assigned to the cued objects survive eye movement so long as the saccade is generated in certain ways (e.g., if the eye is moved to view one of the target objects, but not if it is forced to move to the edge of the screen or to some secondary fixation point [Currie and Pylyshyn 2003]). This means that after the rapid saccade subjects were able to pick out the cued objects even though they were now in a different place on the retina. Having such a mechanism provides the beginnings of an account of how the world retains its apparent stability in the course of the 100,000 or so saccades each day—it does it by maintaining a cross-saccade correspondence on a few significant objects. Studies have shown that we cannot recall more than a few items from one fixation to another, so this mechanism may be all we need (Irwin 1992).

This type of study provides a clear picture of the property of indexes that we have been emphasizing: They provide a mechanism of direct access, rather like the access provided by pointers in computers or demonstratives in language. Certain visual objects can be indexed without appealing to their encoded properties (the indexing being due to such transients as their sudden appearance on the scene), and once indexed, they can be individually examined either in series or in parallel. In other words, one can ask "Is x red?" so long as x is bound to some visual object by an index.

3.2 Assumption 2: Detection of Visual Properties Is the Detection of Properties-of-Objects

When a property is first encoded by the visual system it is encoded not just as a property existing in the visual field, but as the property of an individual, perceived thing in the world. The claim has frequently been made that features are detected as occurring at a location (talk of "feature placing" explicitly assumes that this is what happens). I claim that the visual system does not just detect the presence of redness or circularity in the visual field, or the presence of such properties at some particular location in some frame of reference: It detects that certain individual objects are red or circular or arranged linearly. This, in turn, requires that the individuals be selected first. There are numerous sources of evidence supporting this assumption, most of which were collected in connection with asking somewhat different questions. Some of them are sketched next.

(a) Object-based attention and single-object advantage The first kind of evidence comes from the observation that several properties are most easily extracted from a display when they occur within a single visual object, and therefore that focal attention (which is assumed to be required for encoding conjunctions of properties) is object based (Baylis and Driver 1993). So, for example, if you are asked to judge the relative heights of the two vertices in figure 1.5, you are faster when instructed to view the lighter portion as the object in (a) compared to (b).

Other evidence supporting this conclusion comes from a variety of sources (many of which are reviewed in Scholl 2001), including experiments in which objects move through space or in which they move through feature space. (More examples are discussed in Pylyshyn 2003.) Also, clinical cases of hemispatial visual neglect and Balint syndrome implicate an object-centered frame of reference. Patients with the symptom known as simultanagnosia, who reportedly can see only one object at a time, nonetheless can report properties of two objects if they are somehow linked together. This sort of object-specificity of feature encoding is exactly what would be expected if properties are always detected as belonging to an object. Object-based attention has been widely studied in current vision science, and most of the more impressive evidence comes from cases where objects move so that it is possible to distinguish between objecthood and location.

(b)   The binding problem and detecting conjunctions of properties   Another kind of evidence for the primacy of objecthood comes from the fact that we can distinguish the cooccurrence of features on an individual object from their mere occurrence somewhere in a scene, the aforementioned binding problem or multiple-properties problem. The assumption is that in early vision (or, as some people put it, in sensation) people can distinguish between different displays that consist of redness, greenness, circularity, and squareness. For example, they can distinguish between a display consisting of a red circle and a green triangle from one consisting of a green circle and a red triangle. The usual assumption among psychologists about how the binding problem is solved is that it is done in terms of the common location of the bound properties. This assumption is made

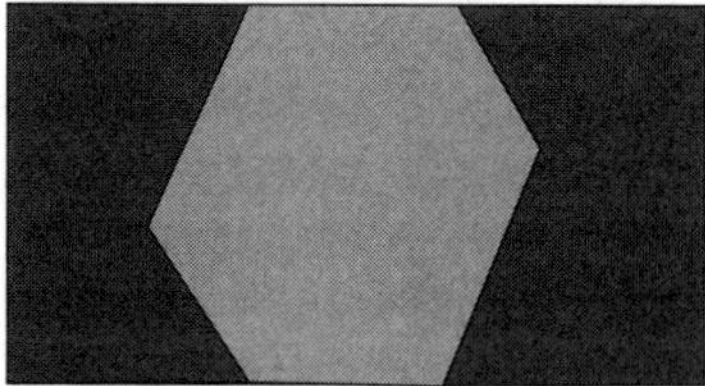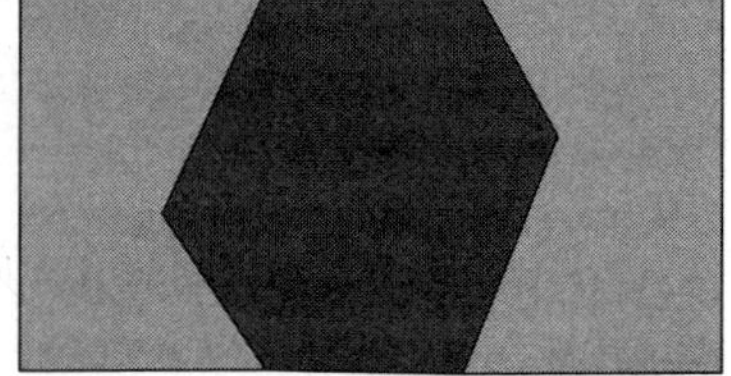

Figure 1.5
Figures used to demonstrate single-object advantage in judging properties of a shape within one figure versus between two figures. (Based on Baylis and Driver 1993.)

in Treisman's feature integration theory (see Treisman and Gelade 1980), in Clark's theory of sentience, in Campbell's analysis of consciousness (see Campbell 2002), and in most psychological theories (see, e.g., Pashler 1998). But this will not work in general; and where it does work, it confounds location and objecthood.

Evidence often cited in support of the assumption that properties are detected in terms of their location is compatible with the view that it is the object with which the property is associated, rather than its location, that is primary. A good example of a study that was explicitly directed at the question of whether location is central is one carried out by Mary-Jo Nissen (Nissen 1985). She argued that in reporting the conjunction of two features, observers must first locate the place in the visual field that has both features. In Nissen's studies this conclusion comes from a comparison of the probability of reporting a stimulus property (e.g., shape, color, or location) or a pair of such properties, given one of the other properties as cue. Nissen found that accuracy for reporting shape and color were statistically independent, but accuracy for reporting shape and location, or for reporting color and location, were not statistically independent. More important, the conditional probabilities conformed to what would be expected if the way observers judge both color and shape is by using the detected (or cued) color to determine a location for that color and then using that location to access the shape. For example, the probability of correctly reporting both the location and the shape of a target, given its color as cue, was equal (within statistical sampling error) to the product of the probability of reporting its location, given its color, and of reporting its shape, given its location. From this, Nissen concluded that detection of location underlies the detection of either the color or shape feature given the other as cue. Similarly, Hal Pashler (Pashler 1998, 97–99) reviewed a number of relevant studies and argued that location is special and is the means by which other information is selected. Note, however, that since the objects in all these studies had fixed locations, these results are equally compatible with the conclusion that detection of properties is mediated by the prior detection of the individuals that bear these properties, rather than detection of their location. If the individuals had been moving in the course of a trial, it might have been possible to disentangle these two alternatives and ascertain whether detection of properties is associated with the instantaneous location of the properties or with the individuals that had those properties.

In contrast, it is clear that detection of objects must precede solving the binding problem because the location that would be required cannot be punctate—one must specify a region that contains both features. But which region? Try specifying the regions that share the dual (conjoined) properties in a figure such as the one in figure 1.6. You can tell these two figures apart even though they contain the same figures and textures and can only be distinguished by which shape has which texture. The rectangular bounding region is the same; so the only way to distinguish these two is to refer the particular texture to the region marked out as the outline of the figure with that texture. But you can only specify this sort of region by having selected the object and used its boundary as the region. Neither texture nor shape has a location apart from the object that has those properties. In addition, empty regions by themselves do not have causal properties and so are incapable of grabbing a FINST index.

(c) Object-specific effects move with moving objects A number of experimental paradigms have used moving objects to explore the question of whether the encoding of properties is associated with individual objects, as opposed to locations. These include the studies on object files (Kahneman, Treisman, and Gibbs 1992) and our own studies using multiple-object tracking (MOT) (see below, as well as Pylyshyn 1994, 1998). Kahneman et al. showed that the priming effect of letters presented briefly in a moving box remains attached to the box in which the letter had appeared, rather than to its location at the time it was presented.

Similarly, related studies by Steven Tipper (Tipper, Driver, and Weaver 1991) showed that the phenomenon known as inhibition of return

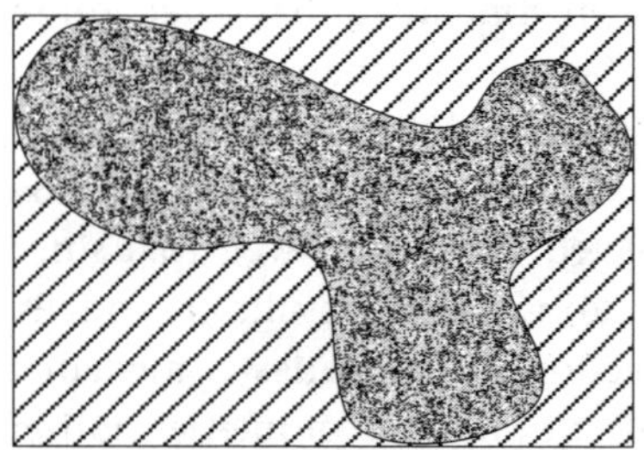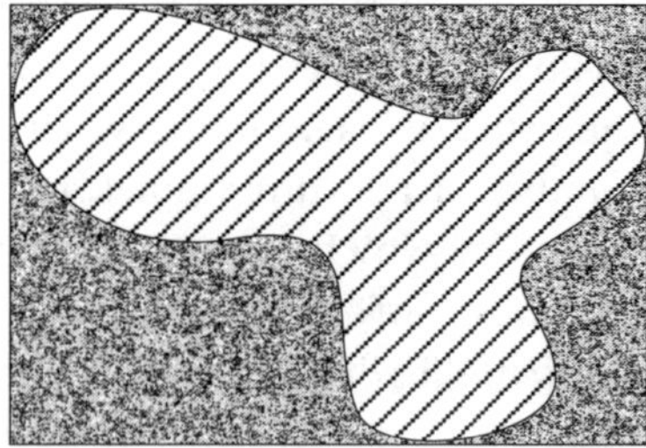

Figure 1.6
To distinguish these two figures you can't simply encode texture, shapes, and their location, as done in feature maps, since they both have the same features and the same centroid (and the same bounding rectangle) location. Instead you have to associate the texture with the region it occupies, and you can't specify that unless you have first picked out the object whose bounds constitute the relevant region.

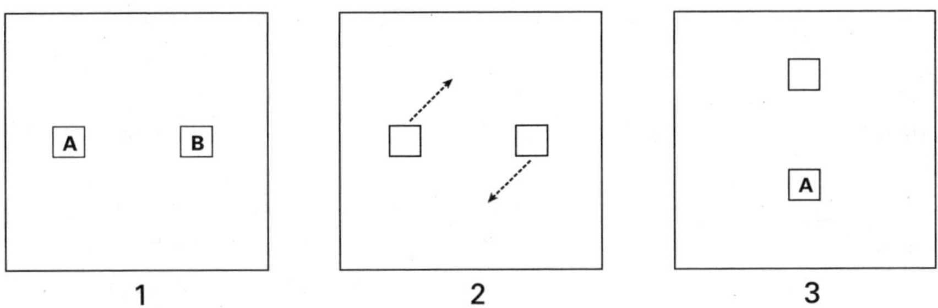

Figure 1.7
Studies showing facilitation of naming a letter (the letter is named faster) when it recurs in the same box as it was in at the start of the trial, even though this was not predictive of which letter it was (since half the time it was the letter that had been in the other, equally distant, box). (Based on Kahneman, Treisman, and Gibbs 1992.)

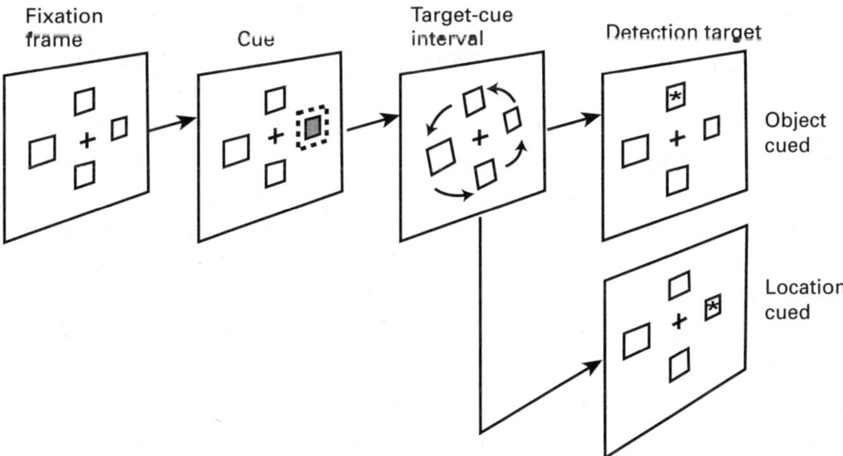

Figure 1.8
Inhibition of return (IOR) is a phenomenon whereby items that are attended and then attention is removed from them become more difficult to reattend during a period of from about 300 ms to 900 ms afterward. It has been shown that what is inhibited in IOR is mostly the individual object that had been attended—IOR travels with the object as it moves.

(whereby the latency for switching attention to an object increases if the object has been attended in the past 300 ms to about 900 ms) was specific to particular objects rather than particular locations within the visual field (though later work by Tipper et al. 1994 suggests that location-specific IOR also occurs).

Although there is evidence that unitary focal attention (sometimes referred to as the *spotlight of attention*) may be moved through space (but see Sperling and Weichselgarter 1995 for an alternative explanation of the apparent attention movement phenomena) and appears to spread away from its central spatial locus, many other attention phenomena appear to be attached to objects with little evidence of spreading to points outside the objects in question. For example, Egly, Driver, and Rafal (1994) showed that attention seems to spread throughout regions defined by contours, but only if those contours are perceived to be the contours of a single object.

3.3   Assumption 3: Visual Representations Are Constructed Incrementally

Another empirical finding is that our visual representation of a scene is not arrived at in one step, but rather is built up incrementally. This finding has strong theoretical support as well. A number of theoretical analyses (e.g., Tsotsos 1988; Ullman 1984) have provided good reasons for believing that some relational properties that hold between visual elements, such as the property of being inside or on the same contour, must be computed serially by scanning a beam of attention over certain parts of a display. We also know from empirical studies that percepts are generally built up by scanning attention and/or one's gaze. Even when attention is not scanned, there is evidence that the achievement of simple percepts occurs in stages over a period of time (e.g., Calis, Sterenborg, and Maarse 1984; Reynolds 1978b; Sekuler and Palmer 1992). If that is so then the following problem immediately arises. If the representation is built up incrementally, we need a mechanism for determining the correspondence between representations of individual objects across different stages of construction of the representation or across different periods of time. As we elaborate the representation by uncovering new properties of a dynamic scene, we need to know which individual objects in the current representation should be associated with the new information. In other words we need to know when a certain token in the existing representation should be taken as corresponding to the same individual object as a particular token in the new representation. We need that so that we can attribute newly noticed properties to the representation of the appropriate individual objects.

A general requirement for adding information to a representation is that we be able to relate the newly discovered properties to particular objects in the existing representation of the figure. If you notice, say, that a certain property or feature is present in the scene, you need to add this information to the current representation. How do you know which represented item is the relevant one, so that you can add the information to the appropriate item? Or how do you know whether a particular object is a new object or one you have seen and represented before? If you don't solve this correspondence problem correctly you will end up with a cacophony of duplicated objects in the representation of a scene. The world does not come with every object conveniently labeled. What constraints on the form and content of an adequate representation are imposed by the need to pick out individual objects?

It might seem that in principle it is possible to pick out an individual object by using an encoded description of its properties. All you need is a description that is unique to the individual in question, say, "the object α with property P" where P happens to uniquely pick out a particular object. But consider how this would have to work. If you want to add to a representation the newly noticed property Q (which, by assumption, is a property of a particular object, say, object α), you must first locate the representation of object α in the current representation. Assuming that individuals are represented as expressions or individual nodes in some conceptual network, you might detect that the object that you just noticed as having property Q also had property P which uniquely identifies it. You might then assume that it had been previously stored as an object with property P. So you find an object in the current representation that is described as having P and conjoin the property Q to it (or use an identity statement to assert that the object with property P is identical to the object with property Q). There are many ways to accomplish this, depending on exactly what form the representation takes. But whatever the details of such an augmentation process, it must be able to locate the representation of a particular individual in order to update the representation properly. Yet this may well be too much to ask of a general procedure for updating representations. It requires working backward from a particular individual in the scene to its previously unique representation. There is no reason to think that locating a previous representation of an individual is even a well-defined function, given that representations are highly partial and schematic (and indeed, the representation of a particular object may not even exist in the current representation) and an individual object may change any of its properties over time while continuing to be the same object.

In fact, the rapidly growing literature on change blindness would suggest that unless objects are attended they may change many of their very obvious properties without their representation being updated (Rensink 2000; Rensink, O'Regan, and Clark 1997, 2000; Simons 1996; Simons and Levin 1997). The alternative to this unwieldy method for locating a representation of a particular individual is to allow the descriptive apparatus to make use of a name or demonstrative reference. If we had such a mechanism, then adding newly noticed information would consist in adding the predicate $Q(\alpha)$ to the representation of a particular object α, where α is the object directly picked out by this demonstrative indexing mechanism. By hypothesis, the visual system's Q-detectors recognize instances of property Q as a property of a particular visual object (in this case of α), so being able to refer to α provides the most natural way to view the introduction of new visual properties by the sensorium.[4] In order to introduce new properties into a representation in that way, however, there would have to be a non-descriptive way of picking out the unique object in question. In the following section I examine experimental evidence suggesting that such a mechanism is needed for independent reasons—and in fact was proposed some time ago in order to account for certain empirical findings.

4 Multiple-Object Tracking (MOT)

I have argued that the visual system must have a mechanism to individuate and keep track of particular individuals in a scene in a way that does not require appeal to any of their properties (including their locations). Thus what we need is a way to realize the following two functions: (a) picking out or individuate visual objects, and (b) providing a means for referring to each individual object as if each individual object had a unique label or proper name. Although (as I will argue later) I believe these two functions to be distinct, I have proposed that they are both realized by a primitive mechanism called a FINST, some of the details of which will be sketched later. In this section I illustrate the claim that there is a primitive mechanism that picks out and maintains the identity of visual objects, by describing an experimental paradigm we have been using to explore the nature of such a mechanism. It is called multiple-object tracking (MOT) and is illustrated in figure 1.9.

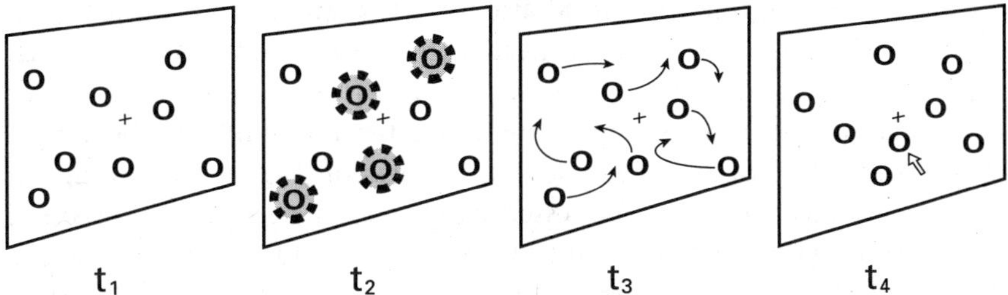

Figure 1.9

Illustration of a typical multiple-object tracking experiment. A number of identical objects are shown, then a subset (the "targets") is selected by making them blink, after which the objects move in unpredictable ways (with or without self-occlusion) for about ten seconds. At the end of the trial the observer has to pick out all the targets using a pointing device. (From Pylyshyn 2003; demonstrations of this and other MOT displays can be viewed at: http://ruccs.rutgers.edu/finstlab/demos. htm.)

In a typical experiment, observers are shown anywhere from 8 to 12 simple identical objects (points, squares, circles, figure-eight shapes). A subset of these objects is briefly rendered distinct (usually by blinking them on and off a few times). Then all the identical objects move about in the display in unpredictable ways. The subject's task is to keep track of this subset of objects (called "targets"). After ten or so seconds of tracking, the objects stop moving and the observer must then indicate which of the objects (all of which are now visually indistinguishable) were the targets by clicking on them using a computer mouse. A large number of experiments, beginning with the studies described in Pylyshyn and Storm 1988, have shown that observers can indeed track up to five independently moving targets within a field of ten identical items.[5] The question we must ask is: How can this be done? What mechanism makes this possible? If it were to be done using some description of each object it would have to be a process that encodes each object's location, since location is the only property that distinguishes one object from the other at a particular point in time. Such a process would have to use focal attention; a reasonable assumption from previous work on attention is that objects must be attended in order for their properties to be encoded. So a possible tracking strategy would be to keep a record of objects' locations and visit them serially to update their location with each iteration until the end of the trial. We have simulated that algorithm on the actual displays we used and have showed that, given very conservative assumptions about location-encoding requiring focal attention that moves at a finite speed, the best performance we could expect is about 30 percent, which is much lower than the observed 87 percent. This means that the moving objects could not have been tracked by using focal attention to update the unique stored description of each figure (i.e., their location). These studies suggest that the early vision system (an essentially encapsulated system, discussed at length in Pylyshyn 1999) is able to individuate and keep track of about five visual objects and does so without using an encoding of any of their visual properties.

The multiple-object tracking task exemplifies what is meant by "tracking" and "maintaining the identity" of objects. It also operationalizes the notion of "visual object" as whatever allows nonconceptual selection and multiple-object tracking (as these things are interdefined with FINSTs, I have sometimes called them FINGs). Of course, it is of interest to discover what sorts of events will in fact count as visual objects from this perspective. We are just beginning to investigate this question. We know from MOT studies that simple figures count as objects and also that certain well-defined clusters of features, such as the endpoints of lines, do not (Scholl, Pylyshyn, and Feldman 2001). Indeed, as we saw earlier, some well-defined visually resolvable features do not allow individuation (see figures 1.2 and 1.3). We also know that the visual system may count as a single persisting individual certain cases where clusters of features disappear and reappear. For example, Scholl and Pylyshyn (1999) showed that if the objects being tracked in the MOT paradigm disappear and reappear in certain ways, they are tracked as though they had a continuous existence. If, for example, they disappear and reappear by deletion and accretion along a fixed contour, the way they would if they were moving behind an occluding surface (even if the edges of the occluder are not invisible), they are successfully tracked. However, performance in the MOT task declines significantly in control conditions where objects suddenly go out of existence and reappear at the appropriate matching time and place, or if they slowly shrink away to a point and then reappear by slowly growing again at exactly the same relative time and place as they had accreted in the occlusion condition. Beyond that, what qualifies as a primitive (potentially indexable) object remains an open empirical question. In fact, more recent evidence (see Blaser, Pylyshyn, and Holcombe 2000) shows that objects can be tracked even though they are not specified by unique spatiotemporal coordinates (e.g., when they share a common spatial locus and move through "feature space" rather than real space).

4.1 How FINSTs Are Used in Multiple-Object Tracking

From the point of view of FINST theory, the way MOT proceeds may be summarized as follows. When a subset of the objects blinks on and off, each individual "target" captures a FINST (so long as there are not more than four or five such blinking objects). Since objects are visually identical, the only current property that distinguishes one object from another is its location on the screen. What distinguishes targets from nontargets is that targets are the visual objects that earlier had been visually distinct in some way (in this case by their blinking)—that is, by their past history. So in order to identify targets as distinct from nontargets it is necessary either to identify them by their location or to trace their provenance or their identity back to the start of the trial, and thereby to ascertain their origin-status as target. In Pylyshyn and Storm 1988, we argued that it is unlikely that observers track the targets by cyclically updating a record of their locations as they move about, and then using this list of target locations at the end of the trial to specify targets. On the basis of that argument we concluded that indexing does not use location information to track objects. Indexes simply attach to objects, and when the objects move they carry the indexes with them (providing that the motion is within certain spatiotemporal bounds). When the objects stop moving, subjects can use the indexes to shift their focal attention and then their gaze to each of the targets in turn. While foveating an indexed object, observers can move the mouse and click on it, then shift their gaze to the next indexed object and repeat.

A slight variation is needed if target objects are indicated by a property that does not automatically draw indexes (e.g., if the targets are vertical lines while nontargets are horizontal lines). We have evidence that in that case a "spotlight of attention" has to visit each of these cued targets in turn and "drop off" an index (Pylyshyn and Annan 2006). There are other findings that tell us more about the nature of these FINST indexes. For example, people do not notice when targets (or nontargets) change color or shape; they have a great deal of trouble recalling which target was which when they are identified with names or numbers at the start of the trial; and they are able to track targets even when the targets disappear briefly but completely behind occluding barriers. When observers make errors these tend to consist in swapping the identity of one object with that of another object that is close to it, and the chances of such swaps is higher between pairs of targets than between targets and nontargets. There is also evidence that the reason for this asymmetry in swapping errors between target-target pairs and target-nontarget pairs is that nontargets are inhibited during a tracking trial (Pylyshyn 2004, 2006).

The story of how basic MOT is carried out in terms of FINST theory is extremely simple, partly because the MOT task was designed to reflect the FINST hypothesis in a fairly direct way. But there are other findings that are not accounted for without some finer-grained assumptions about how FINSTs work. Moreover, there is more to FINST indexing than is revealed in the above story. We assume that FINSTs constitute a very general mechanism that not only is used for tracking simple elements moving on a screen, but that also functions to allow people to keep track of things in the world. The ability to track things has long been recognized as an essential ingredient in identifying individual things, and so the question of what our visual system treats as a thing (an individual or an object in some sense) is extremely important. Thus some of the assumptions we have made about FINSTs have extremely far-reaching implications for how our visual system deals with individuals, properties, and other aspects of the contact between mind and world. What I have found over the last several years of trying to explain to psychologists and philosophers what I think is going on is that finding the right way to describe the empirical phenomena and explaining what they mean in a more general framework is far from an easy task. What I will do very briefly in the next section is present a version of the story that suggests what the FINST idea might mean for the connection between mind and world. Because quite a few pieces of this puzzle are still missing I will have to go out on a limb now and then and engage in some speculation.

5   Viewing FINSTs as Nonconceptual Links between Mind and World

The basic motivation for postulating indexes is that, as we saw at the beginning of this essay, there are both empirical and theoretical reasons for assuming that a small number of individual objects in the field of view must first be picked out from the rest of the visual field and that the identity of these objects qua individuals (sometimes called their numerical identity) must be maintained or tracked despite changes in the individuals' properties, including their location in the visual field. The FINST hypothesis claims that this is done primitively by the FINST mechanism of the early vision system, without identifying the object through a unique descriptor. In other words it is done without cognitive or conceptual intervention. In assigning indexes, some cluster of visual features must first be segregated from the background or picked out as a unit (the gestalt notion of making a figure–ground distinction is closely related to this sort of "picking out," although it carries with it other implications that we do not need to assume in the present context—e.g., that bounding contours are designated as belonging to one of the possible resulting figures). Until some part of the visual field is segregated in this way, no visual operation can be applied to it since it does not exist as something distinct from the entire field.

But segregating a region of visual space is not the only thing that is required. In addition, what is needed is a way for the cognitive system to refer to that particular individual or visual object as distinct from other individuals. It must be possible to bind one of a small number (four or five) of internal symbols or parts of a visual representation to objects in the world by a mechanism that binds them to individual clusters. Moreover, the clusters must be such that the representation can continue to refer to the objects as the same individuals despite changes in their location or any other property (subject to certain constraints, which need to be empirically determined). The existence of such a capacity would make it possible, under certain conditions, to pick out a small number of individual visual objects and also to keep track of them as individuals over time. We are beginning to map out some of the conditions under which such individuation and tracking can occur; they include, for example, spatiotemporal continuity of motion, or else discontinuity in the presence of local occlusion cues such as those mentioned above in discussing the Yantis (1998) and Scholl (Scholl and Pylyshyn 1999) results. They also include the requirement that the object being tracked be a perceptual whole as opposed to some arbitrary but well-defined set of features (Scholl, Pylyshyn, and Feldman 2001).

FINST theory is described in several publications cited earlier and will not be described in detail here beyond the sketch given above. The essential assumptions may be summarized as follows: (1) early visual processes segment the visual field into feature-clusters that tend to be reliable proximal counterparts of distinct individual objects in the distal scene; (2) recently activated clusters compete for a pool of four or five FINST indexes; (3) index assignment is primarily stimulus driven, although cognitive factors, such as scanning focal attention until an object is encountered that activates an index, may have a limited effect; (4) indexes keep being bound to the same individual visual objects as the latter change their properties and locations (which is what makes them perceptually the same objects), within certain as-yet-unknown constraints; and (5) only indexed objects can enter into subsequent cognitive processes, such as recognizing their individual or relational properties, or shifting focal attention or gaze or making other motor gestures toward them.

The basic idea of the FINST indexing and binding mechanism is illustrated in figure 1.10. Certain proximal events (e.g., the appearance of a new visual object) cause an index to be grabbed (since there is only a small pool of such indexes, this may sometimes result in an existing binding being lost). As new properties of the inducing object are detected, they are associated with the index that points to that object. This, in effect, provides a mechanism for connecting objects of an evolving representation with objects in the world (stored temporarily in the object files mentioned earlier). By virtue of this causal connection, the cognitive system can refer to any of a small number of primitive visual objects. The sense of reference I have in mind here is one that appears in computer science when we speak of pointers or when variables are assigned values. To have this sense of reference is to be able to access the referents in certain ways: to interrogate them in order to determine some of their properties, to evaluate multiplace predicates over them, to move focal attention to them, and in general to bind cognitive arguments to them, as would have to be done in order to execute a motor command toward them. What is important to note here is that the inward arrows are purely causal and are instantiated by the nonconceptual apparatus which, following the terminology suggested by Marr (1982), I refer to as *early vision* (Pylyshyn 1999). The indexing system latches onto certain kinds of spatiotemporal objects because it is "wired"

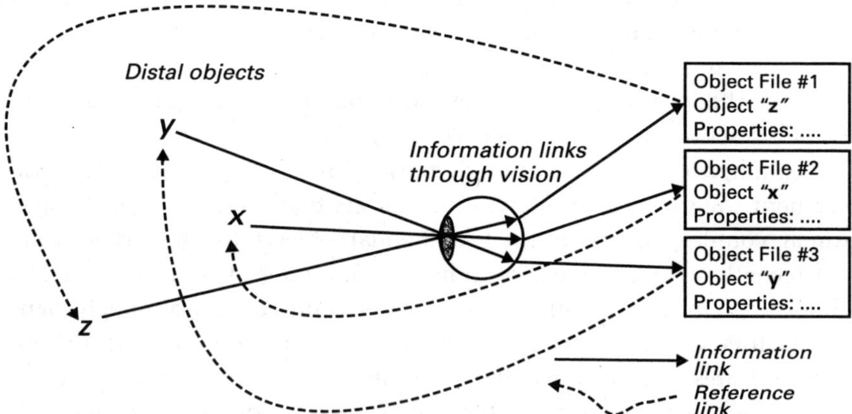

Figure 1.10
Sketch of the types of connections established by FINST indexes between the visual objects and parts of conceptual structures, depicted here as object files. Such a mechanism would clearly have applicability to everyday tasks such as monitoring players in team sports.

to do so, or because it is in the nature of its functional architecture to do so, not because those entities satisfy a certain cognitive predicate—that is, not because they fall under a certain concept. This sort of causal connection between a perceptual system and a visual object in a scene is quite different from a representational or intentional or conceptual connection. For one thing, there can be no question of the object being misrepresented, since it is simply not represented.

The indexing notion that I am describing is extremely simple; it only seems complicated because ordinary language fails to respect certain distinctions (such as the distinction between individuating and recognizing, or between indexing and knowing where something is, and so on). In fact a very simple network, such as the one described by Koch and Ullman (1985), can implement such a function (the application of the Koch and Ullman network to FINST index theory has been explored in Acton 1993; Pylyshyn and Eagleson 1994). Another implementation uses an oscillatory neural network (and uses separate layers for each object; see Kazanovich and Borisyuk 2006). All that is required is some form of winner-take-all circuit whose convergence on a certain active place on a spatiotopic map enables a signal to be sent to that place, thus allowing it to be probed for the presence of specific properties (a simple sketch of such a system is given in appendix 5A of Pylyshyn 2003). What is important about such a network, which makes its indexing function essentially preconceptual, is that the process that sends the probe signal to a particular place uses no encoding of properties of that place, not even its location. Being able to probe a certain place depends only on its being the most active by some measure (such as the activation measures assumed in many theories of visual search, like those of Treisman and Gelade 1980; or Wolfe, Cave, and Franzel 1989). What makes this system object based, rather than location based, is that there are certain provisions in the network that ensure that a smoothly moving object is tracked as the same object (e.g., this can be done by lowering the threshold of the units closest to the selected unit), which results in the FINST moving along with the selected visual object (for details, see Koch and Ullman 1985; Pylyshyn 2003, chapter 5).

What I have described is a mechanism for picking out, tracking, and providing cognitive access to what I have been calling visual objects. The notion of an object is ubiquitous in cognitive science, not only in vision but much more widely. It is also a foundational concern in metaphysics. But for present purposes I will take for granted that the world consists of physical objects. The view I have been proposing assumes that the visual system (or at least that part of it that we refer to as early vision) is encapsulated, that it is a module that works autonomously and independently of cognition. The view also relies on the many studies that have shown that attention (and hence information access to the visual world) is allocated primarily, though not exclusively, to individual visual objects rather than to properties or to unfilled locations. The latter conclusion is also supported by evidence from clinical neuroscience, where it has been shown that deficits such as unilateral neglect (Driver and Halligan 1991) or Balint's syndrome (Robertson et al. 1997) apply over frames of reference that are object based, wherein what is neglected appears to be specified with respect to individual objects. From this initial idea I have sought to analyze the process of attention into distinct stages. One of these involves the detection and tracking of primitive visual objects. This stage allows attention and other more cognitive processes to access and operate on these primitive visual objects.

My focus has been on visual objects—objects that are selected by the visual system without benefit of concepts and knowledge. Although I have mentioned psychophysical experiments, including multiple-object tracking, there are numerous findings in cognitive development that are relevant to our notion of object and index. For example, the notion of object has played an important role in the work by Leslie et al. (1998); Spelke, Gutheil, and Van de Walle (1995); and Xu and Carey (1996). These researchers have explicitly recognized the close relation between this notion of object and the one that is involved in our theory of FINST indexes. Typical experiments show that in certain situations, 8-month-old infants are sensitive to the cardinality of a set of (one or two) objects even before they use the properties of the individual objects in predicting what will happen in certain situations where objects are placed behind a screen and then the screen is removed. For example, Alan Leslie (see Leslie et al. 1998) describes a number of studies in which one or two objects are placed behind a screen and the screen is then lowered. Infants exhibit longer looking times (relative to a baseline) when the number of objects revealed is different from the number that the infant sees being placed behind the screen, but not when the objects have different visual properties. This has been taken to suggest that registering the individuality of objects developmentally precedes recognizing objects by their properties in tasks involving objects' disappearance and reappearance.

Though it is tempting to identify these empirical phenomena as involving the same notion of "object," it is unclear whether all these uses of the term "object" in psychology in fact do mean the same thing. My present use of the term is inextricably connected with the theoretical mechanism of FINST indexing, and therefore to the phenomena of individuation and tracking, and assumes that such objects are picked out in a nonconceptual manner. If the sense of "object" that is needed in other contexts entails that individuating and tracking must appeal to a conceptual category, defined in terms of how the observer represents it or what the observer takes it to be, then it will not help us to ground our concepts, nor will it help with the problem of keeping track of individuals during incremental construction of a percept. In the case of the multiple-object tracking examples, the notion of primitive visual object I have introduced does fulfill these functions. But of course this leaves open the question of what the connection is between the primitive visual object so defined and the more usual notion of physical object, and in particular with the notion of object often appealed to in infant studies. In those studies, an object is defined by Elizabeth Spelke and others as a "bounded, coherent, three-dimensional physical object that moves as a whole" (Spelke 1990). Are such "Spelke Objects" different from what we have been calling primitive visual objects?

My provisional answer to the question of the relation between these two notions of object is that in most natural settings, both primitive visual objects and Spelke Objects correspond to real physical objects. According to this view, the visual system is so structured that it detects visual patterns that in our kind of world tend to be reliably associated with entities that meet the criteria for being an object (or perhaps for being a Spelke object, which is a subset of physical objects). If that is the case, then it suggests that, contrary to claims made by developmental psychologists (Spelke, Gutheil, and Van de Walle 1995; Xu 1997), the concept of an object is not involved in picking out these visual objects, just as no concept (i.e., no description) plays a role in multiple-object tracking. Despite this speculative suggestion, it is less clear whether a concept is involved in all the cases discussed in the developmental literature. From the sorts of considerations raised here, it seems likely that a direct demonstrative reference or index is involved at least in some of the phenomena (see Leslie et al. 1998). However, there also appear to be cases in which clusters of features that one would expect would be perfectly good objects from the perspective of their visual properties may nonetheless fail to be tracked as objects by 8-month-old infants. Chiang and Wynn (2000) have argued that if the infants are given evidence that the things that look like individual objects are actually collections of objects then they do not keep track of them in the studies involving placing objects behind a screen, despite the fact that they do track the visually identical collections when this evidence is not provided. For example, if infants see the apparent objects being disassembled and reassembled, or if they see the them come into existence by being poured from a beaker (Carey 1999) they fail to track them as individual objects. This could mean that whether or not something is treated as an object depends on prior knowledge (which would make them conceptual), or it may just mean that certain aspects of the recent visual history of the objects affect whether or not the visual system treats them as individual objects. What makes the latter at least a possibility is that the ability to track things in psychophysical experiments is also sensitive to the way they appear and disappear, as well as the pattern by which they move.

Several studies have shown that the precise manner in which objects disappear and reappear matters to whether or not they continue to be tracked (Scholl and Pylyshyn 1999). In particular, if their disappearance is by a pattern of deletion and accretion such as occurs when the object goes behind an occluding surface and reappears in a complementary manner (by disocclusion), then it continues to be tracked in a multiple-object tracking paradigm. But the effect of recent visual history is quite plausibly subsumed under the operation of a nonconceptual mechanism of the early vision system. This is consistent with the story I have been telling about how objects are selected and tracked, for I have not said what the time frame or temporal window is within which object properties are effective either in index-grabbing or in tracking; the immediate history (of being put down or being poured) may well be part of what determines whether the thing qualifies as a visual object (for other examples of what appear on the surface as knowledge-based phenomena but which can be understood as the consequence of a nonconceptual mechanism, see Pylyshyn 1999).

The central role that objects play in vision has another, perhaps deeper, consequence. The primacy of objects as the focus through which properties are encoded suggests a rather different way to view the role of objects in visual perception and cognition. Just as it is natural to think that we detect properties such as color and shape as properties of objects, so has it also been natural to think that we recognize and encode objects as a kind of property that particular places have. In other words, we usually think of the matrix of space-time as being primary and of objects as being occupants of places and times. Yet the ideas I have been discussing suggest an alternative and rather intriguing possibility: the notion that the primitive visual object is the primary and more primitive category of early (nonconceptual) vision. It may be that we detect objecthood first and then determine location the way we might determine color or shape—as a property associated with the detected objects. If this is true it raises some interesting possibilities concerning the nature of the mechanisms of early vision. In particular, it adds further credence to what I have argued is needed for independent reasons—some way of referring directly to primitive visual objects without using a unique description which that object satisfies. Perhaps this function can be served in part by the mechanism I referred to as a FINST index or a visual demonstrative (or a FINST).

Notice that what I have been describing is not the full concept of an individual physical object. The usual notion of a physical object, such as a particular table or chair or a particular individual person, does require concepts (in particular, it requires what are called *sortal concepts*) in order to establish criteria of identity, as many philosophers have argued (e.g., Hirsch 1982). The individual items that are picked out by the visual system and tracked primitively are something less than full-blooded individual objects. Yet because they are what our visual system gives us through a brute causal mechanism (because that is its nature), and also because what are picked out in this way are typically real objects in our kind of world, indexes may serve as the basis for real individuation of physical objects. While it is clear that you cannot individuate objects in the full-blooded sense without a conceptual apparatus, it is also clear that you cannot individuate them with only a conceptual apparatus. Sooner or later concepts must be grounded in a primitive causal connection between thoughts and things. The project of grounding concepts in sense data has not fared well and has been abandoned in cognitive science. However, the principle of grounding concepts in perception remains an essential requirement if we are not to succumb to an infinite regress. FINST indexes provide the needed grounding for basic objects—the individuals to which perceptual predicates apply, and hence about which cognitive judgments and plans of action are made. Without such a nonconceptual grounding our percepts and our thoughts would be disconnected from the real-world objects of those thoughts. With indexes we can think about things (I am sometimes tempted to call them FINGs since they are the *things* selected by FINSTs) without having any concepts of them: One might say that we can have demonstrative thoughts. We can think thoughts about *this* object without any description under which the object of that thought falls: You can pick out one speck among countless identical specks on a beach, for example. And because you can pick out that individual you can move your gaze to it or you can reach for it—your motor system cannot be commanded to reach for a red thing, only to reach for a particular individual (of course, the motor system eventually needs coordinates, but that function is established further downstream, rather than being part of the command issued by the cognitive system).

Needless to say, there are some details to be worked out, so this is a work in progress. But there are real problems to be solved in connecting visual representations to the world in the right way, and whatever the solution eventually turns out to be, it will have to respect a collection of facts, some of which are sketched here. Moreover, any visual or attentional mechanism that might be hypothesized for this purpose will have far-reaching implications, not only for theories of situated vision, but also for grounding the content of visual representations and perhaps for grounding perceptual concepts in general.

6 Addendum: Alternative Explanation of Multiple-Object Tracking

In his chapter in this volume Brian Scholl raises an objection to my account of the multiple-object tracking experiment. The alterative proposal is that tracking utilizes split attention, so no visual indexes or FINSTs are needed. Because this is a common view (e.g., Cavanagh and Alvarez 2005) I thought it might be worthwhile briefly to address this alternative proposal.

6.1 Tracking Objects and Tracking Sets

It is of course obvious that you don't have to remember a target's history going back to the beginning of a trial in order to track it. But even though we need not encode the history, the decision to call a particular object token a target connects to its role in the preceding instant, and that, in turn, connects through a chain of individuals and sets to the initial state. Having inferred that a particular object token is a target we can then "flush" the basis for that inference and move on to the next instant in time, just as Brian Scholl says.

But that leaves a puzzle: How do you know whether a particular object had been a target in the immediately preceding instant without tracking it? According to the alternative account this is done by determining whether it was a member of the *set of targets*. Here everyone seems to assume that we can keep track of a set without keeping track of its individual members. But how can we do that? Sometimes there are ways to do this because the set has properties that the individual members do not have, either because they are aggregate properties or they are relational properties. For example, we might be able to identify the targets by first identifying the set to which they belonged if all the targets were in the top right quadrant of the screen, or if they traveled in a rigid configuration. The most popular account of this sort is due to Steve Yantis (1992), who proposed that we could treat the set as a whole by imagining the targets being connected by an elastic band that forms a polygon—then we could track a single *distorting polygon* rather than the individual targets that form its vertices.

The trouble with polygon-tracking and related methods is that they only work if at each instant you *already know* (i.e., have some way to distinguish) which objects are the targets and therefore constitute the vertices of the polygon. The imagined elastic does not automatically wrap around the targets as it would if it were a real elastic attached to real objects; it only does so if you know which objects are the targets and wrap them accordingly. But the objects in MOT move in unpredictable independent trajectories, so in order to keep the elastic wrapped around the targets rather than be taken over by identical moving nontargets, we would have to first distinguish the individual targets from the nontargets. Although Brian may not wish to subscribe to that particular model of MOT, his view does require a similar sort of mechanism that keeps track of the targets as a set, rather than tracking the individual objects that constitute the set. It is this desideratum that leads him to propose that tracking is purely a phenomenon of divided attention. You place an attention beam on each target so each target is tracked individually. But if you now add the novel (and gratuitous) assumption that attention beams are indistinguishable, you get tracking-by-sets without access to individual targets. (It is not clear, by the way, why one couldn't add the same indistinguishability assumption to the FINST version, but it's not one that has an independent motivation).

6.2 Failing to Recall a Name Associated with a Target

As Brian points out, the set-tracking hypothesis fits well with our own data (Pylyshyn 2004) showing that recalling a particular identifier (e.g., a number or name) that had been associated with a target is much harder than simply recalling that it had been a target. In examining this finding we found evidence that the chances of attributing a particular target identifier to the wrong target were significantly higher than attributing it to a nontarget. We postulated that this asymmetry was due to the inhibition of nontargets—a hypothesis for which we subsequently found independent evidence. I now believe that there is very likely more going on in this surprising phenomenon than just index-switching. But as Brian points out, such identification errors seem natural on the account of MOT that assumes that we track sets through split (and unmarked) beams of attention and thus fail to distinguish among members of the set. However, we pay a heavy price for this naturalness since any set-tracking option not only fails to distinguish among the targets, it also skirts the notion of individual entirely and so cannot account for the wide range of empirical phenomena I have discussed here (as well as in chapters 4 and 5 of Pylyshyn 2003). In addition, since one of the main functions of focal attention is to allocate resources in order to facilitate property detection, one would not expect the tracking task to be so insensitive to object properties (as reported in Bahrami 2003; Scholl, Pylyshyn, and Franconeri 1999b).

Recall the many purposes for which FINST indexes were postulated— including distinguishing parts in recognizing patterns (using visual routines) and solving the binding problem (i.e., determining when several visual features are features of the same object). If you cannot distinguish the different attention beams you cannot associate a property with a particular object (as in the study of object-specific priming; see Noles, Scholl, and Mitroff 2005). Such faceless attention beams appear to be little more than FINSTs without token distinctiveness or the pointer function. If you allow them to have these functions then you have FINSTs by another name—a name that, unfortunately, merges them with focal attention and so misses the special feature of FINSTs, such as their failure to encode object properties and their important nonconceptual nature. Though many psychologists may not care about the latter, it is an issue that has been preoccupying me more in recent years (and which I address in Pylyshyn 2007). It's also the sort of issue that cognitive science, as an interdisciplinary pursuit, was intended to address.

Notes

1. I use the term "element" when referring to a graphical unit such as used in experiments. Otherwise when speaking informally I use the term "thing," on the grounds that nobody would mistake that term for a technical theoretical construct. Eventually I end up calling them "visual objects" to conform to usage in psychology.

2. Even visual concepts, like perceived shape, cannot be specified in terms of transducer outputs (see Pylyshyn 2003, chapter 1). Julian Hochberg spent years searching for the geometrical basis of pattern complexity but gave up on the grounds that it was the form of the representation and not the form of the objective stimulus that mattered (see Hochberg 1968).

3. For details see Pylyshyn 2003 and the experimental reports cited there or in more recent reports such as: Pylyshyn 2004, 2006; Pylyshyn and Annan 2002.

4. The reader will have noticed that this way of putting it makes the reference mechanism appear to be a *name* (in fact the name "α"). What I have in mind is very like a proper name insofar as it allows reference to a particular individual. However, this reference relation is less general than a name since it ceases to exist when the referent is no longer in view. In that respect it functions like a demonstrative, which is why I continue to call it that, even as I use examples involving names like α.

5. There have been well over a hundred studies in our laboratory alone (Annan and Pylyshyn 2002; Blaser, Pylyshyn, and Holcombe 2000; Keane and Pylyshyn 2006; Pylyshyn 2004, 2006; Pylyshyn and Annan 2002; Scholl, Pylyshyn, and Feldman 2001), as well as in other laboratories (Allen et al. 2004; Alvarez et al. 2005; Alvarez and Scholl 2005; Bahrami 2003; Cavanagh 1992; Cavanagh and Alvarez 2005; Chiang and Wynn 2000; Horowitz et al. 2006; Liu et al. 2005; Ogawa and Yagi 2002; O'Hearn, Landau, and Hoffman 2005; Oksama and Hyona 2004; Suganuma and Yokosawa 2002; Trick, Perl, and Sethi 2005; vanMarle and Scholl 2003; Viswanathan and Mingolla 2002; Yantis 1992), that have replicated these multiple-object tracking results using a variety of methods, confirming that observers can successfully track around 4 or 5 independently moving objects. In a set of unpublished studies (Scholl, Pylyshyn, and Franconeri 1999a) we showed that observers do not notice and cannot report changes of color or shape of objects they are tracking when the change occurs while they are behind an occluder or during a short period of blank screen, thus lending credence to the view that properties are ignored during tracking. This was confirmed independently by Bahrami (2003), who showed that observers cannot detect changes in color or shape of either nontargets or targets while tracking.

References

Acton, B. (1993). A network model of visual indexing and attention. Unpublished MSc. thesis, University of Western Ontario, London, Ontario, Canada.

Allen, R., P. McGeorge, D. Pearson, and A. B. Milne (2004). Attention and expertise in multiple target tracking. *Applied Cognitive Psychology* 18: 337–347.

Alvarez, G. A., H. C. Arsenio, T. S. Horowitz, and J. M. Wolfe (2005). Are multielement visual tracking and visual search mutually exclusive? *Journal of Experimental Psychology: Human Perception and Performance* 31(4): 643–667.

Alvarez, G. A., and B. J. Scholl (2005). How does attention select and track spatially extended objects? New effects of attentional concentration and amplification. *Journal of Experimental Psychology: General* 134(4): 461–476.

Annan, V., and Z. W. Pylyshyn (2002). Can indexes be voluntarily assigned in multiple object tracking? *Journal of Vision* 2(7): 243a.

Bahrami, B. (2003). Object property encoding and change blindness in multiple object tracking. *Visual Cognition* 10(8): 949–963.

Ballard, D. H., M. M. Hayhoe, P. K. Pook, and R. P. N. Rao (1997). Deictic codes for the embodiment of cognition. *Behavioral and Brain Sciences* 20(4): 723–767.

Baylis, G. C., and J. Driver (1993). Visual attention and objects: Evidence for hierarchical coding of location. *Journal of Experimental Psychology: Human Perception and Performance* 19: 451–470.

Blaser, E., Z. W. Pylyshyn, and A. O. Holcombe (2000). Tracking an object through feature-space. *Nature* 408(Nov. 9): 196–199.

Brooks, R. A. (1991). Intelligence without representation. *Artificial Intelligence* 47: 139–159.

Burkell, J., and Z. W. Pylyshyn (1997). Searching through subsets: A test of the visual indexing hypothesis. *Spatial Vision* 11(2): 225–258.

Calis, G. J., J. Sterenborg, and F. Maarse (1984). Initial microgenetic steps in single-glance face recognition. *Acta Psychologica* 55(3): 215–230.

Campbell, J. (2002). *Reference and Consciousness*. New York: Oxford University Press.

Campbell, J. (2004). Reference as attention. *Philosophical Studies* 120: 265–276.

Carey, S. (1999). Establishing representations of new individuals: New infant results and old studies by Michotte. Paper presented at Object Cognition: Underlying Mechanisms and Their Origins (May 20–21), Rutgers University, New Brunswick, New Jersey.

Cavanagh, P. (1992). Attention-based motion perception. *Science* 257: 1563–1565.

Cavanagh, P., and G. A. Alvarez (2005). Tracking multiple targets with multifocal attention. *Trends in Cognitive Sciences* 9(7): 349–354.

Chiang, W.-C., and K. Wynn (2000). Infants' tracking of objects and collections. *Cognition* 75: 1–27.

Clark, A. (1999). An embodied cognitive science? *Trends in Cognitive Sciences* 3(9): 345–351.

Clark, A. (2000). *A Theory of Sentience*. New York: Oxford University Press.

Currie, C. B., and Z. W. Pylyshyn (2003). *Maintenance of FINSTs across Eye Movements*. Unpublished ms available at http://ruccs.rutgers.edu/~zenon/ccurrie/TitlePage .html.

Driver, J., and P. Halligan (1991). Can visual neglect operate in object-centered coordinates? An affirmative single case study. *Cognitive Neuropsychology* 8: 475–494.

Egly, R., J. Driver, and R. D. Rafal (1994). Shifting visual attention between objects and locations: Evidence from normal and parietal lesion subjects. *Journal of Experimental Psychology: General* 123(2): 161–177.

Frohlich, W. D., and L. Laux (1969). Sequential perception, microgenesis, integration of information, and orienting reactions: I. Actual genetic model and orientation reaction. *Zeitschrift für Experimentelle und Angewandte Psychologie* 16(2): 250–277.

He, S., P. Cavanagh, and J. Intriligator (1997). Attentional resolution. *Trends in Cognitive Sciences* 1(3): 115–121.

Hirsch, E. (1982). *The Concept of Identity*. Oxford: Oxford University Press.

Hochberg, J. (1968). In the mind's eye. In *Contemporary Theory and Research in Visual Perception*, ed. R. N. Haber, 309–331. New York: Holt, Rinehart, and Winston.

Horowitz, T. S., R. S. Birnkrant, D. E. Fencsik, L. Tran, and J. M. Wolfe (2006). How do we track invisible objects? *Psychonomic Bulletin and Review*.

Intriligator, J., and P. Cavanagh (2001). The spatial resolution of attention. *Cognitive Psychology* 4(3): 171–216.

Irwin, D. E. (1992). Memory for position and identity across eye movements. *Journal of Experimental Psychology: Learning, Memory, and Cognition* 18(2): 307–317.

Jackson, F. (1997). *Perception: A Representative Theory*. Cambridge: Cambridge University Press.

Kahneman, D., A. Treisman, and B. J. Gibbs (1992). The reviewing of object files: Object-specific integration of information. *Cognitive Psychology* 24(2): 175–219.

Kazanovich, Y., and R. Borisyuk (2006). An oscillatory neural model of multiple object tracking. *Neural Computation* 18(6): 1413–1440.

Keane, B. P., and Z. W. Pylyshyn (2006). Is motion extrapolation employed in multiple object tracking? Tracking as a low-level, non-predictive function. *Cognitive Psychology* 52(4): 346–368.

Kimchi, R. (2000). The perceptual organization of visual objects: A microgenetic analysis. *Vision Research* 40(10–12): 1333–1347.

Koch, C., and S. Ullman (1985). Shifts in selective visual attention: Towards the underlying neural circuitry. *Human Neurobiology* 4: 219–227.

Lepore, E., and K. Ludwig (2000). The semantics and pragmatics of complex demonstratives. *Mind* 109: 199–240.

Leslie, A. M., F. Xu, P. D. Tremolet, and B. J. Scholl (1998). Indexing and the object concept: Developing "what" and "where" systems. *Trends in Cognitive Sciences* 2(1): 10–18.

Liu, G., E. L. Austen, K. S. Booth, B. D. Fisher, R. Argue, M. I. Rempel, and J. T. Enns (2005). Multiple-object tracking is based on scene, not retinal, coordinates. *Journal of Experimental Psychology: Human Perception and Performance* 31(2): 235–247.

Marr, D. (1982). *Vision: A Computational Investigation into the Human Representation and Processing of Visual Information*. San Francisco: W. H. Freeman.

McDowell, J. (1994). *Mind and World*. Cambridge, Mass.: Harvard University Press.

Nakatani, K. (1995). Microgenesis of the length perception of paired lines. *Psychological Research* 58(2): 75–82.

Navon, D. (1977). Forest before trees: The precedence of global features in visual perception. *Cognitive Psychology* 9: 353–383.

Nesmith, R., and A. S. Rodwan (1967). Effect of duration of viewing on form and size judgments. *Journal of Experimental Psychology* 74(1): 26–30.

Nissen, M. J. (1985). Accessing features and objects: Is location special? In *Attention and Performance XI*, ed. M. I. Posner and O. S. Marin, 205–219. Hillsdale, N.J.: Lawrence Erlbaum.

Noles, N. S., B. J. Scholl, and S. R. Mitroff (2005). The persistence of object file representations. *Perception and Psychophysics* 67(2): 324–334.

Ogawa, H., and A. Yagi (2002). The effect of information of untracked objects on multiple object tracking. *Japanese Journal of Psychonomic Science* 22(1): 49–50.

O'Hearn, K., B. Landau, and J. E. Hoffman (2005). Multiple object tracking in people with Williams syndrome and in normally developing children. *Psychological Science* 16(11): 905–912.

Oksama, L., and J. Hyona (2004). Is multiple object tracking carried out automatically by an early vision mechanism independent of higher-order cognition? An individual difference approach. *Visual Cognition* 11(5): 631–671.

Parks, T. E. (1995). The microgenesis of illusory figures: Evidence for visual hypothesis testing. *Perception* 24(6): 681–684.

Pashler, H. E. (1998). *The Psychology of Attention*. Cambridge, Mass.: MIT Press/A Bradford Book.

Perry, J. (1979). The problem of the essential indexical. *Noûs* 13: 3–21.

Pylyshyn, Z. W. (1984). *Computation and Cognition: Toward a Foundation for Cognitive Science*. Cambridge, Mass.: MIT Press.

Pylyshyn, Z. W. (1989). The role of location indexes in spatial perception: A sketch of the FINST spatial-index model. *Cognition* 32: 65–97.

Pylyshyn, Z. W. (1994). Some primitive mechanisms of spatial attention. *Cognition* 50: 363–384.

Pylyshyn, Z. W. (1998). Visual indexes in spatial vision and imagery. In *Visual Attention*, ed. R. D. Wright, 215–231. New York: Oxford University Press.

Pylyshyn, Z. W. (1999). Is vision continuous with cognition? The case for cognitive impenetrability of visual perception. *Behavioral and Brain Sciences* 22(3): 341–423.

Pylyshyn, Z. W. (2003). *Seeing and Visualizing: It's Not What You Think.* Cambridge, Mass.: MIT Press/A Bradford Book.

Pylyshyn, Z. W. (2004). Some puzzling findings in multiple object tracking (MOT): I. Tracking without keeping track of object identities. *Visual Cognition* 11(7): 801–822.

Pylyshyn, Z. W. (2006). Some puzzling findings in multiple object tracking (MOT): II. Inhibition of moving nontargets. *Visual Cognition* 14(2): 175–198.

Pylyshyn, Z. W. (2007). *Things and Places: How the Mind Connects with the World.* Cambridge, Mass.: MIT Press/A Bradford Book.

Pylyshyn, Z. W., and V. J. Annan (in press). Dynamics of target selection in multiple object tracking (MOT). *Spatial Vision.*

Pylyshyn, Z. W., and R. A. Eagleson (1994). Developing a network model of multiple visual indexing (abstract). *Investigative Ophthalmology and Visual Science* 35(4): 2007.

Pylyshyn, Z. W., E. W. Elcock, M. Marmor, and P. Sander (1978). Explorations in visual-motor spaces. Paper presented at the Second International Conference of the Canadian Society for Computational Studies of Intelligence, University of Toronto.

Pylyshyn, Z. W., and R. W. Storm (1988). Tracking multiple independent targets: Evidence for a parallel tracking mechanism. *Spatial Vision* 3(3): 1–19.

Rensink, R. A. (2000). Visual search for change: A probe into the nature of attentional processing. *Visual Cognition* 7: 345–376.

Rensink, R. A., J. K. O'Regan, and J. J. Clark (1997). To see or not to see: The need for attention to perceive changes in scenes. *Psychological Science* 8(5): 368–373.

Rensink, R. A., J. K. O'Regan, and J. J. Clark (2000). On the failure to detect changes in scenes across brief interruptions. *Visual Cognition* 7: 127–145.

Reynolds, R. I. (1978). The microgenetic development of the Ponzo and Zoellner illusions. *Perception and Psychophysics* 23(3): 231–236.

Robertson, L., A. Treisman, S. Friedman-Hill, and M. Grabowecky (1997). The interaction of spatial and object pathways: Evidence from Balint's syndrome. *Journal of Cognitive Neuroscience* 9(3): 295–317.

Schlottman, A., and D. R. Shanks (1992). Evidence for a distinction between judged and perceived causality. *Quarterly Journal of Experimental Psychology A*, 2: 321–342.

Scholl, B. J. (2001). Objects and attention: The state of the art. *Cognition* 80(1/2): 1–46.

Scholl, B. J., and Z. W. Pylyshyn (1999). Tracking multiple items through occlusion: Clues to visual objecthood. *Cognitive Psychology* 38(2): 259–290.

Scholl, B. J., Z. W. Pylyshyn, and J. Feldman (2001). What is a visual object: Evidence from target-merging in multiple-object tracking. *Cognition* 80: 159–177.

Scholl, B. J., Z. W. Pylyshyn, and S. L. Franconeri (1999a). *The Relationship between Property-Encoding and Object-Based Attention: Evidence from Multiple-Object Tracking.* Unpublished manuscript.

Scholl, B. J., Z. W. Pylyshyn, and S. L. Franconeri (1999b). When are featural and spatiotemporal properties encoded as a result of attentional allocation? *Investigative Ophthalmology and Visual Science* 40(4): 4195.

Sekuler, A. B., and S. E. Palmer (1992). Visual completion of partly occluded objects: A microgenetic analysis. *Journal of Experimental Psychology: General* 121: 95–111.

Simons, D. J. (1996). In sight, out of mind: When object representations fail. *Psychological Science* 7(5): 301–305.

Simons, D. J., and D. T. Levin (1997). Change blindness. *Trends in Cognitive Sciences* 1: 261–267.

Spelke, E. S. (1990). Principles of object perception. *Cognitive Science* 14: 29–56.

Spelke, E. S., G. Gutheil, and G. Van de Walle (1995). The development of object perception. In *Visual Cognition*, second ed., ed. S. M. Kosslyn and D. N. Osherson, vol. 2, 297–330. Cambridge, Mass.: MIT Press.

Sperling, G., and E. Weichselgarter (1995). Episodic theory of the dynamics of spatial attention. *Psychological Review* 102(3): 503–532.

Strawson, P. F. (1963). *Individuals: An Essay in Descriptive Metaphysics.* New York: Anchor Books.

Suganuma, M., and K. Yokosawa (2002). Is multiple object tracking affected by three-dimensional rigidity? Paper presented at the Vision Sciences Society, Sarasota, Florida.

Tipper, S., J. Driver, and B. Weaver (1991). Object-centered inhibition of return of visual attention. *Quarterly Journal of Experimental Psychology A,* 43: 289–298.

Tipper, S. P., B. Weaver, L. M. Jerreat, and A. L. Burak (1994). Object-based and environment-based inhibition of return of selective attention. *Journal of Experimental Psychology: Human Perception and Performance* 20: 478–499.

Treisman, A. (1995). Modularity and attention: Is the binding problem real? In *Visual Selective Attention,* ed. C. Bundesen and H. Shibuya. Hillsdale, N.J.: Lawrence Erlbaum.

Treisman, A., and G. Gelade (1980). A feature integration theory of attention. *Cognitive Psychology* 12: 97–136.

Trick, L. M., T. Perl, and N. Sethi (2005). Age-related differences in multiple-object tracking. *Journals of Gerontology: Series B: Psychological Sciences and Social Sciences* 2: 102.

Trick, L. M., and Z. W. Pylyshyn (1994). Why are small and large numbers enumerated differently? A limited capacity preattentive stage in vision. *Psychological Review* 101(1): 80–102.

Tsotsos, J. K. (1988). How does human vision beat the computational complexity of visual perception. In *Computational Processes in Human Vision: An Interdisciplinary Perspective,* ed. Z. W. Pylyshyn, 286–340. Norwood, N.J.: Ablex Publishing.

Tucker, V., and K. D. Broota (1985). Effect of exposure duration on perceived size. *Psychological Studies* 30(1): 49–52.

Ullman, S. (1984). Visual routines. *Cognition* 18: 97–159.

vanMarle, K., and B. J. Scholl (2003). Attentive tracking of objects versus substances. *Psychological Science* 14(4): 498–504.

Viswanathan, L., and E. Mingolla (2002). Dynamics of attention in depth: Evidence from multi-element tracking. *Perception* 31(12): 1415–1437.

Watson, D. G., and G. W. Humphreys (1997). Visual marking: Prioritizing selection for new objects by top-down attentional inhibition of old objects. *Psychological Review* 104(1): 90–122.

Wolfe, J. M., K. R. Cave, and S. L. Franzel (1989). Guided search: An alternative to the feature integration model for visual search. *Journal of Experimental Psychology: Human Perception and Performance* 15(3): 419–433.

Xu, F. (1997). From Lot's wife to a pillar of salt: Evidence that *physical object* is a sortal concept. *Mind and language* 12: 365–392.

Xu, F., and S. Carey (1996). Infants' metaphysics: The case of numerical identity. *Cognitive Psychology* 30: 111–153.

Yantis, S. (1992). Multielement visual tracking: Attention and perceptual organization. *Cognitive Psychology* 24: 295–340.

Yantis, S. (1998). Objects, attention, and perceptual experience. In *Visual Attention*, ed. R. Wright, 187–214. Oxford: Oxford University Press.

Yantis, S., and E. Jones (1991). Mechanisms of attentional selection: Temporally modulated priority tags. *Perception and Psychophysics* 50(2): 166–178.

2 What Have We Learned about Attention from Multiple-Object Tracking (and Vice Versa)?

Brian J. Scholl

1 Introduction

If you weren't paying attention, you could be forgiven for thinking that this chapter was part of a collection assembled in honor of several people named Zenon Pylyshyn: the philosopher of psychology who has helped define the relation between mind and world; the computer scientist who has characterized the power of computation in the study of cognition; the cognitive psychologist whose imagery research is in every introductory textbook; and the vision scientist whose ideas and experimental paradigms form a foundation for work in visual cognition. (When I first learned of "Zenon Pylyshyn" in college, I figured that this couldn't really be someone's name, and given the breadth and importance of his contributions I figured that "he" must be some sort of research collective—a Nicolas Bourbaki of cognitive science. I was lucky to have been able to study later with this excellent research collective in graduate school, though I discovered that it was housed in one head.)

This chapter is about the last of the Zenons noted above: the vision scientist. In the study of visual cognition, his lasting influence has stemmed in part from the way that he has bucked one of the most dangerous trends in experimental research: whereas most of us too easily fall into the trap of constructing theoretical questions to fit our experimental paradigms, Zenon has consistently managed the reverse. And there is perhaps no better example of this than his development of the multiple-object tracking (henceforth MOT) paradigm. This chapter focuses on the nature of MOT, with three interrelated goals: (1) to explore what makes MOT unique—and uniquely useful—as a tool for studying visual cognition; (2) to characterize the relationship between attention and MOT; and (3) to highlight some of the important things we've learned about attention from the study of MOT—and vice versa.

2 Multiple-Object Tracking

Perhaps the most active area in visual cognition research in the last few decades has been the study of attention. Attention seems to involve a perceptual resource that can both intentionally and automatically select—and be effortfully sustained on—particular stimuli or activities. The core aspects of attention comprise three phenomena (Pashler 1998): (1) the fact that we can process some incoming stimuli more so than others (selectivity), (2) an apparent limitation on the ability to carry out simultaneous processing (capacity-limitation), and (3) the fact that sustained processing of visual stimuli seems to involve a sense of exertion (effort).

There is no paradigm that more viscerally illustrates these three components of attention than MOT (Pylyshyn and Storm 1988). One of the appeals of MOT is that at root it is a very simple task. In a typical experiment (see figure 2.1), observers initially see a number of identical objects. A subset of these are then flashed to indicate their status as targets, after which all of the (again identical) objects begin moving independently and unpredictably about the display. When they stop moving, observers must indicate which of the objects are the original targets.

2.1 What Makes MOT Special?

This procedure contrasts with most other paradigms that have been used to study attention in several ways. First, MOT requires continuous sus-

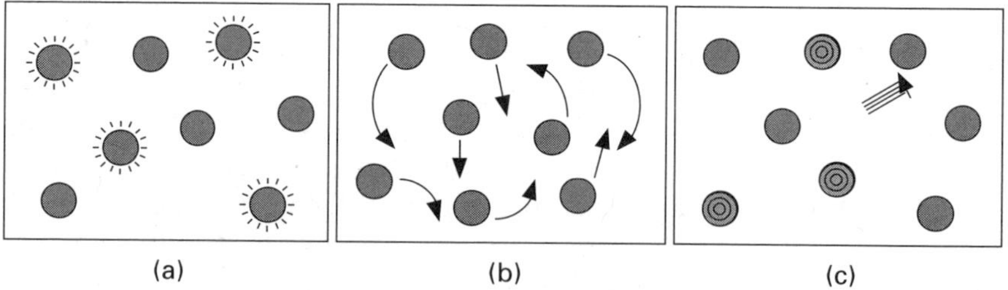

Figure 2.1

A schematic depiction of multiple object tracking. (a) Four items are initially flashed to indicate their status as targets. (b) All items then begin moving independently and unpredictably around the display. (c) At the end of the motion phase, the subject must move the cursor about the screen to highlight the four targets—here the subject has just highlighted three of the targets, and is moving the mouse to the fourth. Animations of many different variants of this task—including those of all the figures in this chapter—can be viewed at or downloaded from http://www.yale.edu/perception/.

tained attention over time rather than brief attentional shifts (as in spatial cueing studies). Second, MOT involves attention to multiple objects rather than focal attention to only a single object at a time (as in most attentional capture studies). Third, MOT is an inherently active task, rather than requiring mere passive vigilance (e.g., when waiting for a target to appear). Fourth, the magnitude of the attentional demands in MOT can be directly manipulated in terms of the underlying tracking load, rather than via indirect temporal manipulations (e.g., as used in the brief masked displays of many divided-attention experiments). Moreover, it is worth noting that each of these features is characteristic of real-world visual cognition: day-to-day experience is filled with situations—driving, hunting, sports, or even just trying to cross a street—that call for sustained attention to multiple objects over time and motion. As such, MOT has proven to be one of the most useful tools in the study of attention. (As a bonus, MOT typically yields relatively large and robust effects, making it ideal for studies that need to distinguish several different levels of performance, beyond simply demonstrating that various attentional effects do or do not exist.)

Perhaps the most central result in the study of MOT is simply that it is possible in the first place. As Pylyshyn and his colleagues have noted, this was not a foregone conclusion, given that classical theories of attention tended to assume a single unitary "spotlight" of selection. Since targets and distractors are spatially interleaved in MOT, though, the only natural way for a unitary spotlight to succeed would be if it cycled repeatedly from target to target, storing and updating their "last known addresses." This possibility seems implausible on its face, given the phenomenology of MOT: It certainly does not seem as if one's attention is constantly cycling around to different targets (though of course it is possible to attend to each of the objects independently or to consider them as a single global deform-ing shape—e.g., as a deforming polygon with targets at the corners; Yantis 1992). This is not an entirely empty point, perhaps, given the tight relation between attention and awareness (see Most et al. 2005): In most situations, you are at least somewhat aware of how and where you are attending. At the same time, however, phenomenology is often a poor guide to the underlying nature of the mind, and so that alone cannot definitively rule out a "single roving spotlight" explanation.

The initial report of MOT, however, effectively ruled out single-spotlight explanations via additional computational modeling results (Pylyshyn and Storm 1988). This model focused on how well a particular single-spotlight model could do when faced with actual MOT trajectories, when the spot-light was constrained to move at physiologically plausible speeds. Even given very generous assumptions about such speeds, the central result of this modeling project was that single-spotlight performance could never match actual human tracking abilities for those same trajectories. Moreover, this is true even when the spotlight is made as intelligent as we can think to make it—for example, employing subtle heuristics that involve extrapolating objects' trajectories over multiple temporal scales, and prioritizing objects in locally dense regions of the display from moment to moment (Chan et al., in preparation). The reasonable conclusion is that the underlying architecture of MOT must involve parallel selection and tracking—perhaps including up to four separate loci of attention, which might then directly explain the fact that tracking suffers beyond this number of targets (see Hulleman 2005).

2.2 MOT as a Phenomenon and a Paradigm

Since its introduction, MOT has been used in many different studies of visual cognition. Some of this work has focused on MOT as a phenomenon in its own right, exploring its constraints and underlying processes. For example, research has characterized how the ability to track multiple objects is influenced by the number of targets (Oksama and Hyönä 2004), their speeds (Liu et al. 2005), their relative depths (Viswanathan and Mingolla 2002), the reference frame in which they move (Liu et al. 2005), individual differences (Oksama and Hyönä 2004), and various higher-level strategies that may be employed by observers (Yantis 1992). Perhaps the most surprising result to come out of this larger research project is the discovery that MOT seems not to involve much extrapolation of objects' trajectories: with only a few exceptions (Franconeri, Pylyshyn, and Scholl 2006), observers are better at MOT when objects that have disappeared reappear at their last known addresses, rather than where they "should" be had their motions continued (Fencsik et al. 2007; Franconeri, Pylyshyn, and Scholl 2006; Keane and Pylyshyn 2006).

MOT has also been frequently used as a tool with which to study other aspects of visual cognition. This work may not necessarily depend on any details about how MOT does or does not work, but simply employs it to manipulate attention in the study of other topics, such as working memory (Fougnie and Marois 2006; Postle, D'Esposito, and Corkin 2005), task switching (Alvarez et al. 2005), spatial resolution (Intriligator and Cavanagh 2001), occlusion (Flombaum, Scholl, and Pylyshyn 2008; Scholl and Pylyshyn 1999), dual-task interference (Allen et al. 2004, 2006; Fougnie and Marois 2006; Trick, Guindon, and Vallis 2006), or even self-regulation (Oaten and Cheng 2006). More generally, MOT has been used as a tool to study the operation of attention in many different populations, including young children (O'Hearn, Landau, and Hoffman 2005), older adults (Trick, Guindon, and Vallis 2005; Sekuler, McLaughlin, and Yotsumoto 2008), special populations (O'Hearn, Landau, and Hoffman 2005), and visual experts such as radar operators (Allen et al. 2004) and videogame players (Green and Bavelier 2006).

The goal of this chapter is to explore MOT as both a phenomenon and a paradigm, focusing on how it interacts with visual attention.

3 The Relationship between MOT and Attention

A common assumption is that MOT is an illustration of the dynamics of attention; indeed, it is sometimes even referred to as simply "attentive tracking" (e.g., Fougnie and Marois 2006; vanMarle and Scholl 2003) or "multifocal attention" (e.g., Cavanagh and Alvarez 2005). However, the relationship between MOT and attention in Pylyshyn's own work is more subtle.

3.1 Visual Indexing

As noted in the introduction to this chapter, Pylyshyn initially created (discovered?) MOT for a specific theoretical purpose. In order to detect even simple geometrical properties among the elements of a visual scene (e.g., being collinear, or being "inside"), he argues, the visual system must be able to simultaneously reference—or "index"—multiple objects in parallel, and to maintain that referential contact over time. This indexing is even necessary to shift attention to an object, since you can't shift attention *to* anything unless you are already referencing it. Pylyshyn noted that this visual indexing theory (e.g., Pylyshyn 1989, 1994, 2001, 2003, 2007) predicted that something like MOT should be possible, and so he created the paradigm in order to test this prediction.

In Pylyshyn's theory, visual indexes (or "FINSTs," for FINgers of INSTantiation, by analogy to pointing fingers) are independently assigned to various items in the visual field on the basis of bottom-up salience cues, and the indexes serve as a means of access to those items for the higher-level processes that allocate focal attention. In this regard, they function like pointers in a computer data structure: They reference certain items in the visual field (identifying them as distinct objects), without themselves encoding any properties of those objects. Indexes are thought to be assigned to objects in the visual field regardless of their spatial contiguity (in contrast with spotlight models), but with the restriction that the architecture of the visual system provides only a limited number of indexes (roughly four). Furthermore, the indexes are sticky: If an indexed item in the visual field moves, the index moves with it, maintaining the referential connection.

3.2 Indexing and Attention

A key assumption of the indexing theory has been that (at least part of) the assignment and maintenance of indexes—that is, the selection of targets and the actual tracking in MOT—is preattentive, automatic, and data driven. This is a key assumption because it underlies the entire reason for indexing in the first place. This aspect of the proposal serves to link visual processing up with the world, providing an exit from the regress in which various representational systems are explained in terms of other representational systems. If a significant portion of the indexing process is truly data driven, then indexing might serve as a sort of interface between the world and the mind, and could underlie higher-level types of object-based processing. In the words of Fodor (this volume), indexing "is where the intentional gets its grip on the physical; it's where psychology starts to get 'naturalized'" (xiii). But indexing can't serve this function unless it operates at least in part at a lower level than attention.

The strongest form of this assumption would be that MOT is entirely preattentive, but this view is clearly wrong. For example, without other assumptions this view is inconsistent with the basic finding that MOT decays with longer tracking durations, and that it is subject to large individual differences that correlate with other aspects of attention (Oksama and Hyönä 2004). However, this no-attention view is (and always has been) a straw man. The initial presentation of MOT (Pylyshyn and Storm 1988) did explicitly suggest that some of the actual tracking was preattentive: the "stage . . . that maintains the identity of a visual feature as it moves about in the visual field" can be "shown to have more than one independent locus and may thus actually be a 'preattentive' stage" (180). However, even this initial report noted that attention was likely to be involved in MOT in other ways—that indexing "is a preattentive operation, although the selection of some subset of these automatically indexed places for . . . tracking may involve deliberate cognitive intervention" (181).

Later discussions of visual indexing have helped to clarify this view (Pylyshyn 1994, 2001, 2003, 2007). These discussions have maintained the view that the actual tracking is in part an automatic and preattentive function (such that tracking is "primitive" and a part of "early vision"), but they have noted that MOT may nevertheless be effortful and attentionally demanding, since indexes may have to "be periodically refreshed" to prevent decay (Pylyshyn 1994, 369), since the task "requires effort inasmuch as it involves warding off competing events" (Pylyshyn et al. 1994, 266) and since observers must also employ an "error recovery stage" to rescue "lost" objects during motion (Sears and Pylyshyn 2000). In sum, "more is going on in tracking tasks than the mere invocation of an automatic tracking mechanism" (Pylyshyn et al. 1994, 266). This is also true for the initial assignment of the indexes in MOT: while the theory has always maintained—as it must, given its purpose—that indexes can be assigned in an automatic and data-driven manner, this is not exclusive, and indexes can also be assigned deliberately via focused attention (Pylyshyn and Annan 2006). In some ways, of course, this has to be true, since the MOT task is a task, and like all tasks it involves central executive resources involving goal maintenance, response selection, and performance monitoring. This can also be easily demonstrated experimentally, since for example MOT interferes with even very general tasks involving auditory tone-monitoring (e.g., Alvarez et al. 2005).

3.3 What If There's No More to MOT Than Attention? How Could We Tell? And Where Is the Burden of Proof?

In noting the ways in which attention may influence MOT, Pylyshyn has commented that "it is clear that more is going on in MOT experiments than just tracking based on data-driven index maintenance" (Pylyshyn 2001, 149). Here I would like to turn this question around: given that the role of attention in MOT is so salient (even phenomenologically), is it clear that there is any more going on in MOT experiments than the application of attention itself? Put more bluntly, is there reason to think that there is any data-driven index maintenance in MOT? (Note that Pylyshyn has employed compelling conceptual arguments to suggest that there must be some data-driven system of visual demonstrative reference in the mind in order to "get vision off the ground," but of course that doesn't mean that it plays a role in this particular task, despite its provenance.) Here I propose that although MOT may have taught us several important things about attention, there may be nothing to MOT beyond attention.

This is a difficult view to defend, simply because it is not obvious how one could falsify the possibility that data-driven index maintenance is involved at some stage. After all, any example of an attentional effect on MOT can be easily (and perhaps too easily) deflected to some other aspect of the global "task"—for example, to an "error recovery stage" or to response selection—without any preexisting constraints on when and how such stages should and should not operate. Moreover, since MOT is "interruptible"—you can do other things for up to at least several hundred milliseconds while you ignore tracking (Alvarez et al. 2005)—any attentional effects during MOT could also always be argued to reflect additional processing that simply occurred "in between" periods of data-driven tracking. In short, to borrow a phrase from a recent study of individual differences in this task (Oksama and Hyönä 2004), in order to evaluate whether MOT involves anything other than attention, one would need a measure of "pure tracking"—but such a measure has never been developed.

Where this leaves us depends on where one thinks the "burden of proof" lies. Pylyshyn has always been clear on this issue: because the indexing theory is a bold attempt to "ground" cognition in a type of brute demonstrative reference (see especially Pylyshyn 2001, 2007), it is worth taking seriously. The view that there is some "pure tracking," in other words, is "the more interesting hypothesis to pursue, pending evidence to the contrary" (Pylyshyn 2001, 149). However, though it may be true that this is a good reason for "pursuing" the hypothesis, I question whether this is a good reason for (even provisionally) accepting the hypothesis. We already know that attention exists from countless studies, that it can be "split" under several circumstances (e.g., Cassidy, Sheremata, and Somers 2007; Castiello and Umiltà 1992; Driver and Baylis 1989; Kramer and Hahn 1995; McMains and Somers 2004), and that it can move (e.g., Cavanagh 1992; Driver and Baylis 1989; Verstraten, Cavanagh, and Labianca 2000). Meanwhile, the visual indexing view proposes an entirely new mechanism of mind—one without a large body of independent supporting evidence, and without any independent evidence for involvement in MOT. So, I suggest, we should prefer the attentional theory of MOT simply on the grounds of parsimony, without some positive evidence for the involvement of a novel "extra" mechanism. Of course, on this view it may still be important to pursue the possibility that visual indexing exists and is involved in MOT, but we should not start from that position without such evidence.

3.4 Is There Evidence against Indexing in MOT? Tracking Individuals versus Sets

As noted above, it is not clear how the hypothesis that indexing is involved in MOT could be directly tested and potentially refuted. As such, in this section I will argue against the involvement of indexing in MOT (and thus argue indirectly for the view that MOT is realized only by attentional tracking) in a different way, by emphasizing an aspect of MOT that seems inconsistent with the purpose of indexing. And, in what is perhaps an unorthodox move, I will make this argument based on one of Pylyshyn's own recent discoveries about MOT.

One of the key assumptions about MOT since its initial discovery has been that each target object is being tracked as a distinct individual: during tracking one is keeping track of *this* target, *that* target, and *that* target as each moves about the display. Recently, however, Pylyshyn (2004) noted an apparent challenge to this view. This challenge can be readily appreciated by any observer in the following way: During the initial target phase, internally name each of the targets. (If you try this using online MOT movies, you can also simply pause the movie during this phase.) For example, if you must track four of eight objects, think of the four targets as A, B, C, and D. Then, at the end of the tracking interval (when your task would normally be to indicate the four targets), give yourself the additional following task: Identify which is which. What you will find is that this is extremely difficult—and is certainly much more challenging than the basic MOT task. Indeed, when you've accurately tracked the four targets, it can be exceedingly difficult even to identify one of them in this way (e.g., which one is B, or which one started out in the upper right quadrant). Pylyshyn (2004) experimentally confirmed the extreme difficulty of keeping track of "which is which" during MOT, and showed that this difficulty is not due to any general dual-task interference (since there is no such deficit when the "labels" on static objects must be remembered, even through a separate tracking interval with additional objects).

This result is exactly what you would expect if targets are maintained during MOT simply by split foci of object-based attention. Under this view, there is nothing that makes one focus of attention different from another: They simply enhance processing on (and as a result, help us keep track of) each of the targets, as a set, not as individuals. As such, the attentional tracking view provides a ready mechanism for keeping targets separate from distractors, but not for keeping any of the targets distinct from each other.

In contrast, I suggest that this result is potentially a much greater challenge for the visual indexing view than Pylyshyn realizes. The reason is simply that this inability eliminates that one part of MOT that most directly supports the purpose of visual indexing in the first place: the ability to keep referring to an individual over time and motion such that its properties can be probed, or attention can be shifted to it. This is clearly not possible if you never know which target you are indexing: Any reliance on visual indexes as a foundation for attentional shifts, for example, would lead you to frequently shift attention to the wrong target. This problem can perhaps be most easily appreciated by harkening back to the initial analogy of visual indexes with pointers in computer data structures: Such pointers are of no use (or worse) if different pointers can frequently end up swapping their referents! Similarly, this result undercuts the analogy with pointing fingers—the idea that "the access that the finger [or visual index] contact gives makes it inherently possible to track a particular token, that is, to keep referring to what is . . . the same object" (Pylyshyn 1989, 68). The inability to do just this in Pylyshyn's experiments is essentially equivalent to tracking two objects by continually pointing to one with each index finger, but then later having no idea which object you were initially pointing to with your left index finger!

In Pylyshyn's article, the inability to track individuals per se is ultimately explained away by appeal to the idea that during tracking some targets are mistakenly "swapped" with other targets—and that target-target swaps are more frequent than target-distractor swaps. Such data are reported in a final experiment, showing that errors when attempting to track individuals are more likely to be errors of mistakenly "ID-ing" other targets: For example, when asked which object was target B, you'll mistakenly select target C more often than you'll select one of the distractors. I suggest that this interpretation is not convincing, however, for three reasons. First, it does not really help to salvage a link between MOT and indexing, since even under this interpretation the frequent target-target swaps would still frustrate any automatic target maintenance via indexing. Indexing, in other words, would still not be especially *useful* for MOT. Second, note that these experiments do not actually provide data that directly support this view; rather, they are merely consistent with it. For again, these results are exactly what you would expect if target maintenance is due solely to attention maintained on the targets as a set. Under this scenario, what Pylyshyn calls "target-target swaps" are nothing of the sort: There is nothing to swap, because there is nothing distinguishable about individual targets in the first place. In other words, the response that is being interpreted as a target-target swap is really just a guess: Observers know which items are the targets, but they have no idea which is which, and so during forced-choice responses they frequently ID the wrong target.

The third argument against Pylyshyn's interpretation, I suggest, is that it clearly doesn't apply in all of the cases where it would have to apply. Even when there is no special danger of targets being "swapped" during tracking—say, because they never come near each other—you still have essentially no idea which is which! This can be readily appreciated by viewing any MOT display in which two of the targets never approach each other. Here you can readily discern at the end of the motion that they are both targets, but you will have no idea which is which.

On balance, then, I suggest that what Pylyshyn's (2004) experiments show is exactly what they intuitively seem to show: We can keep track of the targets in MOT, but not which one is which. This undercuts any reason to suggest that data-driven index maintenance is playing any role in MOT, though, since the only way to modify the functioning of indexes to match these results would be to strip them of the one property they must have in order to fulfill the purpose for which they are theorized to exist in the first place. But again, all of this seems easily explained—and perhaps even necessarily predicted—by the view that MOT is simply realized by split object-based attention to the MOT targets as a set.

3.5　MOT = Tracking In the Present

One reason that Pylyshyn (2004) thinks that there must still be some bona fide tracking of individuals going on in MOT, despite the results discussed above, is that he thinks this is conceptually necessary. He calls this the *discrete reference principle*, and suggests that "a critical part of determining whether some object is a target is being able to trace its individuality . . . back over time to the start of each trial. . . . [T]he only way to determine that a particular individual object belonged to the target set in the previous instant is by knowing which particular individual in the target set it had been" (804, 805).

This logic seems mistaken. In order to identify an object as a target, you need only know that it was a target an instant ago—and everything that came before that moment can be "flushed" from the system without any cost. This is, in fact, what I think occurs during MOT: We are continually tracking *in the present*, without necessarily storing and using some sort of spatiotemporal trace back to the start of the trial. (This is not to say that such implicit memories are not possible, just that we don't use them during tracking.) During the very first frame of motion, you may indeed have a representation that demonstratively IDs each target, but the very next moment that information is gone, and all you know is that it was *a* target a moment ago, and so it must still be *a* target now.

I think this view—that tracking does not require a spatiotemporal trace back to the start of a trial—can be appreciated empirically as well as logically. One way to highlight this is to explore the ways in which tracking can be interrupted and resumed. Dual-task studies of MOT and visual search, for example, have convincingly shown that observers can switch back and forth between these two tasks in sequence, picking up the tracking from where it left off (Alvarez et al. 2005). This seems mysterious according to the indexing view, however, since presumably the indexes would also be required to help implement the search task: Given that search proceeds via the movement of attention, each shift of attention to a potential search target would by hypothesis have to be preceded by the assignment of an index to that object. But given the limited number of available indexes, this means that the indexes would have to be removed from the MOT targets during these "search interruptions," with no data-driven means to later reassign them to the targets. Nevertheless, tracking is not impaired. Why? Because you don't need to trace each target back to its origin in order to succeed in tracking through interruptions: All you need to know is where the targets are as a set in order to recover them, without any need to know which is which. This is also what happens, I suggest, from moment to moment during MOT even without any extrinsic interruptions: We track only in the present, knowing that the tracked objects are the targets, but without any necessary memory trace of how or where they initially acquired that status. (Indeed, note that two people could even "hand off" the tracking tack back and forth to each other, if the display paused at the right moments: the first person could simply describe to the second person where the four targets are, so that the second person could continue the tracking when the motion restarts. In this case, there would obviously be no possibility of maintaining an explicit tag back to the start of the trial, since the second person might not even have been present at the start of the trial!)

4 What Have We Learned about Attention from MOT?

For the remainder of this chapter I will assume—based on the arguments presented above—that MOT just is attentional tracking of multiple objects. In fact, the only initial difference between MOT and focal attentional tracking—though this is an important difference indeed!—is that attention is (necessarily) split during MOT. Other paradigms have also been used to demonstrate the ability to split attention (e.g., Cassidy, Sheremata, and Somers 2007; Castiello and Umiltà 1992; Driver and Baylis 1989; Kramer and Hahn 1995; McMains and Somers 2004) and for attention to track motion (Cavanagh 1992; Driver and Baylis 1989; Verstraten, Cavanagh, and Labianca 2000), but no paradigm has ever illustrated either of these features of attention more powerfully than MOT—or shown how they can be combined.

The goal of this penultimate section is to emphasize that this "attentional tracking" view is in no way deflationary. MOT may not interact with theories of visual indexing in this view, but it has nevertheless allowed us to make several important discoveries about the nature of attention—including several that would not likely have been possible without MOT. This section briefly reviews five such examples from our laboratory.

4.1 Attention Is (Sometimes Necessarily) Object Based

One key question about any cognitive or perceptual process concerns the units over which it operates. As noted earlier, most traditional theories of attention either assumed or explicitly argued that attention was fundamentally spatial, as in metaphors based on spotlights or zoom lenses (for a review, see Cave and Bichot 1999). Such spatial models inherently ignored the structure of the attended information: The process of selection was based on an extrinsic filter, and as a result you could attend to an object, multiple objects, only parts of objects, or even nothing at all—whatever fell within the spotlight. More recent models of attention, in contrast, have stressed the complex interplay between attention and the structure of the attended information (see Ben-Shahar, Scholl, and Zucker 2007). For example, many studies of object-based attention have demonstrated that the underlying units of attention are often discrete visual objects: Rather than spreading uniformly through a spatially defined region, attention flows more readily through individual objects—or alternately, attention is constrained by their boundaries (for a review, see Scholl 2001a).

The possibility of MOT in the first place demonstrates, as do many other paradigms, that attention can be object based in at least one sense, since the targets and distractors are frequently spatially interleaved. But MOT is still consistent with the possibility that attention is simply split into multiple spatial spotlights. Additional experiments using MOT, however, confirm that in some cases attention is necessarily directed only to discrete objects. For example, observers in one experiment still attempted to track multiple independently and unpredictably moving items, but the nature of these items was altered so that target-distractor pairs were perceived as single objects—with a target at one end and a distractor at the other end (Scholl, Pylyshyn, and Feldman 2001). Such a pair might be drawn as a simple line segment connecting the two points, as in figure 2.2b. Crucially, each end of a pair still moved completely independently. Tracking was greatly impaired in such conditions, despite the use of identical sets of trajectories and target selections: Observers could track individual objects, but not individual ends of uniform objects. This result is readily explained

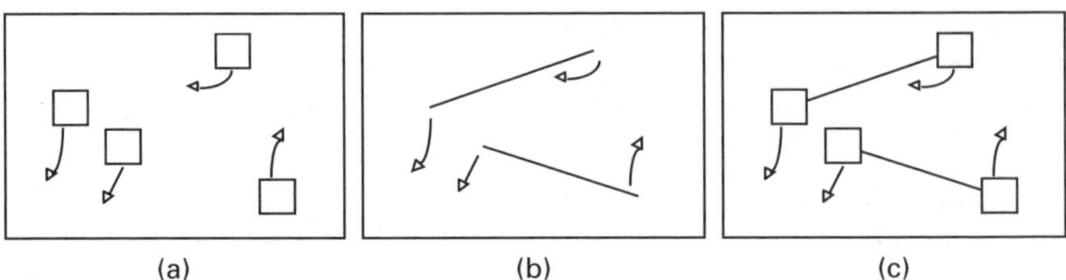

Figure 2.2
Sample "target merging" displays from Scholl et al. 2001. Each display shows four items, each of which always moves independently from all other items. (Actual displays had eight items total.) (a) A control condition, where observers must track punctate objects and perform as in most MOT tasks. (b) Items are merged into pairs, with each pair always consisting of a target and a distractor. Observers are greatly impaired when trying to track one end of each line, though they move through the same trajectories as in (a). (c) When curvature discontinuities are added to the ends of the lines by redrawing the boxes, tracking is better than with the lines alone, but worse than with the boxes alone.

in terms of object-based attention: Selection spreads uniformly throughout the lines, causing observers to lose track of which end was the target.

This demonstration of object-based attention has two advantages over similar demonstrations using paradigms of divided attention and spatial cueing (among others). First, these results demonstrate that object-based attention is in some cases a necessary "mode" of attention that cannot be avoided even when observers have specific task goals to the contrary. In contrast, object-based attention in most other paradigms is heavily influenced by task goals and various other details (e.g., the specific types of cues used and their probabilistic structure). Second, these results indicate that object-based attention can in some cases have a phenomenological component: When trying to track the undifferentiated ends of the lines in this paradigm, you can *feel* object-based attention in action.

Further manipulations of the precise ways in which the targets were connected in such displays indicated how MOT can be used to explore subtler aspects of object-based attention. For example, when observers had to track ends of "dumbbells" as in figure 2.2c, performance was worse than with boxes alone (figure 2.2a), but better than with lines alone (figure 2.2b). This indicates that object-based attention is not an all-or-nothing phenomenon (see also Marino and Scholl 2005), but can be independently affected by multiple cues including connectedness and curvature discontinuities.

4.2 Dynamic Object-Based Attention Requires Cohesive Objects

Objects are most commonly contrasted with spatial areas (as in section 4.1) or visual surface features such as color and shape (see section 4.5). But another contrast that is common from the study of objects in developmental psychology is that of objects versus nonsolid substances. In the study of infant cognition, for example, one of the most powerful principles of "core knowledge" is that of cohesion: An object must maintain a single bounded contour over time (see, e.g., Spelke 1990, 1994). Indeed, this principle may be uniquely important in that it helps define what counts as an object in the first place. If you want to know what an object is, just "grab some and pull"; the stuff that comes with your hand is the object, and the stuff that doesn't (and thereby fails to maintain a single unified boundary with the stuff that moved with your hand) is not. This has led some theorists to claim that cohesion is perhaps the single most important principle of what it means to be an object (e.g., Bloom 2000; Pinker 1997). And, correspondingly, infants' object-tracking abilities are greatly impaired by simple cohesion violations (Cheries et al. 2008; Huntley-Fenner, Carey, and Solimando 2002).

Using MOT, we were able to demonstrate that object-based attention in adult visual cognition is also constrained by cohesion. For example, observers can be asked to track spatially extended objects that move repeatedly in a particular type of noncohesive motion (figure 2.3): Each object began as a small square, but then split into many smaller units and moved in a nonrigid manner—essentially "pouring" from one location to another, as would a nonsolid substance. This manipulation greatly impaired tracking, despite the fact that the "objects" still followed the same trajectories as in typical MOT control conditions (vanMarle and Scholl 2003). We argue that this was due to the fact that each object's location could no longer be characterized by a single point, so that there was no unambiguous location for attention to select on this shrinking and growing extended object.

4.3 Beyond Object-Based Attention: Nonuniform Attention to Uniform Objects

The distinction between object-based and space-based attention need not always be a dichotomy: These views can interact, such that attention can be both spatially oriented and object based, in different ways but at the same time. This is the conclusion drawn from another recent study of MOT that used spatially extended objects. The first study of mandatory object-based attention using MOT, described in section 4.1 (Scholl, Pylyshyn, and Feldman 2001) assumed that attention was spreading equally throughout

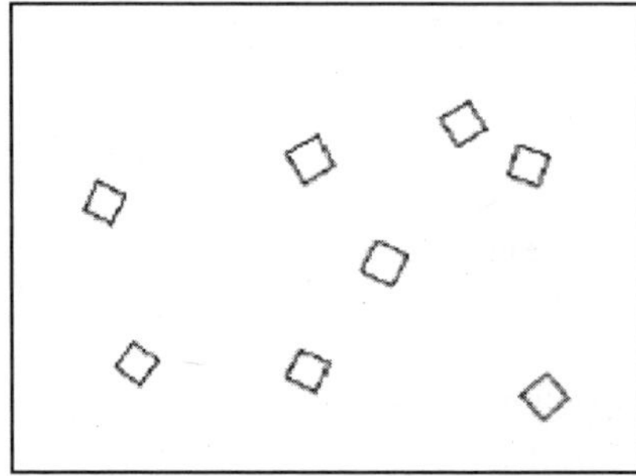

(a) : Object condition

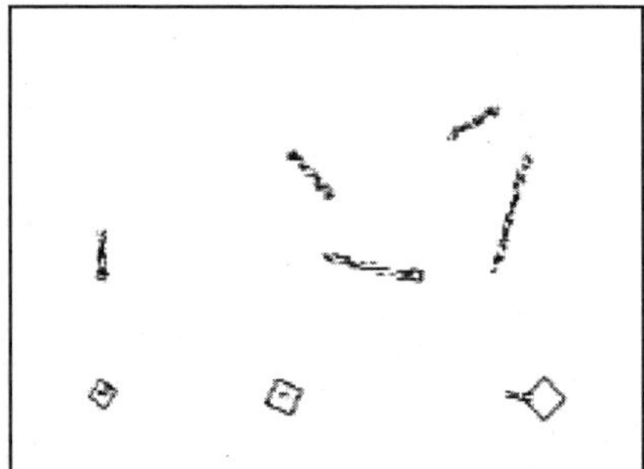

(b) : Substance condition

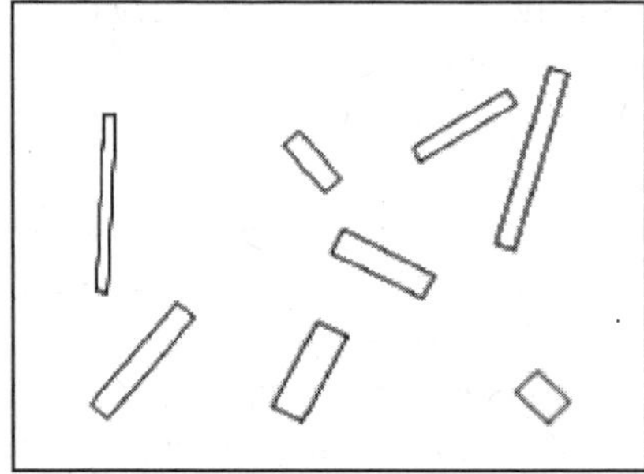

(c) : Morphing condition

Figure 2.3

Sample midtrial screenshots for studies of cohesion and spatial extent in MOT (vanMarle and Scholl 2003). (a) With punctate objects, tracking is accurate. (b) The "objects" move through the same trajectories but split into multiple units during their motion, as if they were liquids being "poured" from one location to another—a manipulation that greatly disrupts tracking. (c) Tracking is also disrupted when each square simply "stretches" its leading edge to its new location (becoming a long thin rectangle), then shrinks its trailing edge, as if it were a caterpillar. Tracking is also greatly disrupted here, perhaps because there is no unambiguous point on the object for attention to select.

the spatially extended lines. But it turns out that this is not the case: Though the lines are uniform, the distribution of attention within them is not.

In these experiments (Alvarez and Scholl 2005), observers had to track three of six long lines that moved haphazardly around a display. The lengths of the lines were randomly increased and decreased as the objects moved, since each of the lines' endpoints moved independently. To allow for an assessment of the distribution of attention within these objects, observers performed a simultaneous probe-detection task in which they were required to press a button whenever they detected the appearance of a probe (a small gray circle). Probes could appear at an object's center or near one of its ends, as depicted in figure 2.4. If attention was uniformly distributed over an object during the MOT task, we might expect that probe-detection rates would be similar for both center and end probes. However, this was not the case. Center probes were detected far more accurately than end probes, suggesting that more attentional resources were concentrated on the centers of the lines than near their ends. This effect was termed *attentional concentration*. Furthermore, the attentional concentration effect was modulated by the lengths of the objects being probed: As a line's length increased, center probes were detected increasingly well and end probes were detected increasingly poorly. In other words, the size of the concentration effect was largest for long lines and smallest for short lines, suggesting that the distribution of attention within an object becomes increasingly concentrated on its center as its length increases. This effect was termed *attentional amplification*, to emphasize that the attentional concentration effect was exaggerated or amplified by increased object length. These effects were both extremely robust (with differences in probe-detection accuracy on the order of 25%–50%), and they cannot be explained by differential patterns of eye fixations (Doran, Hoffman, and Scholl, in press). Both of these effects are illustrated schematically in figure 2.5.

These results begin to show how object-based and space-based attention interact, and they complement the other MOT results described above by narrowing in on the constraints that determine how and whether objects can be attentionally tracked. Both attentional concentration and amplification may reflect the difficulty of tracking spatially extended objects in the first place. Whereas such tracking is impossible for spatially extended objects that grow and shrink at especially fast rates (see the study of cohesion in section 4.2), it is possible when the lines' endpoints simply move independently, as in these studies. Because there is no single explicit

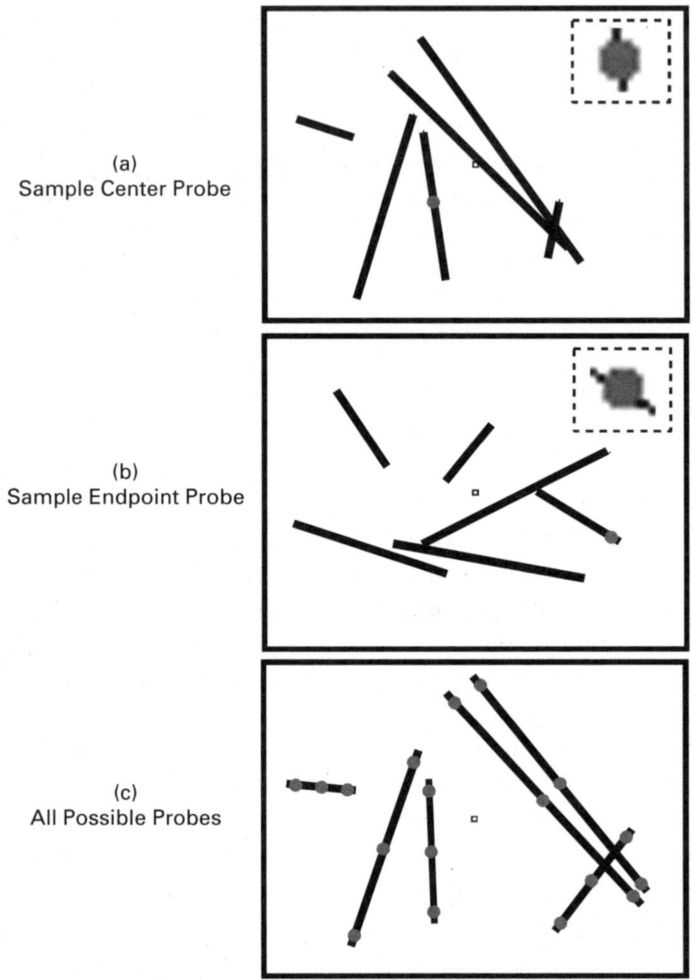

Figure 2.4
The concurrent MOT and probe-detection tasks used to discover the effects of attentional concentration and attentional amplification (Alvarez and Scholl 2005). Observers were required to keep track of three out of six moving lines while concurrently monitoring for the appearance of gray dot probes. (The box near the center of the displays is a fixation marker.) (a) A center probe trial in which a gray dot appears at the center of a line during the tracking task. The inset shows the local contrast of the center probe. (b) An endpoint probe trial in which a gray dot appears near the end of a line during the tracking task. (Note that the local contrast here in the inset is identical to that for center probes.) (c) A single frame of a trial highlighting all possible probe positions within that frame. (Only one probe was presented at a time in the actual experiment.)

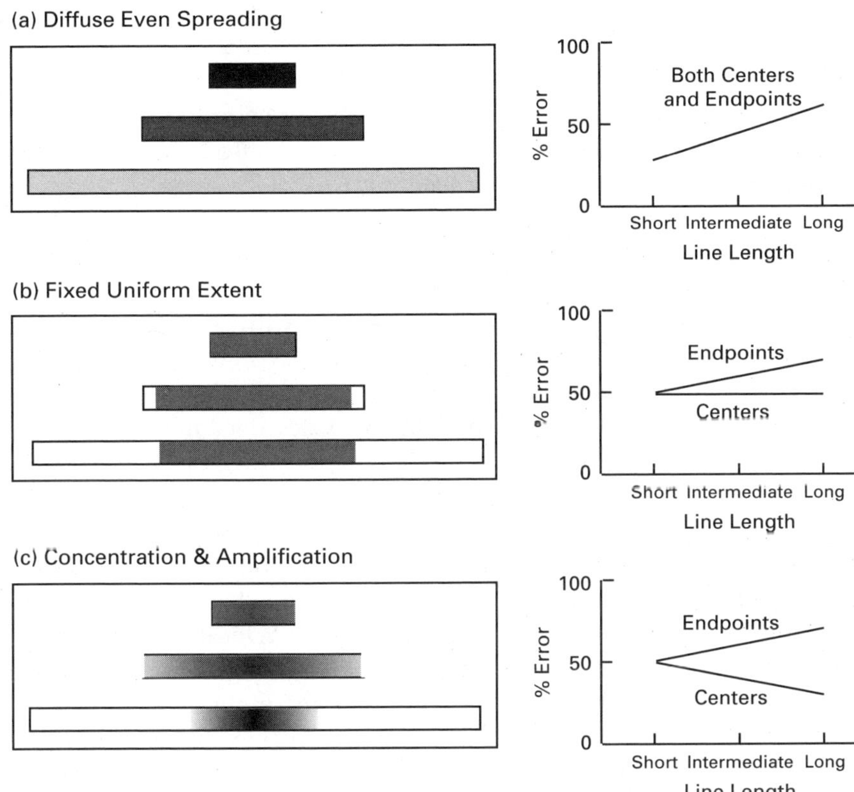

Figure 2.5

Three possible patterns of attentional distribution that could arise from the studies of MOT and probe detection from figure 2.4 (Alvarez and Scholl 2005). Here we depict three possible patterns of attention across the centers and endpoints of both long and short lines. In each case, the color of the line at each point represents the amount of attention (and the likelihood of probe detection), with darker areas indicating more attention, and lighter areas indicating less attention. (a) The performance predicted by a model in which attention always spreads uniformly through entire objects, but becomes more diffuse with increasing spatial extent. (b) The performance predicted by a model in which attention spreads uniformly through as much of a line as is allowed by available capacity. In short lines this yields uniform attention over the whole object, whereas in long lines this yields uniform attention over only a central portion, with little or no attention at the endpoints. (c) A schematic depiction of the actual results, illustrating both concentration and amplification: attention is concentrated at centers compared to endpoints, with centers receiving relatively more attention as line length increases, and endpoints receiving relatively less attention as line length increases.

punctate location for attention to select, a prioritized location may have to be effectively "constructed" via an attentional discontinuity (as in the concentration effect), and the need for such a discontinuity may map onto the degree to which there fails to be a single salient point-location for such objects, which would increase the prevalence of this effect (i.e., attentional amplification) as the lines grow longer.

4.4   Attention Is Influenced by Spatiotemporal Stability

Because MOT is an inherently dynamic paradigm, it allows us to ask questions about attention that would not be possible with paradigms employing only static displays. For example, in one recent study we asked about how attention is influenced by spatiotemporal stability (Alvarez, White, and Scholl, in preparation). In our earlier work described in the preceding section (Alvarez and Scholl 2005), we showed that when tracking spatially extended objects, attention is often concentrated at their centers. The centers of such objects may be important in part because they prove to be the most stable points across various types of motion. To track a person, for example, you would do well to track a point along his torso rather than his hands or feet (which may undergo many spurious local motions). Thus, under conditions of high load, as in MOT, attention might have a tendency to concentrate near the most stable point within an object, as a heuristic to help keep track of it. (This idea may help to explain why the attentional concentration effect exists, but it cannot explain away the effect: It persists even when the subject is comparing probes at endpoints and centers that are matched for velocity; Alvarez and Scholl 2005.)

We recently explored directly whether spatiotemporal stability influences attention by combining the tracking of long lines with probe detection as in our earlier studies, but now using lines that moved in different ways, making some points more stable than others (Alvarez, White, and Scholl in preparation). In these conditions, the "attentional concentration" effect still dominates: Attention is concentrated at lines' centers even when those points are the least stable of all. This was demonstrated by having observers track "bows" as in figure 2.6—long curves whose centers were constantly oscillating in a direction orthogonal to the endpoints' orientation. Probe detection revealed that attention was concentrated at the curves' centers (more so than their endpoints), despite the fact that the centers were always moving faster than the endpoints.

However, an effect of spatiotemporal stability can be observed when stability is not competing with attentional concentration. To demonstrate this, we had observers track "walkers"—long rigid lines where only one

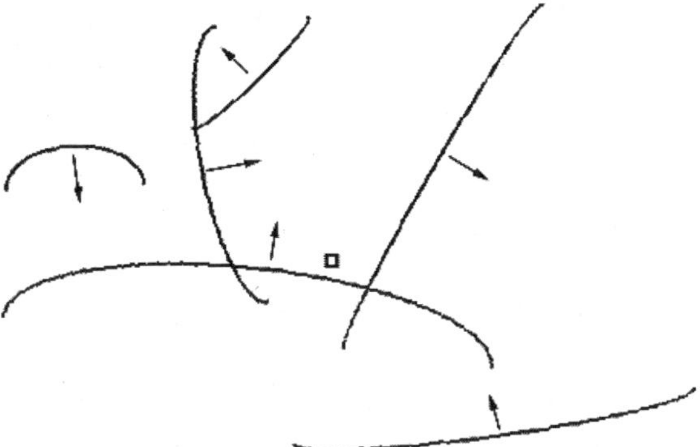

Figure 2.6
Illustration of the "bows" used by Alvarez, White, and Scholl (in preparation). Observers tracked long curves whose centers were constantly oscillating in a direction orthogonal to the endpoints' orientation (as indicated by the arrows, which were not present in the actual displays). Probe detection revealed that attention was concentrated at the curves' centers (compared to near their endpoints), despite the fact that the centers were always moving faster than the endpoints. As described the text, a different stimulus—"walkers"—yielded a different result, wherein there was an advantage in probe detection for more stable positions along tracked objects.

endpoint moves at a time—with the static and moving endpoints frequently swapping. (Static frames of this experiment thus looked just like those in figure 2.4, though now only one endpoint was moving at a time.) Probe detection revealed that attention concentrated at the lines' centers, but also prioritized the lines' stable (unmoving) endpoints over their moving ends. (The same effect obtained for slow vs. fast moving endpoints.) This is, to our knowledge, the first demonstration that spatiotemporal stability influences attention. This phenomenon presumably operates frequently in the real world, but would not be apparent in most experimental paradigms, since (unlike MOT) they are not able to test the distribution of attention on objects that move in such ways over relatively long periods of time.

4.5 Spatiotemporal Priority and Multiple Types of Attention

In section 3, I argued that MOT reflects attentional tracking rather than any special kind of indexing mechanism. However, the way that attention operates during MOT may still be interestingly different than in some other tasks. In particular, there may be different types of attention that support different kinds of visual processing. This was the conclusion drawn from a recent dual-task study that explored the nature of visual memory for natural scenes. Like most aspects of visual processing, the perception of scenes seems quick and effortless, as does the resulting memory for scenes: We can retain accurate memories for thousands of scenes based on only quick presentations (see, e.g., Standing 1973). This may seem to illustrate a type of automaticity, but in such situations observers are typically attending to the scenes that must be encoded, and without attention we often fail to see (much less remember) anything at all (Mack and Rock 1998; Most et al. 2005).

So, does scene memory require attention? This question can be studied via dual-task experiments, exploring the fidelity of both short- and long-terms scene memory when the presentation of the initial scenes occurs while observers are engaged in an attentionally demanding competing task. The results of such studies, however, turn out to depend on the specific types of tasks that are used to engage attention. When attention is engaged by a visual search task during initial scene presentation, for example, the resulting scene memory suffers (Wolfe, Horowitz, and Michod 2007)—and indeed it suffers beyond the baseline impairment produced by combining scene presentation with a generic central executive task such as auditory tone monitoring. When scenes must be encoded during MOT, however, a different picture emerges (no pun intended). In a recent study, observers completed a standard MOT task while several scenes were presented (see figure 2.7), but the resulting impairments of scene memory did

Figure 2.7
A screenshot from an experiment wherein natural scenes were presented in the background of a MOT task (Jungé et al. unpublished). (Gray arrows indicate motion of the discs, and were not present in the actual displays.) Unlike other competing attention tasks such as visual search, MOT did not greatly impair the resulting scene memory. See the text for details.

not exceed those produced by a baseline central executive task (Jungé et al., unpublished).

Why would scene memory be especially disrupted by one attention-demanding task (visual search) but not another (MOT)? We think this is because there are different forms of attention. In particular, many studies indicate that distinct attentional processes may be involved in *identification* (i.e., the processing of what an object is, on the basis of surface features) versus *individuation* over time (i.e., determining how and where objects move, on the basis of spatiotemporal information). Perhaps most famously, these sorts of processes seem to be localized in anatomically distinct cortical streams (e.g., Livingstone and Hubel 1988), with the ventral pathway corresponding to identification, and the dorsal pathway corresponding to individuation. In addition, a variety of behavioral evidence supports this distinction. The surface features of objects (e.g., their colors and shapes), while obviously critical for many visual processes including object recognition, seem to be largely discounted by many other processes (for a review, see Flombaum, Scholl, and Santos in press). For example, surface features play little or no role in determining apparent motion correspondence (Burt and Sperling 1981), identity over time in the tunnel effect (Flombaum et al. 2004; Flombaum and Scholl 2006; Michotte, Thinès, and Crabbé 1964/1991), or object-specific priming (Mitroff and Alvarez 2007).

This distinction can help to explain the scene memory results. In particular, perhaps the two relevant types of attention can be characterized in terms of the distinction between identification and individuation. Visual search (as employed in Wolfe, Horowitz, and Michod 2007) seems chiefly concerned with identifying objects on the basis of what they look like. In contrast, MOT (as employed in Jungé et al. unpublished) seems principally concerned with keeping a set of objects distinct from others over time on the basis of how and where they move (regardless of what they look like.) Thus, though both search and MOT can be highly attentionally demanding, they may do so via demands on partially independent attentional subsystems. In particular, visual search may interfere dramatically with scene encoding because both processes rely heavily on the same underlying ventral identification-based form of attention. In contrast, MOT fails to interfere with scene encoding more than central executive tasks because MOT relies primarily on a different underlying type of visual attention, one that is dorsal and individuation based. This distinction may also help to explain why MOT and search interfere with each other so little (Alvarez et al. 2005): They may both be highly attention-demanding, yet they may draw on fundamentally different forms of attention. Similarly, this may help to explain why observers are relatively poor at encoding surface features of objects in MOT—including those of tracked targets (see, e.g., Bahrami 2003; Ko and Seiffert 2006; Scholl, Pylyshyn, and Franconeri 1999). In sum, MOT may contrast with most other paradigms used to study attention not only in its requirements for attention to multiple objects and for attention to moving objects, but also in the type of attention it invokes.

5 Conclusions

Research on MOT—particularly as a tool with which to study and manipulate attention—is thriving. Indeed, in the last two decades since Pylyshyn's initial report of this phenomenon, the year with the most publications using MOT was 2008 (this year, as of this writing), and the runner-up was 2006. (A frequently updated bibliography of all work employing MOT can be found online at http://www.yale.edu/perception/MOT-Papers/.) The ideas and results discussed in this chapter suggest two reasons for this. First, the special nature of MOT matches key aspects of real-world visual experience: Whereas many or even most paradigms of attention involve unitary attentional shifts to single objects in static displays, real-world perception—and MOT—involves sustained attention to multiple moving objects. Second, these very features of MOT have allowed us to ask and answer questions about attention that we would not otherwise be able to address.

Acknowledgments

For helpful conversation and/or comments on earlier drafts, I thank George Alvarez, Jon Flombaum, and the members of the Yale Perception and Cognition Laboratory. None of our own work as described in this chapter would have been possible without the encouragement and mentorship of Zenon Pylyshyn.

References

Allen, R., P. McGeorge, D. G. Pearson, and A. B. Milne (2004). Attention and expertise in multiple target tracking. *Applied Cognitive Psychology* 18: 337–347.

Allen, R., P. McGeorge, D. G. Pearson, and A. B. Milne (2006). Multiple-target tracking: A role for working memory? *Quarterly Journal of Experimental Psychology* 59: 1101–1116.

Alvarez G. A., T. S. Horowitz, H. C. Arsenio, J. S. Dimase, and J. M. Wolfe (2005). Do multielement visual tracking and visual search draw continuously on the same visual attention resources? *Journal of Experimental Psychology: Human Perception and Performance* 31: 643–667.

Alvarez, G. A., and B. J. Scholl (2005). How does attention select and track spatially extended objects? New effects of attentional concentration and amplification. *Journal of Experimental Psychology: General* 134: 461–476.

Alvarez, G. A., A. White, and B. J. Scholl (in preparation). Attention and spatiotemporal stability. Manuscript in preparation.

Bahrami, B. (2003). Object property encoding and change blindness in multiple object tracking. *Visual Cognition* 10: 949–963.

Ben-Shahar, O., B. J. Scholl, and S. W. Zucker (2007). Attention, segregation, and textons: Bridging the gap between object-based attention and texton-based segregation. *Vision Research* 47: 845–860.

Bloom, P. (2000). *How Children Learn the Meanings of Words*. Cambridge, Mass.: MIT Press.

Burt, P., and G. Sperling (1981). Time, distance, and feature trade-offs in visual apparent motion. *Psychological Review* 88: 171–195.

Cassidy, B. S., S. Sheremata, and D. C. Somers (2007). Spatially specific training effects in multiple spotlight attention [Abstract]. *Journal of Vision* 7(9): 700, 700a.

Castiello, U., and C. Umiltà (1992). Splitting focal attention. *Journal of Experimental Psychology: Human Perception and Performance* 18: 837–848.

Cavanagh, P. (1992). Attention-based motion perception. *Science* 257: 1563–1565.

Cavanagh, P., and G. A. Alvarez (2005). Tracking multiple targets with multifocal attention. *Trends in Cognitive Sciences* 9: 349–354.

Cave, K.R., and N. P. Bichot (1999). Visuospatial attention: Beyond a spotlight model. *Psychonomic Bulletin and Review* 6: 204–223.

Chan, D. T., B. J. Scholl, B. Scassellati, and H. Qian (in preparation). Computational models of heuristic strategies in multiple object tracking. Manuscript in preparation.

Cheries, E. W., S. R. Mitroff, K. Wynn, and B. J. Scholl (2008). Cohesion as a principle of object persistence in infancy. *Developmental Science* 11: 427–432.

Doran, M. M., J. E. Hoffman, and B. J. Scholl (in press). The role of eye fixations in concentration and amplification effects during multiple object tracking. *Visual Cognition*.

Driver, J., and B. Baylis (1989). Movement and visual attention: The spotlight metaphor breaks down. *Journal of Experimental Psychology: Human Perception and Performance* 15: 448–456.

Fencsik, D. E., S. B. Klieger, and T. S. Horowitz (2007). The role of location and motion information in the tracking and recovery of moving objects. *Perception and Psychophysics* 69: 567–577.

Flombaum, J. I., and B. J. Scholl (2006). A temporal same-object advantage in the tunnel effect: Facilitated change detection for persisting objects. *Journal of Experimental Psychology: Human Perception and Performance* 32(4): 840–853.

Flombaum, J. I., S. M. Kundey, L. R. Santos, and B. J. Scholl (2004). Dynamic object individuation in rhesus macaques: A study of the tunnel effect. *Psychological Science* 15(12): 795–800.

Flombaum, J. I., B. J. Scholl, and Z. W. Pylyshyn (2008). Attentional resources in visual tracking through occlusion: The high-beams effect. *Cognition* 107: 904–931.

Flombaum, J. I., B. J. Scholl, and L. R. Santos (in press). Spatiotemporal priority as a fundamental principle of object persistence. In *The Origins of Object Knowledge*, ed. B. Hood and L. Santos. Oxford: Oxford University Press.

Fougnie, D., and R. Marois (2006). Distinct capacity limits for attention and working memory: Evidence from attentive tracking and visual working memory paradigms. *Psychological Science* 17: 526–534.

Franconeri, S. L., Z. W. Pylyshyn, and B. J. Scholl (2006). Spatiotemporal cues for tracking multiple objects through occlusion. *Visual Cognition* 14: 100–103.

Green, C. S., and D. Bavelier (2006). Enumeration versus object tracking: Insights from video game players. *Cognition* 101: 217–245.

Hulleman, J. (2005). The mathematics of multiple object tracking: From proportions correct to number of objects tracked. *Vision Research* 45: 2298–2309.

Huntley-Fenner, G., S. Carey, and A. Solimando (2002). Objects are individuals but stuff doesn't count: Perceived rigidity and cohesiveness influence infants' representations of small groups of distinct entities. *Cognition* 85: 203–221.

Intriligator, J., and P. Cavanagh (2001). The spatial resolution of visual attention. *Cognitive Psychology* 43: 171–216.

Jungé, J. A., J. S. DiMase, B. J. Scholl, M. M. Chun, T. S. Horowitz, and J. M. Wolfe (unpublished). Attentional demands of encoding scenes into memory: Evidence from interference with multiple object tracking.

Keane, B. P., and Z. W. Pylyshyn (2006). Is motion extrapolation employed in multiple object tracking? Tracking as a low-level non-predictive function. *Cognitive Psychology* 52: 346–368.

Ko, P., and A. E. Seiffert (2006). Visual memory for colors of tracked objects [Abstract]. *Journal of Vision* 6(6): 1080, 1080a.

Kramer, A., and S. Hahn (1995). Splitting the beam: Distribution of attention over noncontiguous regions of the visual field. *Psychological Science* 6: 381–386.

Liu, G., E. L. Austen, K. S. Booth, B. D. Fisher, R. Argue, M. I. Rempel, and J. T. Enns (2005). Multiple-object tracking is based on scene, not retinal, coordinates. *Journal of Experimental Psychology: Human Perception and Performance* 31: 235–247.

Livingstone, M. S., and D. H. Hubel (1988). Segregation of form, color, movement, and depth: Anatomy, physiology, and perception. *Science* 6: 740–749.

Mack, A., and I. Rock (1998). *Inattentional Blindness*. Cambridge, Mass.: MIT Press.

Marino, A. C., and B. J. Scholl (2005). The role of closure in defining the "objects" of object-based attention. *Perception and Psychophysics* 67: 1140–1149.

McMains, S., and D. Somers (2004). Multiple spotlights of attentional selection in human visual cortex. *Neuron* 42: 677–686.

Michotte, A., G. Thinès, and G. Crabbé (1964/1991). Les complements amodaux des structures perceptives. In *Studia Psychologica*. Louvain: Publications Universitaires. Reprinted and translated as: Michotte, A., G. Thinès, and G. Crabbé (1991). Amodal completion of perceptual structures. In *Michotte's Experimental Phenomenology of Perception*, ed. G. Thines, A. Costall, and G. Butterworth, 140–167. Hillsdale, N.J.: Lawrence Erlbaum.

Mitroff, S. R., and G. A. Alvarez (2007). Space and time, not surface features, underlie object persistence. *Psychonomic Bulletin and Review* 14: 1199–1204.

Most, S. B., B. J. Scholl, E. Clifford, and D. J. Simons (2005). What you see is what you set: Sustained inattentional blindness and the capture of awareness. *Psychological Review* 112: 217–242.

Oaten, M., and K. Cheng (2006). Longitudinal gains in self-regulation from regular physical exercise. *British Journal of Health Psychology* 11: 717–733.

O'Hearn, K., B. Landau, and J. Hoffman (2005). Multiple object tracking in people with Williams Syndrome and in normally developing children. *Psychological Science* 16: 905–912.

Oksama, L., and J. Hyönä (2004). Is multiple object tracking carried out automatically by an early vision mechanism independent of higher-order cognition? An individual difference approach. *Visual Cognition* 11: 631–671.

Pashler, H. (1998). *The Psychology of Attention*. Cambridge, Mass.: MIT Press.

Pinker, S. (1997). *How the Mind Works*. New York: Norton.

Postle, B. R., M. D'Esposito, and S. Corkin (2005). Effects of verbal and nonverbal interference on spatial and object visual working memory. *Memory and Cognition* 33: 203–212.

Pylyshyn, Z. W. (1989). The role of location indexes in spatial perception: A sketch of the FINST spatial index model. *Cognition* 32: 65–97.

Pylyshyn, Z. W. (1994). Some primitive mechanisms of spatial attention. *Cognition* 50: 363–384.

Pylyshyn, Z. W. (2001). Visual indexes, preconceptual objects, and situated vision. *Cognition* 80: 127–158.

Pylyshyn, Z. W. (2003). *Seeing and Visualizing: It's Not What You Think*. Cambridge, Mass.: MIT Press.

Pylyshyn, Z. W. (2004). Some puzzling findings in multiple object tracking (MOT): I. Tracking without keeping track of object identities. *Visual Cognition* 11: 801–822.

Pylyshyn, Z. W. (2007). *Things and Places: How the Mind Connects with the World*. Cambridge, Mass.: MIT Press.

Pylyshyn, Z. W., and V. Annan (2006). Dynamics of target selection in multiple object tracking (MOT). *Spatial Vision* 19: 485–504.

Pylyshyn, Z. W., and R. W. Storm (1988). Tracking multiple independent targets: Evidence for a parallel tracking mechanism. *Spatial Vision* 3: 179–197.

Pylyshyn, Z. W., J. Burkell, B. Fisher, C. Sears, W. Schmidt, and L. Trick (1994). Multiple parallel access in visual attention. *Canadian Journal of Experimental Psychology* 48: 260–283.

Scholl, B. J. (2001). Objects and attention: The state of the art. *Cognition* 80(1/2): 1–46.

Scholl, B. J., and Z. W. Pylyshyn (1999). Tracking multiple items through occlusion: Clues to visual objecthood. *Cognitive Psychology* 38: 259–290.

Scholl, B. J., Z. W. Pylyshyn, and J. Feldman (2001). What is a visual object? Evidence from target merging in multiple-object tracking. *Cognition* 80(1/2): 159–177.

Scholl, B. J., Z. W. Pylyshyn, and S. Franconeri (1999). When are spatiotemporal and featural properties encoded as a result of attentional allocation? [Abstract]. *Investigative Ophthalmology and Visual Science* 40: S797.

Sekuler, R., C. McLaughlin, and Y. Yotsumoto (2008). Age-related changes in attentional tracking of multiple moving objects. *Perception* 37: 867–876.

Sears, C. R., and Z. W. Pylyshyn (2000). Multiple object tracking and attentional processing. *Canadian Journal of Experimental Psychology* 54: 1–14.

Spelke, E. (1990). Principles of object perception. *Cognitive Science* 14: 29–56.

Spelke, E. (1994). Initial knowledge: Six suggestions. *Cognition* 50: 431–445.

Standing, L. (1973). Learning 10,000 pictures. *Quarterly Journal of Experimental Psychology* 25: 207–222.

Trick, L. M., J. Guindon, and L. Vallis (2006). Sequential tapping interferes selectively with multiple-object tracking: Do finger-tapping and tracking share a common resource? *Quarterly Journal of Experimental Psychology* 59: 1188–1195.

Trick, L., T. Perl, and N. Sethi (2005). Age-related differences in multiple object tracking. *Journal of Gerontology* 60B: P102–P105.

vanMarle, K., and B. J. Scholl (2003). Attentive tracking of objects vs. substances. *Psychological Science* 14(5): 498–504.

Verstraten, F., P. Cavanagh, and A. T. Labianca (2000). Limits of attentive tracking reveal temporal properties of attention. *Vision Research* 40: 3651–3664.

Viswanathan, L., and E. Mingolla (2002). Dynamics of attention in depth: Evidence from multi-element tracking. *Perception* 31: 1415–1437.

Wolfe, J. M., T. S. Horowitz, and K. O. Michod (2007). Is visual attention required for robust picture memory? *Vision Research* 47: 955–964.

Yantis, S. (1992). Multielement visual tracking: Attention and perceptual organization. *Cognitive Psychology* 24: 295–340.

3 Multiple-Object Tracking across the Lifespan: Do Different Factors Contribute to Diminished Performance in Different Age Groups?

Lana Trick, Heather Hollinsworth, and Darlene A. Brodeur

In 1988 Pylyshyn and Storm published the first multiple-object tracking study, and in so doing launched the investigation of a new ability—a new type of attentional selection, one that continues to inspire interest and controversy (see, e.g., Allen et al. 2006; Bahrami 2003; Culham et al. 1998; Fougnie and Marois 2006; Green and Bavelier 2006; Intriligator and Cavanagh 2001; Scholl and Pylyshyn 1999; Sears and Pylyshyn 2000; Yantis 1992). Multiple-object tracking involves keeping track of the positions of a number of independent items (designated targets) as they move among other items that are identical to them (distractors). It differs from standard selective attention tasks in two important ways. First, unlike most selection tasks, in which selection is typically accomplished in less than a second, multiple-object tracking requires sustained processing; selection must be sustained for the duration of item movement (the tracking interval). Though most studies involve tracking intervals of five to ten seconds, one involves tracking intervals of up to ten minutes (Place and Wolfe 2005). Second, tracking requires that participants select and track the positions of multiple independent items at once.

Using the multiple-object tracking task, Pylyshyn and Storm found evidence that young adults could simultaneously track four or five independent targets with great accuracy. This surprising result challenged the dominant view of selective attention at the time, which assumed that selection was accomplished by a single processing focus (the attentional spotlight or zoom lens) that performed detailed perceptual analyses such as combining features or deriving spatial relations (see, e.g., Ericksen and St. James 1986; Posner 1980; Treisman and Gelade 1984) but could only occupy one location at a time. Pylyshyn's results implied a paradox. Multiple-object tracking requires attentional selection (target items are selected among nontargets items that are to be ignored) and exhibits the capacity limitations typical of attentional selection tasks (when there are more than five items to track at once, performance deteriorates markedly)—and yet it does not seem to involve the unitary attentional focus, because people seem to be capable of tracking several independent items at once. In fact, according to Pylyshyn (2001), multiple-object tracking relies on a mechanism that is necessary precondition for *moving* the unitary attentional focus from object to object in a complex dynamic visual scene. It requires the ability to pick out a small number of specific items and refer to them without making reference to their current properties or positions (which may change). This mechanism is important for the construction of mental representations for objects that maintain their integrity despite changes in item positions and properties and it is essential to visual-motor coordination. As such, multiple-object tracking plays a critical role in human perception, cognition, and action, standing at the crossroads between low-level sensory processes, attention, working memory, and motor control.

Given this central role, it is important to know how multiple-object tracking develops and changes with age; it may explain age differences in a variety of daily tasks. At present there is relatively little research on how tracking performance changes across the lifespan. Some argue that multiple-object tracking must emerge in infancy because it is fundamental to the development of the object concept and early numerical cognition (e.g., Carey and Xu 2001; Scholl and Leslie 1999). Nonetheless, most of the studies on infants involve enumeration paradigms rather than tracking tasks, and although it has been argued that there is a relationship between tracking and enumeration (and in particular a form of spatial enumeration called *subitizing*: Trick, Audet, and Dales 2003; Trick and Pylyshyn 1994), these tasks differ in important ways. Most notably, tracking tasks require distractors and item movement whereas enumeration tasks do not. There are several recent studies on preschool- and school-aged children that suggest that tracking performance improves with age to adulthood (Black and Pylyshyn 2004; O'Hearn, Landau, and Hoffman 2005; Trick, Jaspers-Fayer, and Sethi 2005). There is also a single study that indicates that tracking performance may decline in the later years with older adults (Trick, Perl, and Sethi 2005). However, these investigations involve different methodologies and different encoding and retrieval parameters, as well as different age groups, and as a result, direct comparisons between studies are difficult.

Because of these problems, in this study we set out to investigate how tracking develops across the lifespan by using exactly the same tasks to test participants ranging in age from 7 to 75 years. This represents the first study of tracking across the lifespan. The first challenge was to create a version of the tracking task that young children would find meaningful and engaging so that their performance would be good reflection of their true abilities—a task that could also be used with older participants. We used a variant called "Catch the Spies," which was designed for children but produces the same pattern of results as standard tracking tasks when used with young adults (Trick, Jaspers-Fayer, and Sethi 2005). In it, participants monitored the positions of a number of sinister-looking individuals (spies: target items) that had "disguised themselves" to look like other people (happy-face figures: distractor items). A challenging version of the task was employed, with a total of ten moving items in the display and a ten-second tracking interval. Participants were required to track one to four spies at the same time, a range that should be adequate to produce variability in performance. In particular, it was important to look at cases where near-perfect performance was expected (e.g., tracking one item at once) and then see how performance deteriorates with increased tracking load (more items to track at once). At the same time, given that we were not interested in the impact of age on guessing strategies, we wanted to make sure that none of the participants was put in a position where they were guessing more than half of the target positions.

There were other complications. Multiple-object tracking is a complex task, and successful tracking requires a number of other abilities besides tracking, such as seeing individual items as they move, and selecting, encoding, and then reporting multiple target locations after a delay. Given that there may be age differences in each of the component abilities, when age differences do emerge, it is unclear whether they reflect actual differences in tracking or age differences in the other components of the task. In fact, tracking deficits may originate from different component abilities at different points in the lifespan.

For example, an inability to see individual items or item movement would clearly have an impact on tracking performance, and in late adulthood, aging and age-related pathologies produce reductions in visual sensitivity and acuity (see Klein 1991 for a review). The tracking task has some built-in controls for item visibility insofar as the appearances of the tracking displays are the same, regardless of the number of items to be tracked (attended) at once. As a result, if people had difficulty seeing individual items as they moved, this would be manifest even when tracking a single item at once. However, cueing studies suggest that there is a relationship between visual sensitivity and attention: When attentional resources are concentrated on a small area, it improves the visibility of items that appear in that location (see, e.g., Posner 1980). Therefore, it is possible that the effects of diminished sensitivity and acuity may not become evident until attentional resources are strained, as would occur when participants have to track a large number of items distributed over a wide area of the display. To find out whether this was in fact the case, in this study measures of acuity and contrast sensitivity were correlated with tracking performance when the attentional load was high (tracking three or four items at once). This is the first study that has ever directly examined the impact of factors related to visual sensitivity on multiple-object tracking performance.

Similarly, it is impossible to carry out a tracking task without having participants report the final locations of the targets, and there may well be age differences in report. Unfortunately, tracking performance is measured as a function of the number of items to be tracked at once, and the number of items to be tracked is perfectly confounded with the number of items to be reported. Participants might simply forget the target locations before they have a chance to report them. Pylyshyn and Storm (1988) tried to control for this possibility by using the partial report methodology, inspired by the partial report tasks used in studies of iconic memory (Sperling 1960). In partial report tracking tasks, after tracking multiple items, participants decide whether one specific (probed) item is a target or distractor. However, as it turns out, partial report in tracking is not the same as partial report in iconic memory. In Sperling's studies, participants could selectively report items based on a cue that appeared at the end of the display; there was no evidence that participants had to cycle through the items that were not probed before reporting probed stimuli. In contrast, in partial report tracking, the time to decide whether a single item is a target or distractor increases with the total number of items to be tracked at once, as if the participants were obliged to work through the list of target items in order to decide whether a single item was a target or distractor (Pylyshyn and Storm 1988). This may explain why full and partial report studies yield the same estimates of the maximal number of items that can be tracked at once (see Pylyshyn 2001 for a review). In either case, participants may forget the location of the item before they get a chance to report it.

Given that there seems to be no way around the problems associated with having multiple items to report, and given that partial report requires twice as many trials (half the time the probed item is a target and half the time it is a distractor), which is a definite impediment when testing children (who may become bored with the experiment if there are too many trials), there were advantages to the full report procedure. Full report tasks also have better ecological validity. If multiple-object tracking is to be useful in real life, it is important to find out if people can sustain selection long enough so they can react appropriately to *all* of the items. For example, when turning left across traffic when driving, it is important to be able to react correctly to the positions of all of the oncoming cars *and* the cyclists and pedestrians.

To distinguish age differences in report from age differences in tracking, control tasks were used that involved report but no tracking. Tracking performance for four items was compared with performance when participants reported the locations of four static targets either immediately or after a delay as long as the tracking interval (see also O'Hearn, Landau, and Hoffman 2005). Conditions for the report task were identical to those for tracking except for item movement. If report performance has a role in explaining individual differences in tracking, then immediate and delayed report for static displays should predict some of the variability in tracking performance. Of course, if participants used different strategies when reporting target locations for static and dynamic displays then there would be no relationship between report task performance and tracking.

The goal of this study was to determine whether the decrements in tracking performance shown in children and older adults originate from a common source (as might be expected if they reflect the operation of a single cognitive ability that develops slowly and then declines with age) or whether they reflect the effect of different factors for different age groups. In particular, we were interested in the role of age differences in the ability to select and report the locations of multiple targets among distractors (immediately or after a delay) and the impact of the reductions in acuity and contrast-sensitivity typical in older adults. Understanding the effect of these factors is of methodological and theoretical importance for investigations of age differences in tracking.

There were three predictions. The first was that increasing the number of items to track from one to four should produce decreased performance in all age groups, but the decrease should be especially pronounced in younger children and older adults (O'Hearn, Landau, and Hoffman 2005; Trick, Jaspers-Fayer, and Sethi 2005; Trick, Perl, and Sethi 2005). The second was that multiple-object tracking performance should be considerably worse than immediate and delayed report, though delayed report performance may account for some of the variability in tracking in children, who may have more difficulty maintaining selection for extended periods of time. The final prediction was that tracking performance should correlate with contrast sensitivity and acuity when the tracking load is high (there are large numbers to track at once).

Method

Participants

There were 76 participants from six age groups (7, 9, 11, 13, 26, and 75 years old). Children were from grades 1, 3, 5, and 7 and were recruited via consent forms sent to their parents. Participants in the 26-year-old group were students or staff at the University of Guelph whereas participants in the 75-year-old group were healthy, active individuals from a seniors' recreational center. See table 3.1 for details.

Questionnaires were filled out either by the parents of participants (for children) or by participants themselves. Data from individuals with diagnosed learning disabilities were dropped from the analysis. All of the younger participants reported normal or corrected to normal vision. Older adults filled out a more extensive general health questionnaire, covering a number of age-related disorders that might affect cognition, vision, hearing, balance, or motor function. None of the older adults reported cognitive deficits and all of their scores exceeded the minimum requirements for informed consent as measured by the Standardized Mini-Mental State Exam ($M = 28.8$, $SD = 1.23$, maximum score possible $= 30$), and their vision scores fell within the normal range for their age group. Their average Pelli-Robson contrast sensitivity and logmarr Early Treatment of Diabetic Retinopathy acuity scores were 11.71 and 0.12 respectively ($SD = 0.13$ and 0.08).

At the end of the study children were given a small gift. The young and older adults were paid at the rate of $15 an hour for their participation.

Table 3.1
Information about the participants.

Age group	Mean age	SD age	Age range	n	Number of females
7 years old	78.1 months	3.9 months	74–85 months	13	7
9 years old	103.7 months	2.9 months	97–107 months	13	9
11 years old	127.8 months	6.6 months	121–146 months	11	8
13 years old	152.9 months	4.2 months	148–164 months	13	4
26 years old	314.1 months	51.3 months	231–397 months	13	9
75 years old	897.9 months	66.8 months`	722–990 months	13	11

Apparatus and Materials

Testing was conducted using a Macintosh G4 PowerBook with an additional remote keyboard and mouse (for use by the research assistant). The viewing screen on the computer was 21.5 × 32.5 cm. The outer perimeter of the computer screen was light gray but the tracking field (the area in which items moved) was a central black rectangle occupying a 22.96°×17.33° visual angle when viewed from 45 cm. A 0.18° white outline square served as central fixation point. The items were 1.53° black spies and 1.45° blue happy-face figures, each outlined by 0.18° white contours, as shown in figure 3.1. Happy-face figures were the only items that moved. Though these figures could touch one another, they could never occlude (they repelled each other). On encountering tracking field boundaries, the items were programmed to bounce back into the interior of the display so that no item was ever lost from view. For each item, the rate of movement varied randomly from frame to frame, ranging between 0°–9.35°/s (a random 0–3 pixels per frame in each of the horizontal and vertical directions, with each frame requiring 16.5 ms). As well, for each item, and for every frame of motion, there was a random 1/100 chance that the item would change its direction of movement.

Questionnaires were used to gain information about learning disabilities and health problems. The Early Treatment of Diabetic Retinopathy (ETDRS) acuity test, Pelli-Robson Contrast Sensitivity test (Pelli, Robson, and Wilkins, 1988), and the Standardized Mini-Mental State Examination (Molloy, Alemayehu, and Roberts, 1991) were administered to the older adults.

Procedure

The main measure was performance on the Catch the Spies game. The object of the game was to keep track of the position of spies that were

Figure 3.1

Targets (spies) and distractors (civilians) in the Catch the Spies task. In the task spies "disguised themselves" as civilians during the tracking interval.

trying to escape by disguising themselves to look like other people. Each trial had five phases.

1. Initialization   When participants were ready, the "OK" button was pressed to initiate the trial. Ten static happy-face figures were then presented for 1105.5 ms on the screen. These figures appeared in random locations on the tracking field and could touch but never occlude one another.

2. Target acquisition   For 1650 ms a random 1–4 of the 10 happy-face figures alternated between happy-face and spy form (165 ms as happy-face, 165 ms as a spy, for the duration). This was done indicate the items that were spies (i.e., targets). Afterward there was a 495 ms pause in which all 10 items were static and once again in their original happy-face form.

3. Tracking   The 10 happy-faces (including spies disguised as happy faces) began moving. Movement continued for 10 s.

4. Report   Items stopped moving and a (0.5 cm) cursor appeared at a random location in the tracking field to indicate that it was time to report the positions of the targets. Participants pointed at the happy-face figures that they thought were "really spies" (targets). A research assistant seated to the side (and out of view) of the participant used the computer mouse to move the cursor to the items the participant pointed at. Participants were required to select as many items as there were targets.

5. Feedback   The "spies in disguise" revealed themselves by resuming their original spy-form for 165 ms.

For immediate and delayed report tasks, the sequence of events in a trial was identical to that in Catch the Spies except for phase 3 (item movement). For immediate report, the program skipped directly to the report phase after target acquisition, and for delayed report, the report phase was delayed 10 s (the duration of the tracking interval). In both report tasks, participants were required to fixate on the computer screen and report the target positions as soon as they saw the cursor. They reported target positions by pointing at them, as they did in the tracking task. In report tasks there were always 4 targets among 10 items.

Participants were seated 45 cm from the screen with a female research assistant beside them. For immediate and delayed report tasks, participants were given two practice trials before eight experimental trials. For multiple-object tracking, participants did 6 graded practice trials (one for 1 and 2 targets, and two each for 3 and 4 targets, starting with the 1 target condition) before 32 randomly ordered experimental trials (eight at each target numerosity). Delayed report and multiple-object tracking followed immediate report. The experimental session took 30–40 minutes. Participants were encouraged to take breaks between tasks and between trials as needed.

Results

The dependent measure was the percentage of correctly identified targets. Thus, on a given trial, if 3 of 4 targets were identified correctly, accuracy would be 75 percent. There were unavoidable violations of the homogeneity of variance assumption that posed problems when direct comparisons between groups were necessary. One solution to this problem is to convert data to proportions and use the inverse sine transformation to control for differences in variance before performing the analyses (Kirk 1982, p. 83). In this document, analyses for the inverse sine transformed data are reported and the conservative Tamhane's test is used whenever post hoc comparisons are required (Tamhane's test does not assume equal variance; ibid., 120–121). However, another way to circumvent the problem of unequal variances is to compare performance to external criteria (e.g., the criterion of perfect accuracy or the expected accuracy if participants were guessing one item), and this was done whenever possible. To facilitate comparison with external criteria, when data are displayed the percentage of correctly identified targets is graphed instead of the inverse-sine-transformed data.

In what follows, we first consider the results from the multiple-object tracking task and then compare multiple-object tracking, immediate, and delayed report performance. In the final section we document the relationships between the various measures of visual sensitivity and tracking performance in older adults.

Multiple-Object Tracking

In figure 3.2, the percentage of accurately identified targets is plotted as a function of the number of targets to be tracked at once (1–4) and age of the participant (7, 9, 11, 13, 26, and 75 years old). The number of targets had an effect on performance, with performance declining as the number of targets increased—an effect that was exaggerated in children and older adults. Nonetheless, all performed significantly better than would be expected if they could only track one item and guessed the location of the second when tracking two items at once (one sample t-tests against an expected accuracy of 55.6 percent; $p < .001$ for all age groups). The results replicate more standard multiple-object tracking studies insofar as they

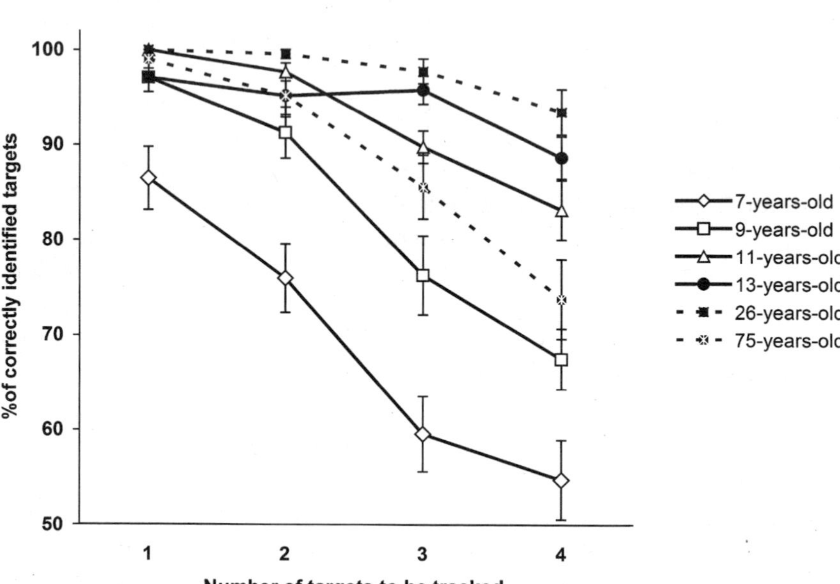

Figure 3.2
Mean percentage of correctly identified targets for 7-, 9-, 11-, 13-, 26-, and 75-year-old participants when tracking one to four spies in a display of ten moving items (standard error bars included).

show that young adults can track up to four targets at once very accurately (M = 93.5 percent accuracy at four), though the number of targets still had an impact for that age group ($F(1.3,15.6) = 4.25$, $MSE = .034$, $p < .05$, $\eta^2 = .26$). When the full transformed dataset was analyzed, effects of age and number of targets emerged, as well as an interaction (Age: $F(5, 70) = 32.93$, $MSE = .108$, $p < .001$, $\eta^2 = .70$; Number of targets: $F(2.5, 174.2) = 86.38$, $MSE = .051$, $p < .001$, $\eta^2 = .55$; Number of targets × Age: $F(12.4, 174.2) = 5.35$, $MSE = .051$, $p < .001$, $\eta^2 = .28$).

Planned comparisons were performed analyzing data for one and four targets in independent analyses. Comparisons of performance when participants were tracking a single item made it possible to ascertain whether participants were having difficulty seeing the items or focusing on the display for the 10-second tracking interval, even when the memory and attentional load was minimal. One sample t-tests indicated that the 7-year-old group was the only one with accuracy significantly below 100 percent when tracking one item at a time (t (12) = −4.07, $p < .05$). Tamhane's test showed that the 7-year-old children performed significantly worse than every other age group except the 13-year-olds. There were no other significant effects ($p > .05$).

Age differences were more pronounced when there were four items. Tamhane's test of means revealed that the 13- and 26-year-old participants performed significantly better than 7-, 9-, and 75-year-old participants when tracking four targets ($p < .05$). Performance for the 11-year-old children was comparable to that of the 75-year-old adults but significantly better than that of either the 7- or 9-year-old children ($p < .05$).

Comparison of Multiple-Object Tracking with Immediate and Delayed Report

Both immediate and delayed report required participants to indicate the positions of four target items, as did the multiple-object tracking task when participants were required to track four targets at once. Analyses were performed comparing transformed accuracies as a function of age and task (immediate report, delayed report, multiple-object tracking for four items). Results are shown in figure 3.3. Task had a significant effect on performance ($F(1.3, 91.32) = 220.95$, $MSE = .048$, $p < .001$, $\eta^2 = .76$). Planned comparisons revealed that for all age groups multiple-object tracking performance for four targets was significantly worse than immediate and delayed report for four targets ($p < .05$). In fact, the differences were dramatic. Multiple-object tracking performance for four items was anywhere from 6.5 percent to 32 percent worse than even the least accurate report task (delayed report). Given that multiple-object tracking performance was so much worse than immediate and delayed report performance, it seems unlikely that limitations in immediate and delayed report were the sole constraint on tracking performance.

Overall, age had a significant effect on performance ($F(5, 70) = 24.76$, $MSE = .049$, $p < .001$, $\eta^2 = .64$) and there was also a significant Age × Task interaction: $F(6.52, 91.32) = 11.24$, $MSE = .048$, $p < .001$, $\eta^2 = .44$. Although all of the children reported the locations of the four target items with 100 percent accuracy on the majority of trials, with young children there were more cases in which the occasional item was missed. Nonetheless, only the 7- and 13-year-old participants had accuracies significantly below 100 percent for immediate report ($t(12) = −5. 46$, $p < .001$; $t(12) = −2.75$, $p < .05$). The 7-year-old children performed significantly worse than every age group except the 13-year-old children on this task. There were no other significant effects ($p > .05$). Delayed report accuracy was significantly below

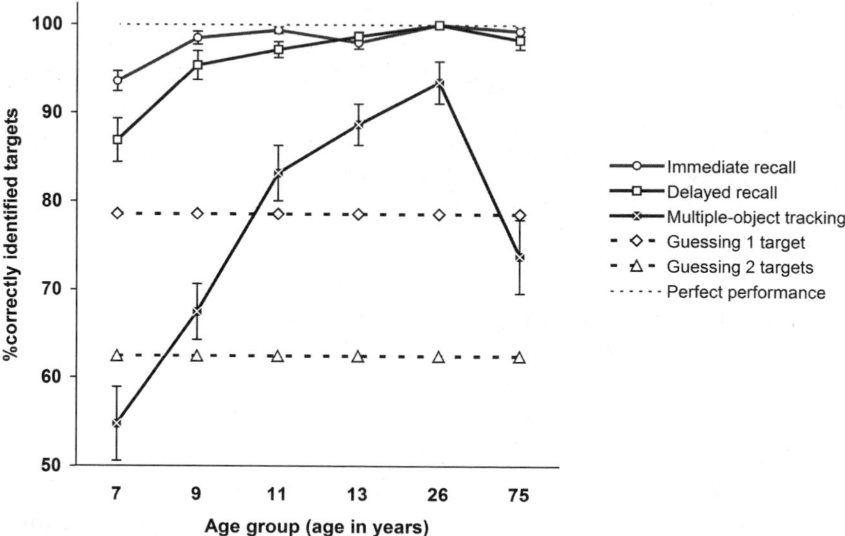

Figure 3.3

Mean percentage of correctly identified targets for 7-, 9-, 11-, 13-, 26-, and 75-year-old participants when reporting the positions of four targets in an immediate or delayed report task or a tracking task in which there are four targets (standard error bars included). Dotted lines indicate expected accuracies if participants were perfectly accurate or if they guessed the positions of one or two of the four targets.

100 percent for all groups of children ($t(12) = -5.36$; $t(12) = -2.78$; $t(10) = -3.13$, $t(12) = -2.61$ for the 7-, 9-, 11-, and 13-year-old participants respectively, $p < .05$), but accuracy did not differ significantly from 100 percent for the 26- and 75-year-old adults. The 7-year-old children performed significantly worse than every group except the 9-year-old children in delayed report. No other effects emerged ($p > .05$).

In contrast, for multiple-object tracking the percentage of correctly identified targets was significantly below 100 percent for all age groups (one sample t, $p < .05$), and exhibited an inverted U-shaped trend across the lifespan, with poorer performance for children and older adults. For purposes of comparison, in figure 3.3 we have plotted the expected outcomes if participants were guessing one or two of the four target locations (Freund 1981, p. 181: expected outcomes for sampling without replacement).

We compared the observed tracking performance with the expected accuracy if participants were tracking three targets and randomly guessing the position of the fourth. One sample t-test revealed that the 13- and 26-year-olds performed significantly better than would be expected if they had tracked three items and guessed the fourth ($t(12) = 4.34$ and $t(12) = 6.25$, respectively, $p < .01$) whereas the 7- and 9-year-old children performed significantly worse ($t(12) = -5.7$ and $t(12) = -3.47$, $p < .01$ for both).

Though limitations in the ability to report target locations were not the sole constraint on tracking performance for any age group (because tracking performance was always significantly worse than report performance), individual differences in delayed report explained some of the variance in tracking performance in children. Analyses of covariance were performed, analyzing multiple-object tracking for one to four items as a function of age, covarying out the effects of immediate and delayed report in the transformed data. Delayed report performance was a significant covariate for the 7- to 13-year old participants ($F(1, 44) = 8.53$, $MSE = .085$, $p < .01$, $\eta^2 = .16$), though there were still robust effects of age ($F(3, 44) = 14.06$, $MSE = .085$, $p < .001$, $\eta^2 = .49$) and an Age × Number of targets interaction ($F(7.8, 114.7) = 2.85$, $MSE = .057$, $p < .01$, $\eta^2 = .16$) when these effects were statistically controlled. Adjusted means are shown in figure 3.4.

Relationship between Acuity and Contrast-Sensitivity and Tracking Performance in Older Adults

The older adults could track one item over the 10-second interval with near perfect accuracy. This suggests that they had no difficulty seeing the items or seeing item motion. However, as the number of targets to be tracked at once increased their performance fell. We correlated tracking performance for three and four items with visual acuity and contrast sensitivity as measured by the ETDRS and Pelli-Robson tests, respectively. There was a marginal correlation between the two measures of visual function ($r = -.54$, $p < .06$) but ETDRS acuity did not correlate significantly with tracking performance for three and four items ($r = -.39$ and $-.17$, $p > .1$, respectively). The magnitudes of the correlations were larger for Pelli-Robson contrast-sensitivity, though the correlation was only significant for tracking three items at once ($r = .57$, $p < .05$ and $r = .42$, $p = .15$ for three and four targets, respectively). This may reflect the fact that there was more variability in performance for tracking three items than four (none of the older adults had 100 percent accuracy when tracking four items at once, for example). There were no other significant correlations between tracking

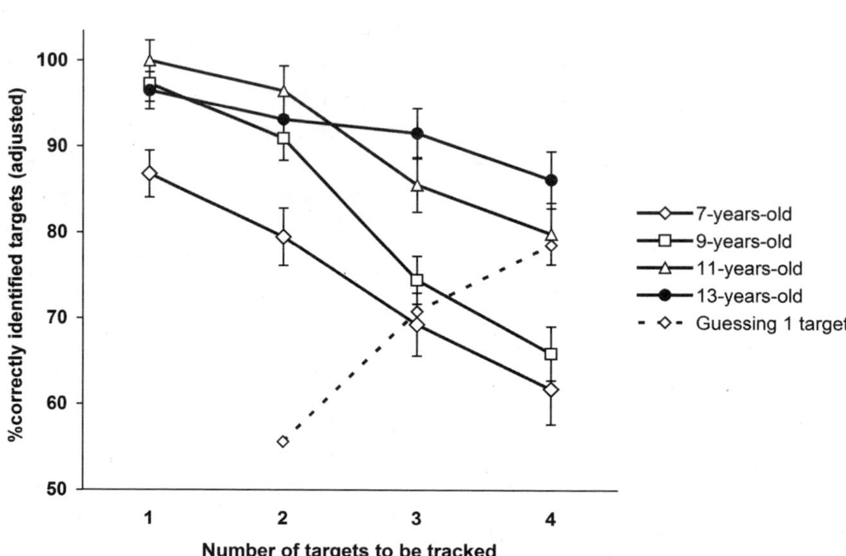

Figure 3.4

Mean adjusted percentage of correctly identified targets for 7-, 9-, 11-, and 13-year-old participants when tracking one to four targets in a display with ten items once the effects of immediate and delayed report have been statistically controlled (standard error bars included). The dotted line indicates the expected accuracy if participants guessed the position of one of the targets they were required to track when tracking two, three, and four targets.

performance and any other measure (including the Standardized Mini-Mental State Examination).

Discussion

As predicted, we found that tracking performance decreased as the number of item increased, but the decreases were especially pronounced in school-aged children and older adults. In particular, when there were three to four items to track at once, tracking performance increased markedly with age to young adulthood. There were corresponding performance decrements in tracking three and four items for the 75-year-old group. This result is of particular importance to the research on driving given that multiple-object tracking is necessary at complex intersections (to keep track of the positions of multiple moving vehicles, pedestrians, and cyclists) and it is there where risk of collision is especially high for older adults (McGwin and Brown 1999). However, when comparing performance in vastly different age groups, it is important to remember that multiple-object tracking is a complex task and deficits in performance may occur for different reasons in different age groups. In this study we found evidence of four different sources of variability in tracking performance.

1. For the 7-year-old children, tracking performance was significantly lower than that for other age groups even when they were only tracking a single item at a time. Given that children appeared to be looking at the screen (they initiated the trial) and there is little reason to suspect diminished visual function in this age group, it seems that this group had difficulty maintaining selection over extended periods of time. As a result, they lost track of the target, even when the tracking load was minimal. To a lesser extent, this difficulty also manifests itself in delayed report for four targets, but the 7-year old children did not differ as much from the other children when dealing with static items. Presumably momentary lapses in selection would be especially deleterious when tracking moving items because the relative positions of targets and distractors might change.

2. For the children (ages 7–13 years), delayed report performance accounted for some of the variability in tracking performance. Report performance was generally very good, with accuracy at 100 percent in the majority of trials for every age group. However, there were more errors in young children and more variability between children of the same age. Thus, although 5-year-old children can report the positions of 10 items in pattern span tasks (Pickering et al. 2001), in this study there were age differences even when participants were only reporting the positions of four items, and individual differences in this ability had an effect on tracking over and above the effects of age.

3. For senior adults, individual differences in contrast-sensitivity predicted tracking performance when tracking three items at once. When tracking only one to two items at once, tracking performance was close to 100 percent accurate in this age group, and as a result there was little evidence of the effects of diminished visual sensitivity or acuity when the tracking load was low.

4. Tracking performance was significantly lower than delayed report performance for all age groups, and moreover, report performance improved with age to young adulthood and remained stable whereas tracking performance increased with age to young adulthood and then declined in old age.

From these four findings, it is apparent that there are three sources of extraneous variance to consider when investigating the effects of age on multiple-object tracking. The first two relate to developmental improvements in the ability to control selection processes consistently from trial to trial. Although the 7-year-old participants could sometimes track with 100 percent accuracy, there were occasional lapses even when all they had to do was maintain selection for a single moving item over the 10-second tracking interval in one trial. For the older children, this showed itself in occasional lapses in immediate and delayed report for four target locations. Older adults did not exhibit these problems, but for that age group contrast-sensitivity predicted some of the variability in tracking performance for larger numbers of items. However, when older adults were tracking a single item their performance was near perfect. As predicted, their problems only revealed themselves when the tracking load is high (and the total area occupied by items is larger), which is a novel finding in the tracking literature.

Although these sources of extraneous variance are important to consider when investigating tracking, it is the fourth and final finding that is of primary importance to those interested in age differences in tracking per se. According to Pylyshyn's FINST theory (Pylyshyn 2001), tracking is possible because people assign spatial reference tokens called FINSTs (FINgers of INSTantiation) to a limited number of selected targets. These reference tokens act like pointer variables in computer languages such as C—they provide information about where the object is without the necessity of referring to the object by its properties or position, which could change from moment to moment. Consequently, FINSTs provide a way of seeing an item as "the same one," even though the item has moved and changed position (e.g., it used to be black and in the center of the screen and now it is blue and in the periphery). They are a necessary precondition for the creation of object files (Kahneman, Treisman, and Burkell 1983). When items move, the positions of the FINSted objects must be updated or the targets will be lost. Based on this account, there are several ways in which age differences might have an impact. One might be age-related change in the number of FINSTs. The theory suggests that there are only a limited number of FINSTs (around four or five in young adults), which makes sense because FINSTs are used for selection: It does not make sense to select everything at once. If there were age-related increases and decreases in the number of FINSTs, this might explain differences in tracking performance. However, this would not explain the pattern of results in this study because FINSTs are supposed to be useful in both static and dynamic displays, and though there are age differences in average report performance for static items, in a majority of the trials, participants of all ages recalled the positions of four items with 100 percent accuracy.

A second possibility is that there are age-related changes in the spatial or temporal resolution of the mechanisms used to update the locations of FINSTed items. Increasing the speed of item motion and the number of distractors reduces tracking performance in young adults (Liu et al. 2005; Trick, Perl, and Sethi 2005, respectively), and consequently there is reason to suspect that there may be limitations to the spatial and temporal resolution of the updating mechanism that provides spatial information for FINSTs. In general, selection processes have a coarser spatial resolution than sensory processes (Intriligator and Cavanagh 2001). As a result, there are situations in which people can see motion of individual items and yet cannot track individual items. If there were age differences in the spatial resolution of the tracking mechanism, or in the rate at which spatial updating occurs, this would explain age differences in tracking. Both of these factors would produce increased spatial uncertainty for target locations, and the probability that this spatial uncertainty would result in confusing the positions of targets and distractors would increase with the number of targets to be tracked at once.

The FINST hypothesis focuses on target selection, but recent investigations have shown that inhibition is associated with distractor items in tracking tasks (Pylyshyn 2006). A number of researchers contend that many age differences in attentional performance originate in problems with inhibition (e.g., Kipp 2005; Kane et al. 1994, for children and older adults respectively), and in particular, there is evidence of differences associated with inhibition as it relates to object-based selection, in which the inhibition is associated with an object (which may move) as opposed to a specific spatial relation in a display. Based on studies of object-based inhibition of return and marking, some maintain that object-based selection relies on cortical mechanisms that develop gradually and decline in old age although spatial selection per se does not (Christ, McCrae, and Abrams 2002; McCrae and Abrams 2001; Watson and Maylor 2002). Thus, it is possible that it is age differences in object-based inhibition that are producing the effects in tracking. At this point, it is unclear whether age differences in tracking are caused by problems in selection or inhibition (or both). Future research will have to disentangle the effects of target selection and distractor inhibition, and accomplishing this may require manipulating the number of targets and distractors separately and going to tasks that do not require distractors though they still involve tracking the positions of multiple moving items at once. Once such task is the enumeration of moving objects (see Trick, Audet, and Dales 2003).

Overall, this chapter makes the following contributions. First, it stands as a warning to future investigators who are interested in studying age differences in multiple-object tracking across the lifespan. Any given experimental task involves a variety of abilities, but the multiple-object tracking task may be more complicated than most in that it involves monitoring the progress of multiple items (which occupy extended areas of visual space) for prolonged periods of time. This poses a challenge for lifespan research, insofar as it makes it necessary to control or quantify the effects of a variety of factors in order to differentiate the impact of age differences in tracking dynamic items from age differences in other components of the task. In this study we found evidence of three sources of extraneous variance: visual sensitivity (for older adults), the ability to consistently maintain selection for a single moving item across period of time (for 7-year-old children), and delayed report (for children 7–13 years of age). Nonetheless, the main finding is that tracking improves in childhood and declines in old age, and these effects are not entirely the product of age differences in visual sensitivity or report.

Acknowledgments

This project was supported by a grant from the Natural Sciences and Engineering Research Council of Canada (238641-01). We would like to thank Julie Famewo, Cheryl Hymmen, and Julie Ojala, who helped in testing.

References

Allen, R., P. McGeorge, D. Pearson, and A. Milne (2006). Multiple-target tracking: A role for working memory? *Quarterly Journal of Experimental Psychology* 59(6): 1101–1116.

Bahrami, B. (2003). Object property encoding and change blindness in multiple object tracking. *Visual Cognition* 10(8): 949–963.

Black, A. K., and Z. W. Pylyshyn (2004). Developmental differences in multiple object tracking [Abstract]. *Journal of Vision* 4(8): 371a, http://journalofvision. org/4/8/371/, doi:10.1167/4.8.371.

Carey, S., and F. Xu (2001). Infants' knowledge of objects: Beyond object files and object tracking. *Cognition* 80: 179–213.

Christ, S.E., C. S. McCrae, and R. A. Abrams (2002). Inhibition of return in static and dynamic displays. *Psychonomic Bulletin and Review* 9: 80–85.

Culham, J. C., S. A. Brandt, P. Cavanagh, N. G. Kanwisher, A. M. Dale, and R. B. Tootell (1998). Cortical fMRI activation produced by attentive tracking of moving targets. *Journal of Neurophysiology* 80(5): 2657–2670.

Ericksen, C., and J. St. James (1986). Visual attention within and around the field of focal attention: A zoom lens model. *Perception and Psychophysics* 40(4): 225–240.

Fougnie, D., and R. Marois (2006). Distinct capacity limits for attention and working memory: Evidence from attentive tracking and visual working memory paradigms. *Psychological Science* 17(6): 526–534.

Freund, J. F. (1981). *Statistics: A First Course*, 3rd ed. Englewood Cliffs, N.J.: Prentice-Hall.

Green, C. S., and D. Bavelier (2006). Enumeration versus multiple object tracking: The case of action video game players. *Cognition* 101: 217–245.

Intriligator, J., and P. Cavanagh (2001). The spatial resolution of visual attention. *Cognitive Psychology* 43(3): 171–216.

Kahneman, D., A. Treisman, and J. Burkell (1983). The cost of visual filtering. *Journal of Experimental Psychology: Human Perception and Performance* 9(4): 510–522.

Kane, M. J., L. Hasher, E. R. Stoltzfus, R. T. Zacks, and S. L. Connelly (1994). Inhibitory attentional mechanisms and aging. *Psychology and Aging* 9(1): 103–112.

Kipp, K. (2005). A developmental perspective on the measurement of cognitive deficits in attention-deficit/hyperactivity disorder. *Biological Psychiatry* 57: 1256–1260.

Kirk, R. E. (1982). *Experimental Design: Procedures for the Behavioral Sciences*, 2nd ed. Belmont, Calif.: Brooks/Cole Publishing.

Klein, R. (1991). Age-related eye disease, visual impairment, and driving in the elderly. *Human Factors* 33: 521–525.

Liu, G., E. Austen, K. Booth, B. Fisher, R. Argue, M. Rempel, and J. T. Enns (2005). Multiple-object tracking is based on scene not retinal coordinates. *Journal of Experimental Psychology: Human Perception and Performance* 31(2): 235–247.

McCrae, C. S., and R. A. Abrams (2001). Age-related differences in object- and location- based inhibition of return of attention. *Psychology and Aging* 16: 437–449.

McGwin, G., and D. Brown (1999). Characteristics of traffic crashes among young, middle-aged, and older drivers. *Accident Analysis and Prevention* 31: 181–189.

Molloy, D. W., E. Alemayehu, and R. Roberts (1991). A Standardized Mini-Mental State Examination (SSMSE): Its reliability compared to the traditional Mini-Mental State Examination (MMSE). *American Journal of Psychiatry* 148: 102–105.

O'Hearn, K., B. Landau, and J. Hoffman (2005). Multiple object tracking in people with Williams syndrome and in normally developing children. *Psychological Science* 16(11): 905–912.

Pelli, D. G., J. F. Robson, and A. J. Wilkins (1988). The design of a new letter chart for measuring contrast sensitivity. *Clinical Vision Sciences* 2: 187–199.

Pickering, S. J., S. E. Gathercole, M. Hall, and S. A. Lloyd (2001). Development of memory for pattern and path: Further evidence for the fractionation of visuo-spatial memory. *Quarterly Journal of Experimental Psychology* 54A(2): 397–420.

Place, S. S., and J. M. Wolfe (2005). Multiple visual object juggling [Abstract]. *Journal of Vision* 5(8): 27a, http://journalofvision.org/5/8/27/, doi:10.1167/5.8.27

Posner, M. I. (1980). Orienting of attention. *Quarterly Journal of Psychology* 32: 3–25.

Pylyshyn, Z. (2001). Visual indexes, preconceptual objects, and situated vision. *Cognition* 80: 127–158.

Pylyshyn, Z. W. (2006). Some puzzling findings in multiple object tracking (MOT): II. Inhibition of moving nontargets. *Visual Cognition* 14(2): 175–198.

Pylyshyn, Z., and R. Storm (1988). Tracking multiple independent targets: Evidence for both serial and parallel stages. *Spatial Vision* 3(3): 179–197.

Scholl, B. J., and A. M. Leslie (1999). Explaining the infant's object concept: Beyond the perception/cognition dichotomy. In *What Is Cognitive Science?*, ed. E. Lepore and Z. Pylyshyn, 26–73. Oxford: Blackwell.

Scholl, B. J., and Z. W. Pylyshyn (1999). Tracking multiple items through occlusion: Clues to objecthood. *Cognitive Psychology* 38(2): 259–290.

Sears, C., and Z. W. Pylyshyn (2000). Multiple-object tracking and attentional processing. *Canadian Journal of Experimental Psychology* 54(1): 1–14.

Sperling, G. (1960). The information available in a brief visual presentation. *Psychological Monographs* 74: 1–29.

Treisman, A., and G. Gelade (1980). A feature integration theory of attention. *Cognitive Psychology* 12: 97–136.

Trick, L., D. Audet, and L. Dales (2003). Age differences in enumerating things that move: Implications for the development of multiple-object tracking. *Memory and Cognition* 31(8): 1229–1237.

Trick, L. M., F. Jaspers-Fayer, and N. Sethi (2005). Multiple-object tracking in children: The "Catch the Spies" task. *Cognitive Development* 20(3): 373–387.

Trick, L. M., T. Perl, and N. Sethi (2005). Age-related differences in multiple-object tracking. *Journals of Gerontology: Series B: Psychological Sciences and Social Science* 60B(2): 102–104.

Trick, L., and Z. W. Pylyshyn (1994). Why are small and large numbers enumerated differently? A limited capacity preattentive stage in vision. *Psychological Review* 101(1): 80–102.

Watson, D. G., and E. A. Maylor (2002). Aging and visual marking: Selective deficits for moving stimuli. *Psychology and Aging* 17: 321–339.

Yantis, S. (1992). Multi-element visual tracking: Attention and perceptual organization. *Cognitive Psychology* 24: 295–340.

4 Vision for Action

Mel Goodale and Marla Wolf

1 Introduction

Ultimately, brains evolved not to enable us to think but to control our movements. Thinking is a mere handmaiden to action. The key word here is "ultimately." To put it bluntly: Without action, we would not be able to compete with others, reproduce, and thus project our genes (and our brains) into the future. This means that a complete account of the "cognitive" functions of the brain will require as much attention to the organization of motor output as to the processing of sensory input.

In the disciplines of physiology and psychology, however, a sharp division is often drawn between sensory and motor systems. The chapters on vision in most undergraduate textbooks, for example, are entirely separate from those devoted to motor control. (In fact, in psychology, motor systems are scarcely discussed at all.) Similar divisions exist in scientific societies—and sometimes within university departments. Although one can find the occasional book that talks about "sensorimotor integration" and the occasional symposium that brings together researchers from both fields, sensory and motor systems have, for the most part, remained two solitudes. Even in the new discipline of cognitive neuroscience, this division remains (Gazzaniga 2004).

As we have already intimated, however, it does not make good biological sense to separate the study of sensory and motor systems in this way. Consider for a moment the relationship between vision and the motor output it controls. Although it might be convenient to talk about visual cortex on the one hand and motor cortex on the other, there is no particular point along the many routes between the retina and the muscles where signals stop being sensory and suddenly become motor. From the moment signals from the photoreceptors enter the central nervous system, the information they convey is on its way to being transformed into motor output.

This is not mere semantics. The artificial division of the brain into visual areas and motor areas, and the mutual isolation of the intellectual traditions that study them, has led to theories of brain function that are quite misleading. Take the case of the "what" versus "where" story, which for many years was the dominant theoretical account of the functional organization of the cortical visual pathways. Over twenty years ago, Ungerleider and Mishkin (1982) identified two "streams of visual processing" arising from early visual areas in the cerebral cortex of the monkey: a ventral stream projecting to inferotemporal cortex, and a dorsal stream projecting to posterior parietal cortex (see figure 4.1). In what was to become one of the most influential theories in behavioral neuroscience, Ungerleider and Mishkin proposed that the ventral stream mediates "object vision," enabling the monkey to identify an object, whereas the dorsal stream mediates "spatial vision," enabling the monkey to locate the object. Notice that the emphasis here is on a difference in sensory processing, with

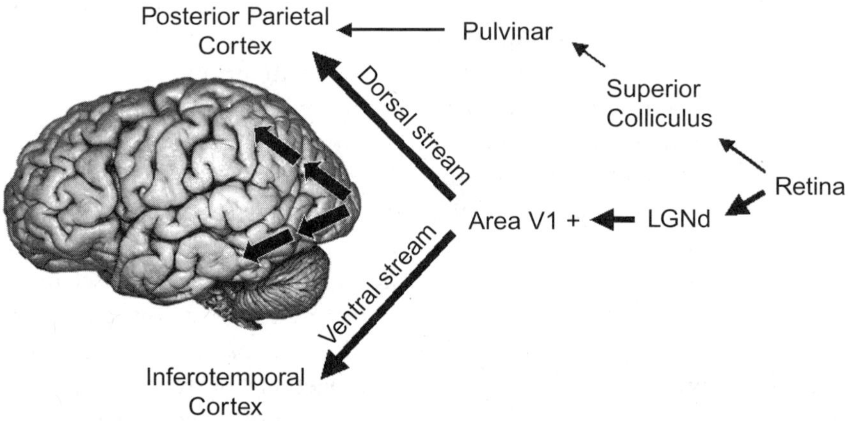

Figure 4.1
Schematic representation of the two streams of visual processing in human cerebral cortex. The retina sends projections to the dorsal part of the lateral geniculate nucleus in the thalamus (LGNd), which projects in turn to primary visual cortex (V1). Within the cerebral cortex, the ventral stream arises from early visual areas (V1+) and projects to regions in the occipito-temporal cortex. The dorsal stream also arises from early visual areas but projects instead to the posterior parietal cortex. The posterior parietal cortex also receives visual input from the superior colliculus via the pulvinar. On the left, the approximate locations of the pathways are shown on an image of the brain. The routes indicated by the arrows involve a series of complex interconnections.

one stream handling information about an object's features (the "what" pathway) and the other handling information about its spatial location (the "where" pathway). This distinction between what and where resonated remarkably not only with psychological accounts of perception, but also with nearly a century of neurological thought about the functions of the temporal and parietal lobes in vision (Schäfer 1888; Brown and Schäfer 1888; Ferrier and Yeo 1884; Holmes 1918).

In the early 1990s, however, the "what versus where" story began to unravel—largely because it treats the dorsal and ventral streams as purely "visual" pathways. New evidence began to accumulate from work with both monkeys and neurological patients, showing that a purely sensory account simply would not work. It soon became apparent that the only way to make sense of these new findings was to consider the different outputs of the two streams—and to work out how visual information is eventually transformed into motor acts.

In 1992, Goodale and Milner proposed a reinterpretation of the Ungerleider and Mishkin account of the two visual streams. According to the Goodale and Milner proposal, the dorsal stream plays a critical role in the real-time control of action, transforming moment-to-moment information about the location and disposition of objects into the coordinate frames of the effectors being used to perform the action. The ventral stream (together with associated cognitive networks) helps to construct the rich and detailed representations of the world that allow us to identify objects and events, attach meaning and significance to them, and establish their causal relations. Such operations are essential for accumulating and accessing a visual knowledge base about the world. Thus, it is the ventral stream that provides the perceptual foundation for the off-line control of action, projecting action into the future and incorporating stored information from the past into the control of current actions.

The ventral stream provides us with our conscious visual experience of the world. But is a matter of some debate as to how much the basic operations of this stream are subject to the influences of our beliefs or knowledge. Pylyshyn (1980, 1999) would almost certainly argue that the early stages of vision-for-perception are "cognitively impenetrable." Others, however, would argue that visual processing in these early stages can be influenced (via recurrent projections) by attention, prior knowledge, and expectations (for reviews, see Friston 2003; Lamme and Spekreijse 2000; Yuille and Kersten 2006). Nevertheless, Pylyshyn (1999) would respond that even though the early and elemental processes that give rise (eventually) to our conscious percepts of the world are subject to top-down influences, their computations are encapsulated from cognition. From Pylyshyn's point of view, the processing of visual information by the dorsal stream is even more cognitively impenetrable. Not only are we denied cognitive access to the underlying visuomotor transformations, but there is nothing remotely "visual" (in the experiential sense) about the products of those transformations. As Goodale and Milner (1992) originally pointed out, processing in the dorsal stream does not generate visual percepts; it generates skilled actions (as part of a network of structures involved in sensorimotor control).

In this chapter, we first introduce some of the original neuropsychological evidence that prompted Goodale and Milner (1992) to propose the distinction between vision-for-perception and vision-for-action. We then move on to examine recent neuroimaging data that complement the earlier neuropsychological observations. Finally, we discuss why such a division of labor should exist in visual processing—and present some new work with pictorial illusions that reveals striking differences in the operating characteristics of vision-for-perception and vision-for-action. This chapter is by no means an exhaustive review of the Goodale and Milner proposal. For more details, readers are directed to Goodale and Milner 2004a and Milner and Goodale 2006.

2 Neuropsychological Evidence

2.1 Optic Ataxia

The original evidence for Goodale and Milner's (1992) proposal came from studies that examined the pattern of deficits and spared visual abilities in neurological patients with selective damage to the dorsal or ventral stream. Patients with lesions in the dorsal stream, particularly in the intraparietal sulcus (IPS) and adjacent regions of the superior posterior parietal cortex, for example, typically have problems reaching toward targets placed in different positions in the visual field, particularly the peripheral visual field (see, e.g., Perenin and Vighetto 1988). This particular deficit is often termed "optic ataxia" (Bálint 1909). But the failure to locate an object with the hand cannot be construed as a problem in spatial vision: Many optic ataxia patients, for example, can describe the relative position of the object in space quite accurately, even though they cannot direct their hand toward it (Perenin and Vighetto 1983). In fact, these patients usually have no difficulty using input from other sensory systems, such as proprioception or audition, to guide their movements. In addition to their deficits in reaching, many patients with damage in the posterior parietal cortex are unable

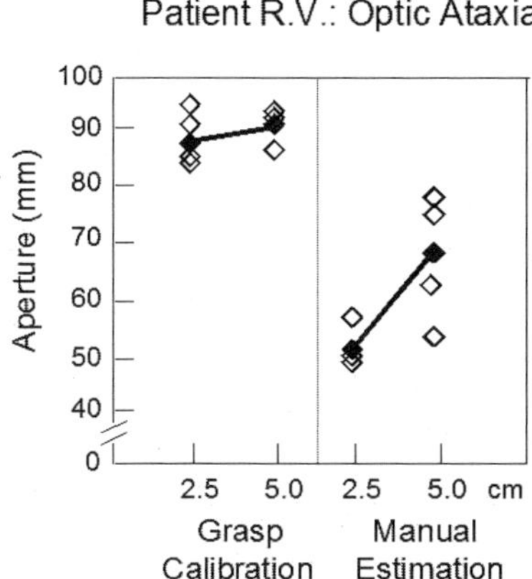

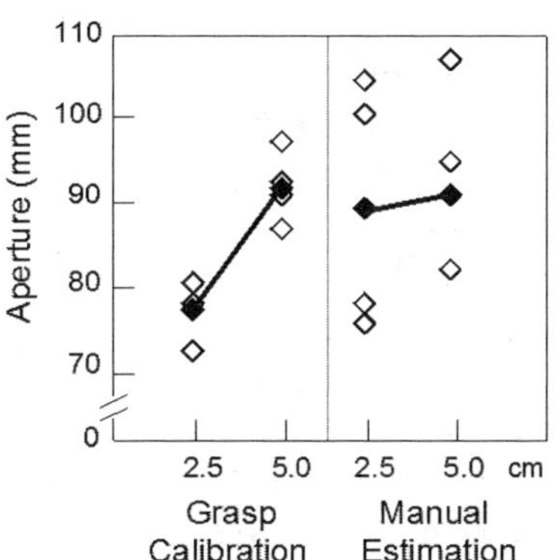

Figure 4.2

Graphs showing the size of the aperture between the index finger and thumb during object-directed grasping and manual estimates of object width for R.V., a patient with optic ataxia, and D.F., a patient with visual form agnosia. R.V. (left) was able to indicate the size of the objects reasonably well (individual trials marked as open diamonds), but her maximum grip aperture in flight was not well tuned. She simply opened her hand as wide as possible on every trial. In contrast, D.F. (right) showed excellent grip scaling, opening her hand wider for the 50 mm-wide object than for the 25-mm wide object. D.F.'s manual estimates of the width of the two objects, however, were grossly inaccurate and showed enormous variability from trial to trial.

to use visual information to rotate their hand, scale their grip, or configure their fingers properly when reaching out to pick up objects (for an example, see figure 4.2), even though they have no difficulty describing the orientation, size, or shape of those objects (Perenin and Vighetto 1983, 1988; Goodale et al. 1994; Jakobson et al. 1991). In addition, they do not take into account the positions of potential obstacles when they are attempting to reach out toward goal objects (Schindler et al. 2004). All of this confirms the critical role that the dorsal stream plays in the visual control of skilled actions. The patients exhibit neither a purely visual nor a purely motor deficit, but instead a specific deficit in visuomotor control.

2.2 Visual Form Agnosia: Patient D.F.

The opposite pattern of deficits and spared abilities has been described in patients with damage to the ventral stream. The best-documented case is patient D.F., a young woman who developed a profound visual form agnosia following carbon monoxide poisoning. Structural MRI showed evidence of diffuse damage consistent with hypoxia, but with specific lesions in ventrolateral regions of the occipital cortex, with primary visual cortex remaining largely spared (Milner et al. 1991; Goodale et al. 1991; Goodale and Milner 2004a; Milner and Goodale 2006). Even though D.F.'s contrast sensitivity and other low-level visual abilities remain reasonably intact, she can no longer recognize everyday objects or the faces of her friends and relatives; nor can she identify or copy line drawings of common objects or even simple geometric shapes. Even though D.F. cannot recognize the shapes of objects, she does appear to perceive their color and texture—and can use these cues to identify an object when those surface cues are diagnostic. It should be emphasized that her inability to perceive the form of objects is largely visual. She has no trouble identifying familiar objects by touch or even by the sounds they make when tapped or placed on a hard surface. Nor does she have any problem recognizing familiar voices. There is some indication, however, that D.F. has difficulty using haptics to identify unfamiliar "nonsense" shapes (James et al. 2006), suggesting that the form vision networks that are presumably damaged in D.F.'s brain may normally play an important role in enabling the haptic system to acquire information about the geometrical structure of new objects.

What is most amazing about D.F, however, is the fact that—despite profound deficits in form vision—she shows strikingly accurate guidance of her hand and finger movements when she attempts to pick up the very objects she cannot identify (Goodale et al. 1991). Thus, when D.F. reaches out to grasp objects, her hand opens wider mid-flight for larger objects than it does for smaller ones, just as it does in someone with normal vision (see figure 4.2). She also takes into account the position of potential obstacles in the immediate vicinity of a goal object to which she is reaching (Rice et al. 2006). In addition, she rotates her hand and wrist quite normally when she reaches out to grasp objects in different orientations (Carey, Harvey, and Milner 1996), and she places her fingers optimally around the edges of objects of different shape (Goodale, Jakobson, and Keillor 1994). At the same time, she is quite unable to distinguish between any of these objects when they are presented to her in simple discrimination tests. As can be seen in figure 4.2, D.F. even fails in manual "matching" tasks in which she is asked to show how wide an object is by opening her index finger and thumb a corresponding amount (Goodale et al. 1991). D.F.'s spared visuomotor skills are not limited to grasping. She can step over obstacles during locomotion as efficiently as controls, even though her perceptual judgments about the height of these obstacles are far from normal (Patla and Goodale 1997). In short, a profound loss of form perception coexists in D.F. with a preserved ability to use information about the form of objects to guide a broad range of actions. The contrast between what D.F. can and cannot do is exactly what one would expect in someone with a damaged ventral "perception" stream but a functionally intact dorsal "action" stream.

To summarize: Even though D.F.'s brain damage has left her unable to perceive the size, shape, and orientation of objects, her visuomotor outputs remain quite sensitive to these same object features. There appears to have been an interruption in the normal flow of shape and contour information into her perceptual system without affecting the processing of shape and contour information by her visuomotor control systems. But where is the damage in D.F.'s brain? If, as was suggested earlier, the perception of objects and events is mediated by the ventral stream of visual projections to inferotemporal cortex, then D.F. should show evidence for damage relatively early in this pathway. As was mentioned earlier, an MRI taken shortly after her accident suggested that there might be bilateral damage in the ventrolateral regions of the occipital lobe. More recent high-resolution anatomical MRIs of D.F.'s brain have confirmed that this is indeed the case (James et al. 2003). In fact, the damage is remarkably localized to the lateral occipital area (LO), part of the lateral occipital complex (LOC), a heterogeneous collection of visual areas that have been implicated in object recognition in a number of functional imaging studies (Grill-Spector, Kourtzi, and Kanwisher 2001; James et al. 2000, 2002; Kourtzi and Kanwisher 2000; Malach et al. 1995). As figure 4.3 shows, the LO lesions are bilateral and do not include that part of LOC extending into the fusiform gyrus on the ventral surface of the brain. It seems likely, then, that it is the lesions in area LO that are responsible for her deficit in form and shape perception.

3 Functional Neuroimaging Studies of D.F.'s Ventral and Dorsal Streams

To test this prediction, James et al. (2003) used functional MRI (fMRI) to examine activation in D.F.'s ventral stream to line drawings of objects, stimuli which D.F. has great difficulty recognizing because the only information about the object is conveyed by form and contour information. Not surprisingly, as can be seen in figure 4.4, D.F. showed no differential activation in her ventral stream (or anywhere else in her brain) to line

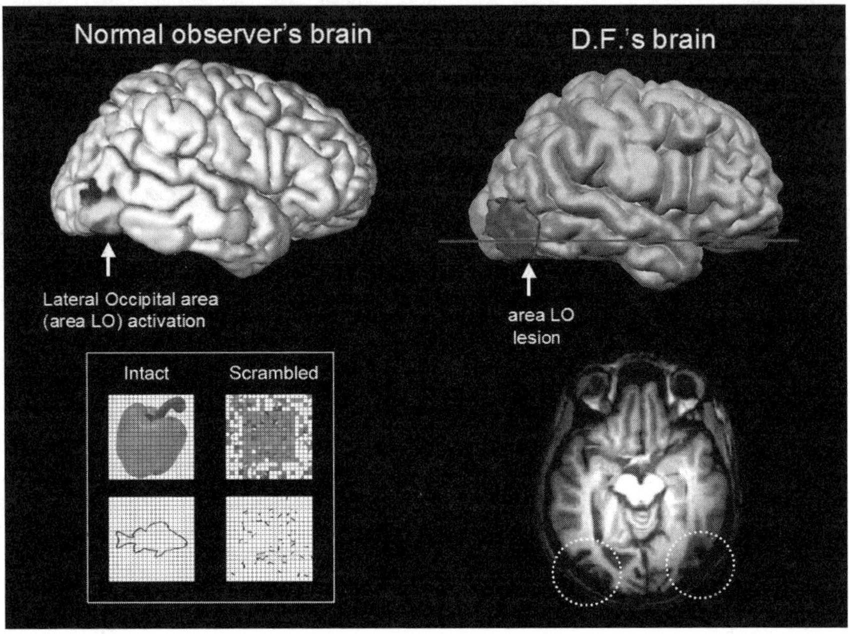

Figure 4.3

Area LO, a ventral-stream area implicated in object recognition (particularly object form), has been localized on the brain of a normal observer by comparing fMRI activation to intact versus scrambled line drawings. Note that the lesion (shaded) on patient D.F.'s right cerebral hemisphere encompasses all of area LO. Area LO in D.F.'s left hemisphere is also completely damaged. Adapted with permission from Goodale and Milner (2004a). (See the book's page at http://mitpress.mit.edu for a color version.)

drawings (as compared to scrambled versions of the same drawings). Neurologically intact observers, of course, showed robust activation to the same stimuli. Indeed, when a normal observer's brain was stereotactically aligned with D.F.'s brain, the differential activation to line drawings fell neatly into D.F.'s area LO lesions (see figure 4.4).

Although D.F. did not show any activation to line drawings of objects, James et al. (2003) predicted that she might show differential activation to images of objects in which color and texture cues were available—since she was often able to identify the material or "stuff" from which objects were made. As figure 4.5 illustrates, when D.F. was tested with such stimuli she showed robust (but somewhat atypical) activation in the fusiform and parahippocampal gyri, anteromedial to the damage in area LO. Normal

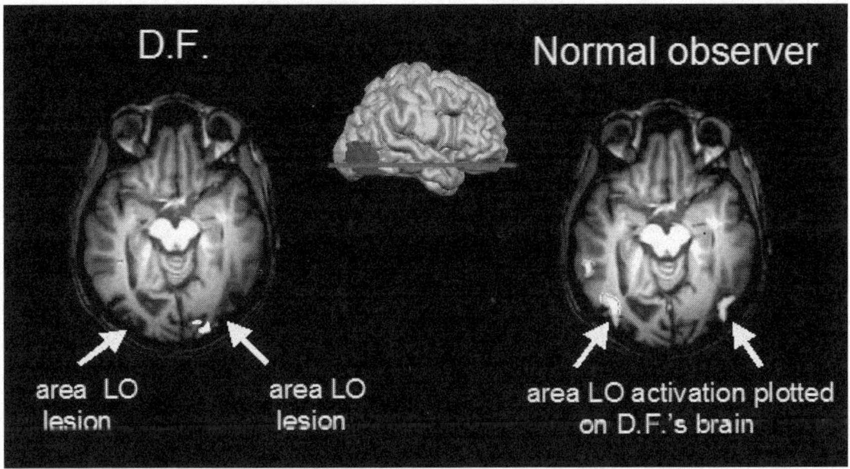

Figure 4.4
A horizontal slice through D.F.'s brain at the level of area LO (see dark line marked on the whole brain). Unlike the normal observer, D.F. showed no difference in fMRI activation with intact as compared to scrambled line drawings. The robust activation seen in the normal observer's brain for the same task has been stereotaxically morphed onto D.F.'s brain. Note that the activation to the line drawings in the normal observer falls neatly into the corresponding LO lesions on both sides of D.F.'s brain. (See the book's page at http://mitpress.mit.edu for a color version.)

observers showed robust activation in area LO, and much less in the neighboring fusiform and parahippocampal gyri (although they did show significant activation in these regions). Interestingly, though, the activation that D.F. showed in the fusiform and parahippocampal gyri was higher for objects that she was able to identify than it was for objects she could not, suggesting that top-down input might have contributed to the observed activation in these anteromedial ventral-stream areas.

So D.F., who has bilateral lesions of area LO, shows no differential activation for line drawings of objects but continues to show robust activation for colored and textured images of objects. These results not only converge nicely with the earlier behavioral findings, but also indicate that area LO may play a special role in processing the geometrical structure of objects whereas more anteromedial regions in the fusiform and parahippocampal gyri might be more involved in processing information about the material properties of objects—the stuff from which they are made. In fact, a more recent fMRI study in normal observers (Cant and Goodale 2007) found

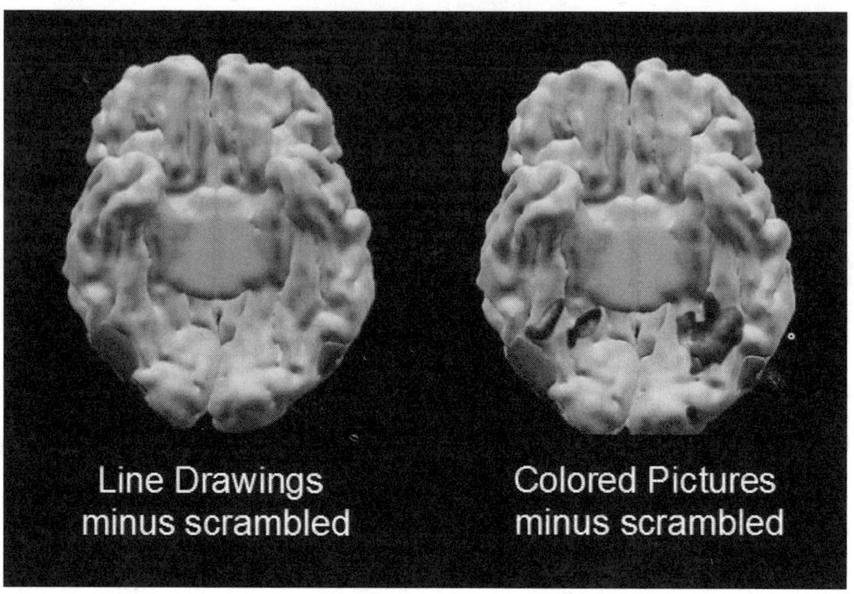

Figure 4.5
Activation for line drawings and colored pictures plotted on the ventral surface of a 3-D rendering of D.F.'s cerebral hemispheres. Note that absence of differential activation for line drawings and the robust (albeit abnormal) activation for colored pictures in the fusiform and parahippocampal regions. The bilateral area LO lesions are shaded. Adapted with permission from James et al. (2003). (See the book's page at http://mitpress.mit.edu for a color version.)

that attention to the form of objects was associated with activation in area LO whereas attention to their surface properties (and by extension their material properties) was associated with activation in the fusiform and parahippocampal gyri—overlapping in part the fusiform face area (FFA) and the parahippocampal place area (PPA). The fact that attention to visual texture and color resulted in activation in the face and place areas underscores the importance of these surface cues in face and scene perception (see, e.g., Vailaya, Jain, and Jiang Shiang 1998; Gegenfurtner and Rieger 2000; Oliva and Schyns 2000; Tarr et al. 2001, 2002). Not surprisingly, D.F. (presumably because of her spared ability to perceive visual texture and color) is able to categorize scenes reasonably well, particularly natural scenes presented in their diagnostic colors (Steeves et al. 2004). In addition, she shows selective activation for scenes in the PPA and (unlike normal observers) shows significantly higher activation for full-color images as opposed to black-and-white renditions (Steeves et al. 2004). Taken together, these fMRI results (and the structural MRI evidence discussed earlier), provide a strong confirmation of Goodale and Milner's (1992) original conjecture, namely that D.F.'s perceptual problems are a consequence of damage to form-processing regions in the ventral stream of visual processing.

But what about the visual control of actions, such as grasping, where D.F. shows relatively normal behavior? What areas of the brain are mediating this behavior? To answer this question, James et al. (2003) carried out an event-related fMRI study of grasping in D.F. It has been known for a long time that neurons in an area in the anterior part of the intraparietal sulcus of the monkey's posterior parietal cortex (area AIP) show activity related to the shape, size, and orientation of objects that are the targets of visually guided grasping movements (Taira et al. 1990). Recent neuroimaging experiments have revealed a human homologue of monkey area AIP that is also activated during visually guided grasping (Binkofski et al. 1998; Culham and Kanwisher 2001; Culham et al. 2003; Culham 2004). When James et al. (2003) asked D.F. to grasp objects in the scanner, they found robust activation in area AIP (figure 4.6). This result, coupled with the observation that area LO is damaged bilaterally in D.F., provides strong support for the argument that the visual control of object-directed grasping does not depend on object form-processing regions in the ventral stream but instead is mediated by object-driven visuomotor systems in the dorsal stream. In fact, it is worth noting that grasp-related activation in area AIP in neurologically intact individuals is also unaccompanied by any differential activation in area LO (Culham et al. 2003).

To sum up: the new MRI and fMRI findings with D.F. provide a striking confirmation of Goodale and Milner's (1992) earlier proposal that visual perception and the visual control of action depend on separate visual pathways in the cerebral cortex, and confirm the respective roles of the ventral and dorsal visual streams in these functions.

4 Acting on Illusions

4.1 Different Metrics and Frames of Reference for Perception and Action

Goodale and Milner (1992, 2004a; Milner and Goodale 2006) argue that the two separate streams of visual processing evolved because perception and action require quite different transformations of the visual signals. For an observer to be able to grasp an object successfully, for example, it is

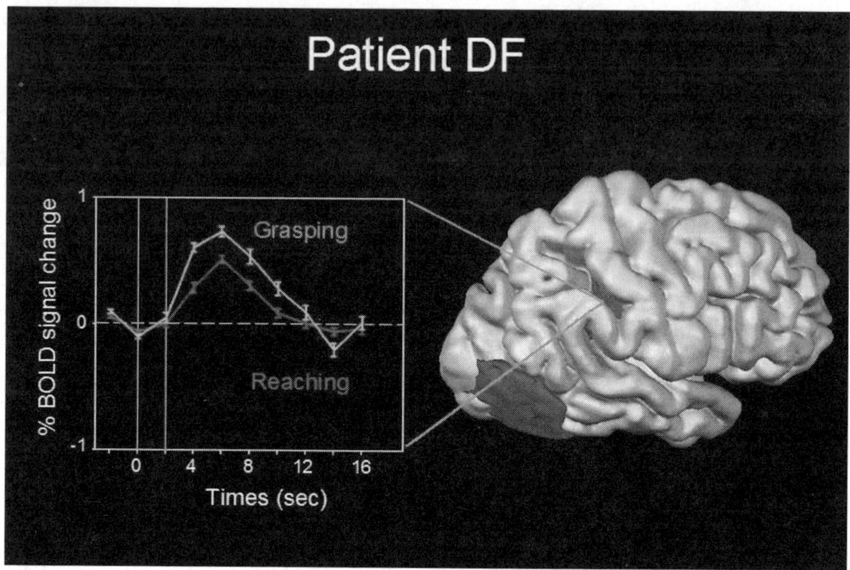

Figure 4.6

Grasp-related fMRI activation in D.F.'s dorsal stream. The task was either to grasp the target using a precision grip, or in a control condition, to simply touch it with the knuckles. Activation associated with grasping is shaded on the 3-D rendered brain on the right. There is activation in area AIP in both hemispheres but it is stronger on the right. The graph on the left shows the average event-related activation in area AIP for grasping and reaching. Adapted with permission from James et al. (2003). (See the book's page at http://mitpress.mit.edu for a color version.)

essential that the brain compute the actual size of the object, and its orientation and position with respect to the observer (i.e., in egocentric coordinates). Moreover, the time at which these computations are performed is also critical. Observers and goal objects rarely stay in a static relationship with one another, and, as a consequence, the egocentric coordinates of a target object can often change dramatically from moment to moment. For this reason, it is essential that the required coordinates for action be computed immediately before the movements are initiated. For the same reason, it would be counterproductive for these coordinates (or the resulting motor programs) to be stored in memory. In short, vision-for-action works very much in an "online" mode.

The requirements of perception are quite different, both in terms of the frames of reference used to construct the percept and the time period over which that percept (or the information it provides) can be accessed. Vision-for-perception appears not to rely on computations about the absolute size of objects or their egocentric locations. Instead, the perceptual system in the ventral stream computes the size, location, shape, and orientation of an object (and its parts) primarily in relation to other objects, object parts, and surfaces in the scene (see, e.g., Ganel and Goodale 2003). Encoding an object in a scene-based frame of reference permits a perceptual representation of the object that preserves the relations between the object and its surroundings without requiring precise information about its absolute size or its exact position with respect to the observer. Indeed, if the perceptual machinery attempted to deliver the real size and distance of all the objects in the visual array, the computational load on the system would be astronomical.

The products of perception also need to be available over a much longer time scale than the visual information used in the control of action. It may be necessary to recognize objects seen minutes, hours, days—or even years before. To achieve this, the coding of the visual information has to be somewhat abstract—transcending particular viewpoint and viewing conditions. By working with perceptual representations that are object or scene based, it is possible to maintain the constancies of size, shape, color, lightness, and relative location, over time and across different viewing conditions. Although there is much debate about the way in which this information is coded, it is clear that it is the identity of the object and its location within the scene, not its disposition with respect to the observer, that is of primary concern to the perceptual system. Thus current perception combined with stored information about previously encountered objects not only facilitates the object recognition but also contributes to the control of goal-directed movements when working in off-line mode (i.e., on the basis of the memory of a goal object and its location in the world). For a more detailed account of these arguments, see Goodale, Westwood, and Milner 2004.

4.2 Grasping in the Context of Pictorial Illusions

As we discussed earlier, some of the most compelling evidence for the two-visual-systems proposal comes from human neuropsychology and neuroimaging (as well as work with nonhuman primates). Nevertheless, differences in the metrics and frames of reference used by vision-for-perception and vision-for-action have also been revealed in studies with normal observers. Pictorial illusions have proved to be a particularly good way of doing this. Aglioti, DeSouza, and Goodale (1995), for example, showed that the scaling of grip aperture in flight was remarkably insensitive to the Ebbinghaus illusion, in which a target disk surrounded by smaller circles appears to be larger than the same disk surrounded by larger circles. In short, maximum grip aperture was scaled to the real and not the apparent size of the target disk (see figure 4.7). This resistance to the illusion was not due to participants simply comparing the amplitude of their grip to the size of the target as they performed the movement; a similar dissociation between grip scaling and perceived size was observed in a later study by Haffenden and Goodale (1998), in which participants were given no visual feedback during the execution of the grasping movement. Although grip scaling escaped the influence of the illusion, Haffenden and Goodale showed that the illusion did affect performance in a manual matching task, a kind of perceptual report, in which participants were asked to open their index finger and thumb to indicate the perceived size of a disk. In other words, the aperture between the finger and thumb was resistant to the illusion when the vision-for-action system was engaged (i.e., when the participant grasped the target) but was sensitive to the illusion when the vision-for-perception system was engaged (i.e., when the participant estimated its size).

In the context of the Goodale and Milner's (1992) two-visual-systems account, a dissociation of this kind between perception and action is not unexpected. The obligatory size-contrast effects that give rise to the illusion (whereby an object that is smaller than its immediate neighbors is assumed to be smaller than a similar object that is larger than its immediate neighbors) normally play a crucial role in scene interpretation (Coren and Girgus 1978), a central function of the perception system. In addition, it is possible that some sort of image distance computation is contributing to the illusion, in which the array of smaller circles is assumed to be more distant than the array of larger circles; as a consequence, the target circle within the array of smaller circles will also be perceived as more distant (and therefore larger) than the target circle of equivalent retinal-image size within the array of larger circles (Gregory 1963). In contrast, the execution of a goal-directed act, such as manual prehension, requires metrical computations that are centered on the target itself. As a consequence, computation of the retinal-image size of the object coupled with an accurate estimate of distance would deliver the true size of the object for calibrating the grip. Such computations would be expected therefore to be quite insensitive to the kinds of pictorial cues that drive our perception of familiar illusions.

The initial demonstrations that grasping is refractory to pictorial illusions triggered a good deal of interest among researchers studying vision

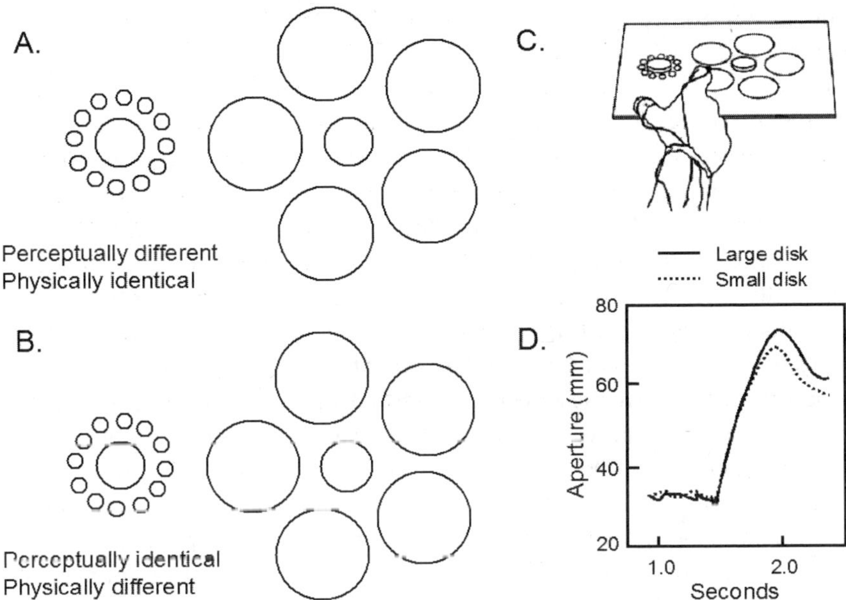

Figure 4.7

The effect of a size-contrast illusion on perception and action. A. The traditional Ebbinghaus illusion in which the central circle in the annulus of larger circles is typically seen as smaller than the central circle in the annulus of smaller circles, even though both central circles are actually the same size. B. The same display, except that the central circle in the annulus of larger circles has been made slightly larger. As a consequence, the two central circles now appear to be the same size. C. A 3-D version of the Ebbinghaus illusion. Participants are instructed to pick up one of the two 3-D disks placed either on the display shown in panel A or the display shown in panel B. D. Two trials with the display shown in panel B, in which the participant picked up the small disk on one trial and the large disk on another. Even though the two central disks were perceived as being the same size, the grip aperture in flight reflected the real not the apparent size of the disks. Adapted with permission from Aglioti et al. (1995).

and motor control. Some investigators have replicated the dissociation between perception and action using different versions of the Ebbinghaus illusion (e.g., Amazeen and DaSilva 2005; Kwok and Braddick 2003; Fischer 2001), as well as other illusions such as the Ponzo illusion (Brenner and Smeets 1996; Jackson and Shaw 2000), the horizontal-vertical illusion (Servos, Carnahan, and Fedwick 2000), the Müller-Lyer illusion (Dewar and Carey 2006), the diagonal illusion (Stöttinger and Perner 2006), and the rod-and-frame illusion (Dyde and Milner 2002). Some have reported that these illusions affect some aspects of motor control but not others (e.g., Gentilucci et al. 1996; Daprati and Gentilucci 1997; van Donkelaar 1999; Glazebrook et al. 2005). And a few investigators have found no dissociation whatsoever between the effects of pictorial illusions on perceptual judgments and the scaling of grip aperture (e.g., Franz et al. 2000; Franz, Bulthoff, and Fahle 2003).

The fact that actions such as grasping are sometimes sensitive to illusory displays is not by itself a refutation of the idea of two visual systems. Indeed, one should not be surprised that perception affects our motor behavior. After all, ultimately, perception has to affect our actions or the brain mechanisms mediating perception would never have evolved. The real surprise (at least for monolithic accounts of vision) is that there are instances where visually guided action is apparently unaffected by perception. But from the standpoint of Goodale and Milner's (1992) proposal, such instances are to be expected. Nevertheless, the fact that action has been found to be affected by pictorial illusions in some experiments has led some authors to argue that the early studies demonstrating a dissociation had not adequately matched action and perception tasks for various input, attentional, and output demands (e.g., Smeets and Brenner 2001; Vishton and Fabre 2003)—and that when these factors were taken into account the apparent differences between perceptual judgments and motor control could be resolved without invoking the idea of two visual systems. Other authors, notably Glover (2004), have argued that action tasks involve multiple stages of processing from purely perceptual to more "automatic" visuomotor control. According to this so-called planning/control model, illusions would be expected to affect the early but not the late stages of a grasping movement (Glover 2004; Glover and Dixon 2001a,b).

Some of these competing accounts, particularly Glover's (2004) planning/control model, are simply modifications of Goodale and Milner's (1992) original proposal. But Glover's model fails to distinguish between planning in the sense of deciding on one course of action rather than another, and planning in the sense of programming the actual constituent movements of an action (Goodale and Milner 2004b). Goodale and Milner would not dispute that the ventral stream is involved in the former kind of planning, but would argue that the other kind of planning (i.e., programming) is mediated by mechanisms in the dorsal stream. In fact, Glover and Dixon's (2002) claim that ventral-stream mechanisms intrude into the early part of motor programming for grasping movements is based on findings that have been difficult to replicate (Danckert et al. 2002). But even so, there are numerous other studies whose results cannot easily be reconciled with the two-visual-systems model, and it remains a real question as to why actions appear to be sensitive to illusions in some experiments but not in others.

One possible explanation for the different findings with the Ebbinghaus illusion has come from work by Haffenden and Goodale (2000) and Haffenden, Schiff, and Goodale (2001), who showed that the two-dimensional arrays of circles surrounding the target disks were sometimes treated as potential obstacles. As a consequence, they argued, the surrounding circles could influence the posture of the fingers during grasping (see also Plodowski and Jackson 2001). In other words, the apparent effect of the illusion on grip scaling in some experiments might simply reflect the operation of visuomotor mechanisms that treat the flanker elements of the visual arrays as obstacles to be avoided. Indeed, recent studies of neurological patients provide convincing evidence that mechanisms in the dorsal stream normally take into account the position of potential obstacles in planning the trajectory of target-directed movements (Schindler et al. 2004; Rice et al. 2006). In addition, de Grave et al. (2005) have recently shown that simply rotating the flanking circles around the target can result in differential effects on maximum grip aperture, presumably because the fingers would be more likely to "collide" with the flankers in some positions than in others. If the direction of this "flanker" effect coincides with the predicted effect of the illusion on grasp aperture, an investigator could erroneously conclude that the visuomotor programming was sensitive to the illusion. In pictorial illusions, such as the Ponzo and diagonal illusions, where the presence of potential "obstacles" is less of a problem, investigators have typically found that grip aperture is quite immune to the effects of the illusion (see, e.g., Brenner and Smeets 1996; Jackson and Shaw 2000; Stöttinger and Perner 2006).

One other variable that might explain the discrepancies in the results is the timing of the grasp with respect to the presentation of the stimuli. Westwood and Goodale (2003) found that when the target was visible during the programming of a grasping movement, maximum grip aperture was not affected by a simple size-contrast illusion, whereas when vision was occluded just before the programming of the movement occurred, a reliable effect of the illusion on grip aperture was observed. A similar dissociation between visually guided and memory-guided grasping was also found with the Müller-Lyer illusion (Westwood, Heath, and Roy 2000). These particular findings not only confirm the dissociation between perception and action, but also provide strong support for the idea, discussed earlier, that the dorsal "action" stream operates in real time and is not normally engaged unless the target object is visible during the programming phase, that is, when (bottom-up) visual information is being converted into the appropriate motor commands. The observation that (top-down) memory-guided grasping is affected by the illusory display is probably due to the fact that the stored information about the target's dimensions was originally derived from a perceptual representation of the scene created moments earlier by mechanisms in the ventral stream (Goodale, Jakobson, and Keillor 1994; Fischer 2001; Hu and Goodale 2000).

But the obstacle argument and differences in timing cannot account for all of the apparent discrepancies in the illusion literature (cf. Carey 2001; Bruno 2001; and Franz 2001). What other factors could be at play? Recent work by Dyde and Milner (2002) suggests that at least some of the differences in the findings could be related to the type of visual illusions that were used. In a particularly clever series of experiments, they showed that the orientation of the grasping hand is sensitive to a simultaneous tilt illusion—similar to the one used by Glover and Dixon (2001a)—but not a rod-and-frame illusion, even though the two visual displays have equivalent effects on judgments of target orientation. Dyde and Milner argue that the simultaneous tilt illusion arises from "early" (i.e., area V1, area V2) stages of visual processing, and thereby influences activity in both the dorsal and ventral visual pathways. The rod-and-frame illusion, in contrast, is thought to arise during later stages of processing (i.e., in inferotemporal cortex) and should consequently not affect action, because the encapsulated visuomotor systems within the dorsal stream do not have direct access to this processing. In a recent review (Milner and Dyde 2003), the authors emphasize that before selecting an illusion to demonstrate possible dissociations between perception and action, one has to take into account the putative brain areas that are involved in generating that illusion. Ideally, then, a definitive test for the dissociation between vision-for-perception and vision-for-action should utilize a visual illusion that is demonstrably "higher-order," that is, one that is dependent on perceptual/cognitive mechanisms in the ventral stream. This is exactly what was done in a recent experiment by Króliczak et al. (2006), who used the hollow-face illusion, a high-order illusion that depends on our knowledge of faces. For a dramatic demonstration of the hollow-face illusion, see http://www. richardgregory.org/experiments/index.htm.

4.3   Unmasking the Hollow-Face Illusion

The hollow-face illusion, in which a hollow mask is perceived (incorrectly) as a normal protruding face, has two important characteristics that make it quite different from other illusions that have been used to examine the possible dissociation between perception and action: (1) unlike traditional illusions of extent or position, the hollow-face illusion involves a reversal of the perceived depth, and (2) under the right testing conditions, the depth reversal can be several centimeters (Gregory 1970). Thus, if action is refractory to the hollow-face illusion, there should be a dramatic difference between the effect of the display on perception and the effect of the same display on visually guided movements directed at targets located on the display. Such a large predicted difference would contrast sharply with what has been predicted in earlier experiments that have used illusions such as the Müller-Lyer and Ebbinghaus illusions, where the differences between the effects on perception and action are typically no more than a few millimeters. Indeed, the very fact that the predicted differences are so small with typical pictorial illusions could also help to explain why the results have sometimes been hard to replicate.

In the Króliczak et al. (2006) study, participants were asked to use their index finger to "flick" off a small target stuck to the surface of an actually hollow, but apparently normal face, or to flick the same target off the surface of a normal protruding face. The task was designed to resemble an ecologically plausible task, quickly flicking a small insect off someone's face. The idea was that the fast flicking movement would engage visuomotor networks in the dorsal stream, and thus would be directed to the veridical rather than the perceived position of the target. On other blocks of trials, participants were asked to indicate the apparent position of the target by pointing to it slowly and on other trials by drawing its position with a pencil on a piece of paper. These latter two tasks provided measures of the perceived position of the targets on the surface of the face. The results conformed almost exactly to what the Goodale and Milner (1992) model would predict. That is, despite the presence of a robust illusory reversal of depth (as measured by the paper-and-pencil measure in particular), the fast flicking movements were directed to the real, not the illusory (perceived) location of the target (see figure 4.8). In slow pointing, the endpoints of

A.

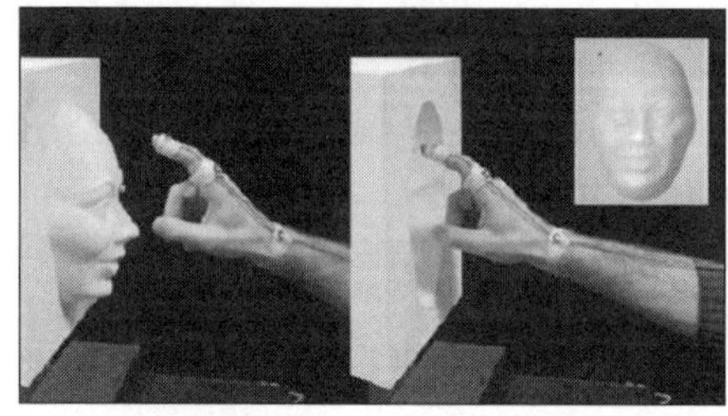

B.

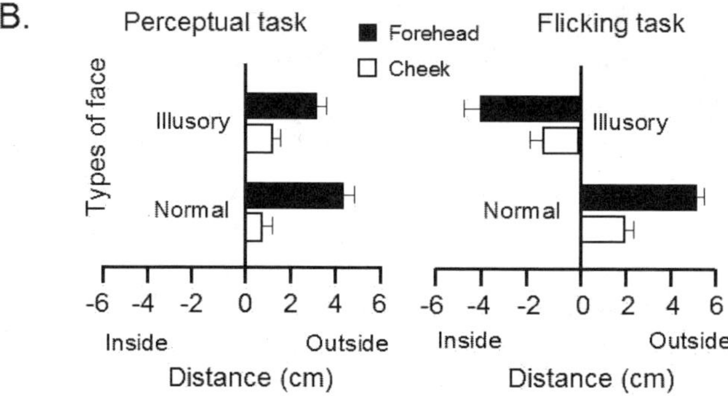

Figure 4.8

Perceptual judgments and visuomotor control with the hollow-face illusion. Panel A. A small magnet was placed on either the cheek or forehead of the normal face (left) or the hollow mask (right). Participants were required either to flick the magnet from the normal or illusory (actually hollow) face or to estimate its distance psychophysically. Inset shows a photograph of bottom-lit hollow face, in which the illusion of a normal convex face is evident. Panel B. Left. The mean psychophysical (paper and pencil) judgments of the apparent position of the magnets on the illusory and normal face with respect to the reference plate from which the two displays either protruded or receded. Note that participants perceived the hollow face as protruding forward like the normal face. Right. The mean distance of the hand at the moment the participant attempted to flick the target off the cheek or forehead of the illusory (actually hollow) or the normal face. In the case of the illusory face, the endpoints of the flicking movements corresponded to the actual distances of the targets, not to consciously perceived distances. Error bars indicate the standard error of the mean. Adapted with permission from Króliczak et al. (2006).

the movements also corresponded to the illusory location of the targets (see figure 4.9), although the performance on this task was more variable. Taken together, the results of this experiment show that the visuomotor system can use bottom-up sensory inputs (perhaps involving vergence) to guide a goal-directed movement to the real location of the target, even when perceived position of the target is influenced, or even reversed, by top-down processing.

In a recent study by Hartung et al. (2005), which also examined the hollow-face illusion but used a 3-D virtual reality display, the claim was made that action and perception were both "fooled" by the illusion. In the Hartung et al. experiment, however, the measure that was used to tap into the visuomotor system was a pointing movement, similar to one used by Króliczak et al. (2006). But such pointing movements, as distinct from the fast target-directed flicking movements that were also used in the Króliczak et al. study, can often reflect cognitive/perceptual judgments about the

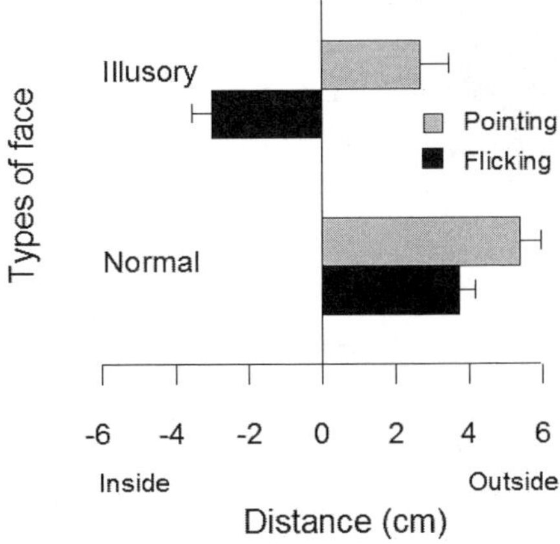

Figure 4.9
The slow pointing and fast flicking responses (the endpoints are averaged over both target positions). For the normal face, both the flicking and slow hand movements were nearly veridical. For the illusory face, the movements were very different. Here, the endpoints of the slow pointing corresponded to the illusory position of the target (in front of the reference plate) whereas the endpoints of the fast flicking movements corresponded to their actual position (behind the reference plate). Error bars indicate standard errors of the mean. Adapted with permission from Króliczak et al. (2006).

location of the target and need not engage the more "automatic" visuomotor mechanisms in the dorsal stream (Bridgeman, Peery, and Anand 1997; Rossetti et al. 2005). Indeed, pointing is a rather anomalous behavior: In some cases, such as rapid target-directed aiming movements, it can be quite automatic; in other cases, such as when one person indicates to another where a particular stimulus is located, it can be much more deliberate and cognitively controlled (Bridgeman et al. 2000).

4.4 The Right Hand and the Left Hemisphere

This distinction between controlled and automatic movements may also help to explain why some researchers (e.g., Franz et al. 2000; Franz, Bulthoff, and Fahle 2003; Radoeva et al. 2005) have found that grip aperture is affected by the Ebbinghaus and other familiar size-contrast illusions. Because the devices that are used to measure grip aperture in these experiments were quite intrusive (see figure 4.10), however, it is possible that

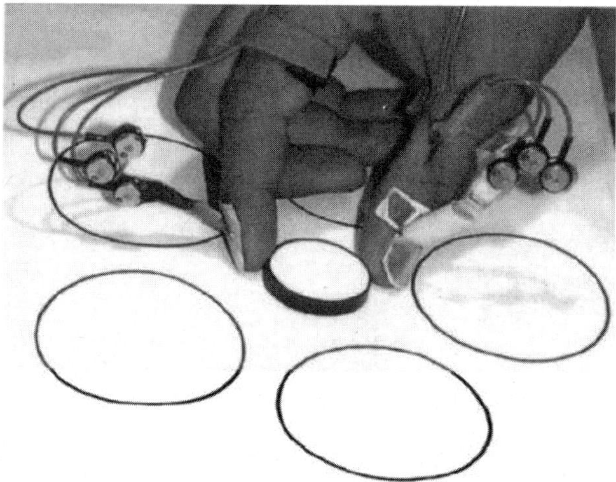

Figure 4.10

Photograph of the optoelectronic sensors used by Franz et al. (2000) to record grip aperture in a study of the effects of the Ebbinghaus illusion on grip scaling. The relatively large extensions on which the three infrared light emitting diodes were mounted (together with the attached wires) on the finger and thumb may well have interfered with normal grasping and thus led participants to use more deliberate control in executing their grasp. As a consequence, the cognitive monitoring of the grasp could have made use of perceptual information derived from ventral-stream processing. Reproduced with permission from Franz et al. (2000).

grasping movements made under these conditions were more controlled than automatic, thus relying more on perceptual information (presumably provided by the ventral stream) than on encapsulated visuomotor control (by the dorsal stream).

Gonzalez, Ganel, and Goodale (2006) explored this possibility by directly comparing the sensitivity of skilled and awkward grasping movements to a pictorial illusion, predicting that the former but not the latter would escape the effects of the illusion. In their experiment, right-handed participants were asked to pick up one of two small rectangular objects with their right hand. On some trials, the objects, which were identical in size, were presented against the backdrop of a Ponzo (railway tracks) illusion. Some participants were required to use a normal precision grip (with the thumb and index finger), while other participants were required to use a much more awkward grip (with the thumb and ring finger). The results were clear and unambiguous. Even though the illusion had no effect on grip scaling in the participants who used a precision grasp, it had a large and significant effect on grip scaling in the participants who used an awkward grasp. This result provides some confirmation of the idea that awkward actions, which require the use of more deliberate cognitive control, are more likely to rely on the same perceptual processing participants use to make conscious judgments about the size of objects in illusory displays. It also suggests that in experiments designed to investigate possible differences between vision-for-action and vision-for-perception, one should be careful to ensure that the recording methods used to measure the actions do not interfere with the "automaticity" of the constituent movements.

But can awkward actions eventually escape cognitive control and become more automatic—and thus less susceptible to visual illusions—as participants gain more experience? To examine this question, Gonzalez et al. (2008) gave participants three days of practice picking up objects using an awkward grasp. By the end of the third day, the grip scaling was no longer sensitive to the illusion. The awkward grasp was no longer awkward—and presumably now engaged the same automatic mechanisms that mediate the familiar precision grip.

If unfamiliar and less practiced actions are more likely to make use of vision-for-perception than vision-for-action, then one might predict that precision grasping with the left hand in right-handers would be much more sensitive to pictorial illusions than precision grasping with the right hand. After all, right-handers are presumably much more skilled at visuomotor tasks with their right hand than they are with their left. In a recent experiment, Gonzalez, Ganel, and Goodale (2006) showed that grasping movements made with the left hand (but not the right) are indeed sensitive to both the Ebbinghaus and the Ponzo illusion (see figure 4.11). In other words, the control of unskilled movements made with the left hand (in right-handers) appears to make use of scene-based perceptual information, whereas the control of skilled movements with the right hand is mediated by encapsulated visuomotor networks that compute the real size of the target objects. But the story becomes more complicated. When Gonzalez and colleagues went on to test a group of left-handers, individuals who by definition favor their left hand, they found that even in this group the left hand was much more sensitive to the illusions than the right (see figure 4.11). In other words, the left-handers behaved just like right-handers. This surprising result suggests that skill is not the only factor affecting sensitivity to pictorial illusions. Indeed, Gonzalez and colleagues suggested that the left hemisphere, which has direct control of the distal musculature of the right hand, has a special role to play in visuomotor control—and that this left-hemisphere specialization is also present in the majority of left-handers. This idea receives support from recent observa-

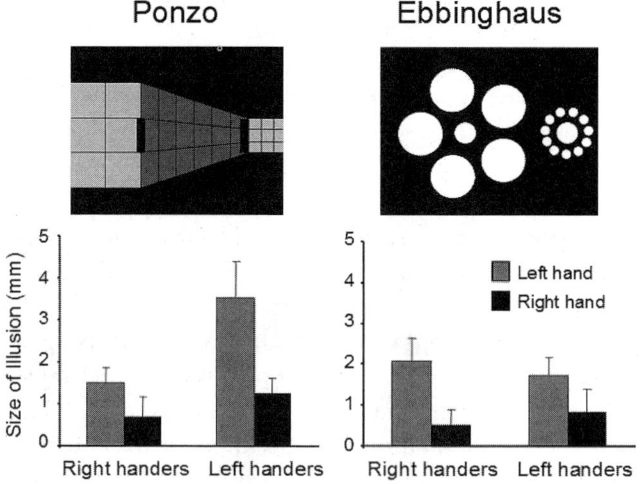

Figure 4.11

The effects of pictorial illusions on grasping with the right and left hand. In both the Ponzo (left) and Ebbinghaus (right) illusions, right-handed subjects showed no effect of the illusion on their handgrip size when using the right hand (black bars), yet showed a strong and significant effect when using the left hand (gray bars). Left-handed subjects also showed a similar effect of larger illusion magnitude when using the left hand. Adapted with permission from Gonzalez et al. (2006).

tions that when left-handers reached out and pick up small objects in a "natural" setting (puzzle pieces and Lego blocks), they do not behave like the mirror image of right-handers (Gonzalez et al. 2007). Thus, although right-handers showed a marked preference for using their dominant (right) hand (78 percent), left-handers did not show this preference and instead used their nondominant (right) hand 52 percent of the time. In fact, some left-handers were more right-handed than some right-handers, at least when it came to employing a precision grip to acquire small objects!

The idea that the left hemisphere is specialized for the visual control of action is consistent with observations of reaching and grasping deficits in patients with unilateral damage to the posterior parietal cortex. As mentioned earlier, such deficits are referred to by neurologists as optic ataxia. Perenin and Vighetto (1988) have reported that target-directed movements with the right hand are more severely impaired in patients with optic ataxia following damage to the left hemisphere than are similar movements with the left hand following damage to the right hemisphere. In other words, patients with left-hemisphere damage show a "hand effect" and have great difficulty reaching toward objects and shaping their grasp appropriately with their right hand anywhere in space. With their left hand, however, the deficit is apparent only in the right visual field, the field contralateral to the lesion. Patients with right-hemisphere damage do not show this hand effect and their deficit in visuomotor control is limited to the field contralateral to their lesion. That is, they can acquire objects successfully when reaching with either hand into the right visual field, but show deficits in both hands when reaching for objects in the left visual field. Similar observations have been made by other investigators working with optic ataxia patients with unilateral damage to the left posterior parietal cortex (e.g., Boller et al. 1975; Ferro 1984). In short, the evidence from the optic ataxia patients, like the work of Gonzalez et al. (2006, 2007, 2008), suggests that the encapsulated visuomotor networks that mediate rapid target-directed movements may have evolved preferentially in the left hemisphere alongside the well-established specialization of this hemisphere for praxis and ultimately speech (for review, see Kimura 1993).

The relationship between the praxis network (which mediates movement selection) and the visuomotor network (which mediates visual control of skilled goal-directed movements) is poorly understood. Patients with apraxia (deficits in movement selection) following damage to the left hemisphere will typically show performance deficits when using either hand—and sometimes with other effectors as well, such as the mouth (Kimura 1982; Koski, Iacoboni, and Mazziotta 2002). Moreover, there is some evidence that left-hemisphere (as opposed to right-hemisphere) damage, even when it does not result in the obvious visuomotor deficits that characterize optic ataxia, will produce subtle deficits in the kinematics of rapid visually guided aiming movements (Fisk and Goodale 1988)—and these deficits are also apparent in the ipsilesional limb (the limb on the same side as the lesion). As was already noted, however, damage to the dorsal-stream visuomotor networks in the left (but not the right) hemisphere results in a contralesional hand effect—and damage to either hemisphere results in a field effect (Perenin and Vighetto 1988).

In neurologically intact individuals, the right-hand advantage that is typically observed in the performance of many skilled tasks, including the use of tools, is thought to reflect the existence of more direct connections between this hand and the left-hemisphere mechanisms involved in the selection and/or control of the constituent movements (Bryden 1982). As figure 4.12 shows, this contralateral advantage for the performance of complex movements is also evident as a right-sided bias in mouth opening during the performance of sequences of different oral postures (Wolf and Goodale 1987). The fact that skilled grasping movements with the right but not the left hand are resistant to the effects of pictorial illusions, as we have already seen, may reflect the special role played by the left hemisphere in visuomotor control (Gonzalez et al. 2006, 2007, 2008). One possible way to link these findings is to suggest that the left-hemisphere praxis system selects the appropriate movements for a precision grasping movement—but because the control of these movements has to be rapidly integrated with incoming visual information, privileged lines of communication have evolved between the praxis networks, dorsal-stream mechanisms in the left hemisphere, and the direct motor control of the right hand. Indeed, this could be one of the driving forces behind the emergence of right-handedness as a population-level trait. Nevertheless, the behavior of left-handers provides some complications for this account. Even though Gonzalez et al. (2006, 2007) found that left-handers are as likely to use their right hand as their left in a precision grip task (and show the same right-hand resistance to pictorial illusions), there is evidence that manual praxis systems tend to be more often lateralized to the right than to the left hemisphere in left-handers (Kimura 1993). This would suggest programming and execution of skilled visually guided movements in left-handers would require much more interhemispheric interaction. In short, it remains unclear at present how the praxis and visuomotor systems interact and how they might have evolved in relation to one another (but see Buxbaum, Kyle, and Menon 2005).

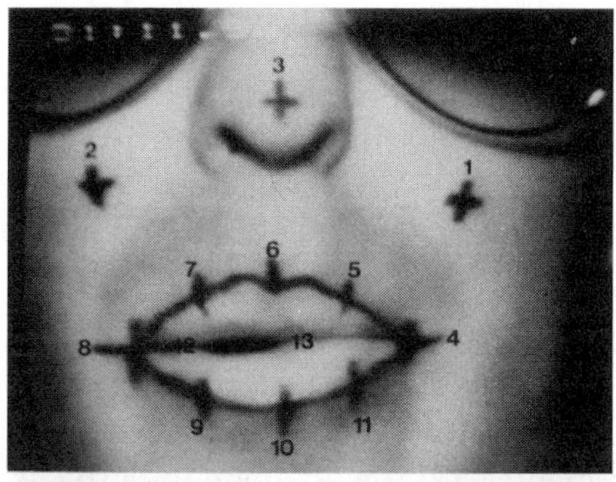
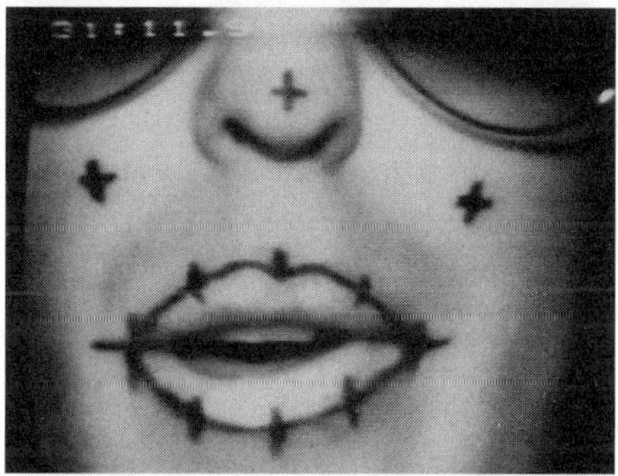
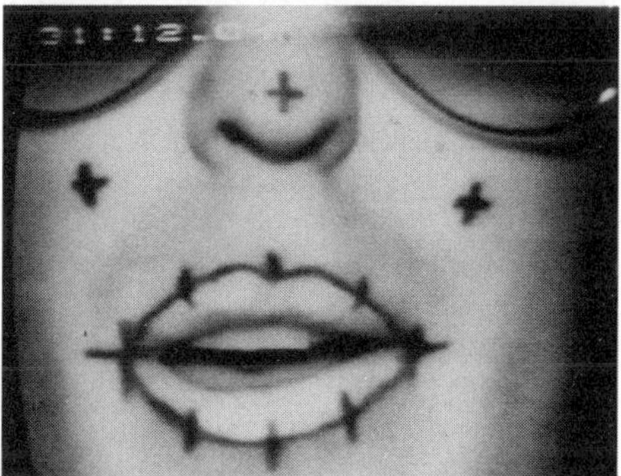

Figure 4.12

Successive video frames illustrating the mouth opening during the production of the syllable "ma" in the sequence "ma-bo-pi." The middle frame shows the mouth 67 ms after the top frame, and the bottom frame shows the mouth 50 ms after the middle frame. The lips were outlined with eyeliner pencil and cream-colored lip paint was applied to the lips. A number of reference points were marked around the perimeter of the lips and on the nose and cheeks. Notice that the right side of the mouth opens before the left. The same asymmetry was evident when participants performed a sequence of different nonverbal mouth movements (opening the mouth, blowing, and retracting the lips). In both verbal and nonverbal tasks, the asymmetry was greater for a sequence of different movements than it was for repetition of the same movements. Adapted with permission from Wolf and Goodale (1987).

Some tentative predictions can nevertheless be made. If the left-hemisphere/right-hand system is adapted for skilled precision grasping (in right-handers, at least), then one might expect that the left hand would be particularly poor at acquiring new visuomotor skills. To test this idea, Gonzalez et al. (2008) trained right-handed participants to pick up targets placed in the Ponzo display using an awkward grasp (thumb and ring finger)—but with their left hand. In contrast to what happened with the right hand, awkward grasping movements with the left hand were still as sensitive to the effects of the Ponzo illusion at the end of the third day of training as they were at the end of the first day. This suggests that the visual control of unskilled grasping movements made with the left hand (unlike those made with the right) cannot be shifted as easily to the encapsulated visuomotor systems in the dorsal stream (in the left hemisphere) that work with the real metrics of the world.

To sum up: The visual control of skilled actions, unlike visual perception, operates in real time and uses the metrics of the real world. As a consequence, many actions such as grasping or target-directed aiming movements are immune to the effects of high-level illusions of size or depth, which by definition affect perceptual judgments. Recent evidence also suggests that some components of the encapsulated visuomotor mechanisms in the dorsal stream that presumably mediate these actions are lateralized to the left hemisphere (at least in the case of grasping)—and that the more skilled the action, the more likely it is that the action will be mediated by these left-hemisphere mechanisms. All of these findings are a cautionary tale for investigators using visual illusions to tease apart the workings of vision-for-action and vision-for-perception in normal observers. Only highly practiced actions with the right hand operating in real time and directed at visible targets presented in the context of high-level illusions are likely to escape the effects of vision-for-perception.

5 Conclusions

We began this chapter by stating that achieving a complete understanding of how vision works will require paying as much attention to the motor outputs vision serves as to the organization of its different inputs. We argued that Goodale and Milner's (1992) proposed division of labor between vision-for-perception and vision-for-action for the ventral and dorsal streams is a useful first step in this direction. According to their model, both streams of visual processing transform visual information into motor output. In the dorsal stream, the transformation is direct:

Visual input and motor output are essentially "isomorphic" with one another. In the ventral stream, however, the transformation is quite indirect, and the construction of a perceptual representation of the world permits a "propositional" relationship between input and output, taking into account previous knowledge and experience. Although both streams process information about the structure of objects and about their spatial locations, they use quite different frames of reference and metrics to deal with this information. The operations carried out by the ventral stream use scene-based frames of reference and relational metrics; those carried out by the dorsal stream use egocentric frames of reference and absolute metrics.

Both streams work together in the production of goal-directed behavior. The ventral stream (together with associated cognitive machinery) identifies goals and plans appropriate actions; the dorsal stream (in conjunction with related circuits in premotor cortex, basal ganglia, and brain stem) programs and controls those actions. This interplay between a "smart" but metrically challenged ventral stream and a "dumb" but metrically accurate dorsal stream is reminiscent of the interaction between the human operator and a semiautonomous robot in what engineers call *teleassistance* (Pook and Ballard 1996; Goodale and Humphrey 1998). A full understanding of the integrated nature of visually guided behavior will require that we specify the nature of the interactions and information exchange that occurs between these two streams of visual processing. This will only happen, however, if we abandon the idea that sensory systems can be studied in isolation from the motor systems they serve.

References

Aglioti, S., J. F. X. DeSouza, and M. A. Goodale (1995). Size-contrast illusions deceive the eye but not the hand. *Current Biology* 5: 679–685.

Amazeen, E. L., and F. DaSilva (2005). Psychophysical test for the independence of perception and action. *Journal of Experimental Psychology: Human Perception and Performance* 31: 170–182.

Bálint, R. (1909). Seelenlämung des "Schauens," optische Ataxie, räumliche Störung der Aufmerksamkeit. *Monatschrift für Psychiatrie und Neurologie* 25: 51–81.

Binkofski, F., C. Dohle, S. Posse, K. M. Stephan, H. Hefter, R. J. Seitz, and H. J. Freund (1998). Human anterior intraparietal area subserves prehension: A combined lesion and functional MRI activation study. *Neurology* 50: 1253–1259.

Boller, F., M. Cole, Y. Kim, J. L. Mack, and C. Patawaran (1975). Optic ataxia: Clinical-radiological correlations with the EMIscan. *Journal of Neurology, Neurosurgery, and Psychiatry* 38: 954–958.

Brenner, E., and J. B. Smeets (1996). Size illusion influences how we lift but not how we grasp an object. *Experimental Brain Research* 111: 473–476.

Bridgeman, B., A. Gemmer, T. Forsman, and V. Huemer (2000). Processing spatial information in the sensorimotor branch of the visual system. *Vision Research* 40: 3539–3552.

Bridgeman, B., S. Peery, and S. Anand (1997). Interaction of cognitive and sensorimotor maps of visual space. *Perception and Psychophysics* 59: 456–469.

Brown, S., and E. A. Schäfer (1888). An investigation into the functions of the occipital and temporal lobes of the monkey's brain. *Philosophical Transactions of the Royal Society of London* 179: 303–327.

Bruno, N. (2001). When does action resist visual illusions? *Trends in Cognitive Science.* 5: 379–382.

Bryden, M. P. (1982). *Laterality: Functional Asymmetry in the Intact Brain.* New York: Academic Press.

Buxbaum, L. J., K. M. Kyle, and R. Menon (2005). On beyond mirror neurons: Internal representations subserving imitation and recognition of skilled object-related actions in humans. *Cognitive Brain Research* 25: 226–239.

Cant, J. S., and M. A. Goodale (2007). Attention to form or surface properties modulates different regions of human occipitotemporal cortex. *Cerebral Cortex* 17: 713–731.

Carey, D. P. (2001). Do action systems resist visual illusions? *Trends Cognitive Science* 5: 109–113.

Carey, D. P., M. Harvey, and A. D. Milner (1996). Visuomotor sensitivity for shape and orientation in a patient with visual form agnosia. *Neuropsychologia* 34: 329–338.

Coren, S., and J. S. Girgus (1978). *Seeing Is Deceiving: The Psychology of Visual Illusions.* Hillsdale, N.J.: Lawrence Erlbaum.

Culham, J. (2004). Human brain imaging reveals a parietal area specialized for grasping. In *Attention and Performance XX. Functional Neuroimaging of Visual Cognition*, ed. N. Kanwisher and J. Duncan, 415–436. Oxford: Oxford University Press.

Culham, J. C., S. L. Danckert, J. F. X. DeSouza, J. S. Gati, R. S. Menon, and M. A. Goodale (2003). Visually-guided grasping produces fMRI activation in dorsal but not ventral stream brain areas. *Experimental Brain Research* 153: 180–189.

Culham, J. C., and N. G. Kanwisher (2001). Neuroimaging of cognitive functions in human parietal cortex. *Current Opinion in Neurobiology* 11: 157–163.

Danckert, J., N. Sharif, A. M. Haffenden, K. C. Schiff, and M. A. Goodale (2002). A temporal analysis of grasping in the Ebbinghaus illusion: Planning versus on-line control. *Experimental Brain Research* 144: 275–280.

Daprati, E., and M. Gentilucci (1997). Grasping an illusion. *Neuropsychologia* 35: 1577–1582.

de Grave, D. D., M. Biegstraaten, J. B. Smeets, and E. Brenner (2005). Effects of the Ebbinghaus figure on grasping are not only due to misjudged size. *Experimental Brain Research* 163: 58–64.

Dewar, M. T., and D. P. Carey (2006). Visuomotor "immunity" to perceptual illusion: A mismatch of attentional demands cannot explain the perception-action dissociation. *Neuropsychologia* 44: 1501–1508.

Dyde, R. T., and A. D. Milner (2002). Two illusions of perceived orientation: One fools all of the people some of the time, the other fools all of the people all of the time. *Experimental Brain Research* 144: 518–527.

Ferrier, D., and G. F. Yeo (1884). A record of experiments on the effects of lesion of different regions of the cerebral hemispheres. *Philosophical Transactions of the Royal Society of London* 175: 479–564.

Ferro, J. M. (1984). Transient inaccuracy in reaching caused by a posterior parietal lobe lesion. *Journal of Neurology, Neurosurgery, and Psychiatry* 47: 1016–1019.

Fischer, M. H. (2001). How sensitive is hand transport to illusory context effects? *Experimental Brain Research* 136: 224–230.

Fisk, J. D., and M. A. Goodale (1988). The effects of unilateral brain damage on visually guided reaching: Hemispheric differences in the nature of the deficit. *Experimental Brain Research* 72: 425–435.

Franz, V. H. (2001). Action does not resist visual illusions. *Trends in Cognitive Science* 5: 457–459.

Franz, V. H., H. H. Bulthoff, and M. Fahle (2003). Grasp effects of the Ebbinghaus illusion: Obstacle avoidance is not the explanation. *Experimental Brain Research* 149: 470–477.

Franz, V. H., K. R. Gegenfurtner, H. H. Bulthoff, and M. Fahle (2000). Grasping visual illusions: No evidence for a dissociation between perception and action. *Psychological Science* 11: 20–25.

Friston, K. (2003). Learning and inference in the brain. *Neural Networks* 6(9): 1325–1352.

Ganel, T., and M. A. Goodale (2003). Visual control of action but not perception requires analytical processing of object shape. *Nature* 426: 664–667.

Gazzaniga, M. S. (2004). *The Cognitive Neurosciences*, third ed. Cambridge, Mass.: MIT Press.

Gegenfurtner, K. R., and J. Rieger (2000). Sensory and cognitive contributions of color to the recognition of natural scenes. *Current Biology* 10: 805–808.

Gentilucci, M., S. Chieffi, E. Daprati, M. C. Saetti, and I. Toni (1996). Visual illusion and action. *Neuropsychologia* 34: 369–376.

Glazebrook, C. M., V. P. Dhillon, K. M. Keetch, J. Lyons, E. Amazeen, D. J. Weeks, and D. Elliott (2005). Perception-action and the Müller-Lyer illusion: Amplitude or endpoint bias? *Experimental Brain Research* 160: 71–78.

Glover, S. (2004). Separate visual representations in the planning and control of action. *Behavioral and Brain Sciences* 27: 3–24; discussion 24–78.

Glover, S., and P. Dixon (2001a). Motor adaptation to an optical illusion. *Experimental Brain Research* 137: 254–258.

Glover, S., and P. Dixon (2001b). The role of vision in the on-line correction of illusion effects on action. *Canadian Journal of Experimental Psychology* 55: 96–103.

Glover, S., and P. Dixon (2002). Dynamic effects of the Ebbinghaus illusion in grasping: Support for a planning/control model of action. *Perception and Psychophysics* 64: 266–278.

Gonzalez, C. L., T. Ganel, and M. A. Goodale (2006). Hemispheric specialization for the visual control of action is independent of handedness. *Journal of Neurophysiology* 95: 3496–3501.

Gonzalez, C. L., T. Ganel, R. L. Whitwell, B. Morrissey, and M. A. Goodale (2008). Practice makes perfect, but only with the right hand: Sensitivity to perceptual illusions with awkward grasps decreases with practice in the right but not the left hand. *Neuropsychologia* 46: 624–631.

Gonzalez, C. L., R. L. Whitwell, B. Morrissey, T. Ganel, and M. A. Goodale (2007). Left handedness does not extend to visually guided precision grasping. *Experimental Brain Research* 182: 275–279.

Goodale, M. A., and G. K. Humphrey (1998). The objects of action and perception. *Cognition* 67: 179–205.

Goodale, M. A., L. S. Jakobson, and J. M. Keillor (1994). Differences in the visual control of pantomimed and natural grasping movements. *Neuropsychologia* 32: 1159–1178.

Goodale, M. A., J. P. Meenan, H. H. Bülthoff, D. A. Nicolle, K. S. Murphy, and C. I. Racicot (1994). Separate neural pathways for the visual analysis of object shape in perception and prehension. *Current Biology* 4: 604–610.

Goodale, M. A., and A. D. Milner (1992). Separate visual pathways for perception and action. *Trends in Neuroscience* 15: 20–25.

Goodale, M. A., and A. D. Milner (2004a). *Sight Unseen: An Exploration of Conscious and Unconscious Vision.* Oxford: Oxford University Press.

Goodale, M. A., and A. D. Milner (2004b). Plans for action. *Behavioral and Brain Sciences* 2: 37–40.

Goodale, M. A., A. D. Milner, L. S. Jakobson, and D. P. Carey (1991). A neurological dissociation between perceiving objects and grasping them. *Nature* 349: 154–115.

Goodale, M. A., D. A. Westwood, and A. D. Milner (2004). Two distinct modes of control for object-directed action. *Progress in Brain Research* 144: 131–144.

Gregory, R. L. (1963). Distortions of visual space as inappropriate constancy scaling. *Nature* 199: 678–680.

Gregory, R. L. (1970). *The Intelligent Eye.* New York: McGraw-Hill.

Grill-Spector, K., Z. Kourtzi, and N. Kanwisher (2001). The lateral occipital complex and its role in object recognition. *Vision Research* 41: 1409–1422.

Haffenden, A. M., and M. A. Goodale (1998). The effect of pictorial illusion on prehension and perception. *Journal of Cognitive Neuroscience* 10: 122–136.

Haffenden, A. M., and M. A. Goodale (2000). Independent effects of pictorial displays on perception and action. *Vision Research* 40: 1597–1607.

Haffenden, A. M., K. C. Schiff, and M. A. Goodale (2001). The dissociation between perception and action in the Ebbinghaus illusion: Nonillusory effects of pictorial cues on grasp. *Current Biology* 11: 177–181.

Hartung, B., P. R. Schrater, H. H. Bulthoff, D. Kersten, and V. H. Franz (2005). Is prior knowledge of object geometry used in visually guided reaching? *Journal of Vision* 5: 504–514.

Holmes, G. (1918). Disturbances of vision by cerebral lesions. *British Journal of Ophthalmology* 2: 353–384.

Hu, Y., and M. A. Goodale (2000). Grasping after a delay shifts size-scaling from absolute to relative metrics. *Journal of Cognitive Neuroscience* 12: 856–868.

Jackson, S. R., and A. Shaw (2000). The Ponzo illusion affects grip-force but not grip-aperture scaling during prehension movements. *Journal of Experimental Psychology: Human Perception and Performance* 26: 418–423.

Jakobson, L. S., Y. M. Archibald, D. P. Carey, and M. A. Goodale (1991). A kinematic analysis of reaching and grasping movements in a patient recovering from optic ataxia. *Neuropsychologia* 29: 803–809.

James, T. W., J. Culham, G. K. Humphrey, A. D. Milner, and M. A. Goodale (2003). Ventral occipital lesions impair object recognition but not object-directed grasping: A fMRI study. *Brain* 126: 2463–2475.

James, T. W., G. K. Humphrey, J. S. Gati, R. S. Menon, and M. A. Goodale (2000). The effects of visual object priming on brain activation before and after recognition. *Current Biology* 10: 1017–1024.

James, T. W., G. K. Humphrey, J. S. Gati, R. S. Menon, and M. A. Goodale (2002). Differential effects of viewpoint on object-driven activation in dorsal and ventral streams. *Neuron* 35: 793–801.

James, T. W., K. Harman James, G. K. Humphrey, and M. A. Goodale (2006). Do visual and tactile object representations share the same neural substrate? In *Touch and Blindness: Psychology and Neuroscience*, 139–155, ed. M. A. Heller and S. Ballesteros. Mahwah, N.J.: Lawrence Erlbaum.

Kimura, D. (1982). Left-hemisphere control of oral and brachial movements and their relation to communication. *Philosophical Transactions of the Royal Society of London* B298: 135–149.

Kimura, D. (1993). *Neuromotor Mechanisms in Human Communication*. New York: Oxford University Press.

Koski, L., M. Iacoboni, and J. C. Mazziotta (2002). Deconstructing apraxia: Understanding disorders of intentional movement after stroke. *Current Opinion in Neurology* 15: 71–77.

Kourtzi, Z., and N. Kanwisher (2000). Cortical regions involved in perceiving shape. *Journal of Neuroscience* 20: 3310–3318.

Króliczak, G., P. Heard, M. A. Goodale, and R. L. Gregory (2006). Dissociation of perception and action unmasked by the hollow-face illusion. *Brain Research* 1080: 9–16.

Kwok, R. M., and O. J. Braddick (2003). When does the Titchener Circles illusion exert an effect on grasping? Two- and three-dimensional targets. *Neuropsychologia* 41: 932–940.

Lamme, V. A., and H. Spekreijse (2000). Modulations of primary visual cortex activity representing attentive and conscious scene perception. *Frontiers in Bioscience* 1;5: D232–243.

Malach, R., J. B. Reppas, R. R. Benson, K. K. Kwong, H. Jiang, W. A. Kennedy, P. J. Ledden, T. J. Brady, B. R. Rosen, and R. B. Tootel (1995). Object related activity revealed by functional magnetic resonance imaging in human occipital cortex. *Proceedings of the National Academy of Sciences USA* 92: 8135–8139.

Milner, A. D., and R. Dyde (2003). Why do some perceptual illusions affect visually guided action, when others don't? *Trends in Cognitive Sciences* 7: 10–11.

Milner, A. D., and M. A. Goodale (2006). *The Visual Brain in Action*, 2nd ed. Oxford: Oxford University Press.

Milner, A. D., D. I. Perrett, R. S. Johnston, P. J. Benson, T. R. Jordan, D. W. Heeley, D. Bettucci, F. Mortara, R. Mutani, E. Terazzi, and D. L. W. Davidson (1991). Perception and action in visual form agnosia. *Brain* 114: 405–428.

Oliva, A., and P. G. Schyns (2000). Diagnostic colors mediate scene recognition. *Cognitive Psychology* 41: 176–210.

Patla A., and M. A. Goodale (1997). Visuomotor transformation required for obstacle avoidance during locomotion is unaffected in a patient with visual form agnosia. *NeuroReport* 8: 165–168.

Perenin, M. T., and A. Vighetto (1983). Optic ataxia: A specific disorder in visuomotor coordination. In *Spatially Oriented Behavior*, ed. A. Hein and M. Jeannerod, 305–326. New York: Springer-Verlag.

Perenin, M. T., and A. Vighetto (1988). Optic ataxia: A specific disruption in visuomotor mechanisms. I. Different aspects of the deficit in reaching for objects. *Brain* 111: 643–674.

Plodowski, A., and S. R. Jackson (2001). Vision: Getting to grips with the Ebbinghaus illusion. *Current Biology* 11: R304–R306.

Pook, P. K., and D. H. Ballard (1996). Deictic human/robot interaction. *Robotics and Autonomous Systems* 18: 259–269.

Pylyshyn, Z. W. (1980). Computation and cognition: Issues in the foundations of cognitive science. *Behavioral and Brain Sciences* 3: 111–169.

Pylyshyn, Z. W. (1999). Is vision continuous with cognition? The case for cognitive impenetrability of visual perception. *Behavioral and Brain Sciences* 22 (3): 341–423.

Radoeva, P. D., J. D. Cohen, P. M. Corballis, T. G. Lukovits, and S. G. Koleva (2005). Hemispheric asymmetry in a dissociation between the visuomotor and visuoperceptual streams. *Neuropsychologia* 43: 1763–1773.

Rice, N. J., R. D. McIntosh, I. Schindler, M. Mon-William, J. F. Demonet, and A. D. Milner (2006). Intact automatic avoidance of obstacles in patients with visual form agnosia. *Experimental Brain Research* 174: 176–188.

Rossetti, Y., P. Revol, R. McIntosh, L. Pisella, G. Rode, J. Danckert, C. Tilikete, H. C. Dijkerman, D. Boisson, A. Vighetto, F. Michel, and A. D. Milner (2005). Visually guided reaching: Bilateral posterior parietal lesions cause a switch from fast visuomotor to slow cognitive control. *Neuropsychologia* 43: 162–177.

Schäfer, E. A. (1888). On electrical excitation of the occipital lobe and adjacent parts of the monkey's brain. *Proceedings of the Royal Society of London* 43: 408–410.

Schindler, I., N. J. Rice, R. D. McIntosh, Y. Rossetti, A. Vighetto, and A. D. Milner (2004). Automatic avoidance of obstacles is a dorsal stream function: Evidence from optic ataxia. *Nature Neuroscience* 7: 779–784.

Servos, P., H. Carnahan, and J. Fedwick (2000). The visuomotor system resists the horizontal-vertical illusion. *Journal of Motor Behavior* 32: 400–404.

Smeets, J. B. J., and E. Brenner (2001). Action beyond our grasp. *Trends in Cognitive Science* 5: 287.

Steeves, J. K. E., G. K. Humphrey, J. C. Culham, R. S. Menon, A. D. Milner, and M. A. Goodale (2004). Behavioral and neuroimaging evidence for a contribution of color and texture information to scene classification in a patient with visual form agnosia. *Journal of Cognitive Neuroscience* 16: 955–965.

Stöttinger, E., and J. Perner (2006). Dissociating size representation for action and for conscious judgment: Grasping visual illusions without apparent obstacles. *Consciousness and Cognition* 15: 269–284.

Taira, M., S. Mine, A. P. Georgopoulos, A. Murata, and H. Sakata (1990). Parietal cortex neurons of the monkey related to the visual guidance of hand movement. *Experimental Brain Research* 83: 29–36.

Tarr, M. J., D. Kersten, Y. Cheng, K. Doerschner, and B. Rossion (2002). Men are from Mars, women are from Venus: Behavioral and neural correlates of face sexing using color. *Journal of Vision* 2: 598a.

Tarr, M. J., D. Kersten, Y. Cheng, and B. Rossion (2001). It's Pat! Sexing faces using only red and green. *Journal of Vision* 1: 337a.

Ungerleider, L. G., and M. Mishkin (1982). Two cortical visual systems. In *Analysis of Visual Behavior*, ed. D. J. Ingle, M. A. Goodale, and R. J. W. Mansfield, 549–586. Cambridge, Mass.: MIT Press.

Vailaya, A., A. Jain, and H. Jiang Shang (1998). On image classification: City images vs. landscapes. *Pattern Recognition* 31: 1921–1935.

van Donkelaar, P. (1999). Pointing movements are affected by size-contrast illusions. *Experimental Brain Research* 125: 517–520.

Vishton, P. M., and E. Fabre (2003). Effects of the Ebbinghaus illusion on different behaviors: One- and two-handed grasping; one- and two-handed manual estimation; metric and comparative judgment. *Spatial Vision* 16: 377–392.

Westwood, D. A., and M. A. Goodale (2003). Perceptual illusion and the real-time control of action. *Spatial Vision* 16: 243–254.

Westwood, D. A., M. Heath, and E. A. Roy (2000). The effect of a pictorial illusion on closed-loop and open-loop prehension. *Experimental Brain Research* 134: 456–463.

Wolf, M. E., and M. A. Goodale (1987). Oral movement asymmetries during verbal and nonverbal tasks. *Neuropsychologia* 25: 375–396.

Yuille, A., and D. Kersten (2006). Vision as Bayesian inference: Analysis by synthesis? *Trends in Cognitive Science* 10(7): 301–308.

5 There's a New Kid in Town: Computational Cognitive Science, Meet Molecular and Cellular Cognition

John Bickle

Most readers of this volume are aware of Zenon Pylyshyn's justifiably famous work on the foundations of cognitive science, in particular his three-level account of cognitive systems. In the preface to his landmark *Computation and Cognition* (1984), Pylyshyn wrote: "Explaining cognitive behavior requires that we advert to three distinct levels of the system: the nature of the mechanism or functional architecture; the nature of the code (that is, the symbol structures); and their semantic content" (p. xviii). Theorists can elaborate and test "regularities" proposed at each level "without concern for the way the regularities are realized at the 'lower' level" (ibid.). In the next sentence Pylyshyn tells us that "elaborating" and "justifying" this view—a view *"implicit in cognitive-science practice"* (my emphasis)—is the book's principal aim.

Pylyshyn is correct in finding this view implicit in the practices of many cognitive scientists. This was especially true in the early 1980s, but it remains prominent even today—not only among cognitive psychologists but also in much cognitive neuroscience. However, a different practice is implicit in a more recent branch of cognitive science—a reductionistic branch that has grown in prominence with applications of techniques from cellular physiology and molecular biology. My goal in this essay is to "elaborate and justify" its view, and to juxtapose it with Pylyshyn's. I'll start with a detailed example of research from this field on an aspect of cognition that has long interested Pylyshyn (visual attention) and use that example to illustrate some of the field's general features. Pylyshyn's type of cognitive science, even supplemented with cognitive neuroscience, is no longer the only game in town. We now confront competing accounts of cognitive-scientific practice. Ultimately we, as individual cognitive scientists, must decide which approach best warrants our time and efforts. And politically, we must decide which warrants our research dollars.

In this time of tolerance and diversity, many may wonder why I'm making this fuss. Can't we all just get along—and pursue whichever features of cognition interest us, with whatever tools and approaches we think will be fruitful? Of course we can, do, and will continue to. The issue here is neither coercion nor conversion. But this challenge misses a crucial point. The two approaches to cognitive scientific practice to be juxtaposed here are *irreconcilably contradictory*. The contradiction is subtle, however, and is easily missed or misconstrued. Both Pylyshyn-style cognitive science and the approach I champion ascribe a status to the discoveries of the other approach, and it is these ascriptions that contradict. What is irreconcilable across these two practices is *not* that they investigate the same phenomenon at different levels; rather it is the role they ascribe to the investigations pursued by the other approach.

Pylyshyn's account does not disparage neurobiology. Late in his book he wrote:

Giving a full account [of a cognitive system] of course requires discussing not only goals and beliefs but the functional (or symbol-processing) mechanisms that encode these goals and beliefs and make the system work, as well as the way in which such mechanisms are realized in brain tissue. In short the other two levels of the account [the functional and the neurobiological] must be given. (1984, 211)

He even suggests the proper places to look for neurophysiological constraints on cognitive models. Concerning his famous story of the pedestrian who, having witnessed an automobile accident, rushes to a nearby telephone and dials a "9" and a "1," he insists that "I do not deny that the minute muscular movements that unfold . . . are governed precisely by physical and biological laws, and thereby are, in principle, predictable from the current physical and neurophysiological state of the person making the movement" (1984, 10).

But, according to Pylyshyn, demanding such predictions from cognitive science asks for both too much and too little. It is too strong a demand because it yields a prediction of every microscopic event that occurs in the person, not just the ones relevant for the behavior. It also requires complete microscopic descriptions of both the subject's neurophysiology and the causally efficacious states of the environment that partly determined the specific movement. It is too weak a demand because it will miss regularities across physiologically distinct behaviors—for example, people responding to perceived disasters—that are there for the finding if we avail ourselves of semantic-level explanatory resources. Neurobiology is an important cohort of cognitive science for some aspects of the endeavor.

But with regard to "representation-governed regularities," it can only tell us how the system's goals, beliefs, and functional architecture are "realized in brain tissue" on that specific behavioral occasion. That is not "what we really need to know" (ibid., 11).

I will now contrast Pylyshyn's picture with a competitor that has grown in prominence over the past decade. That competitor has been dubbed *molecular and cellular cognition*. Its practitioners set their approach in opposition even to cognitive neuroscience, in addition to the computational cognitive psychology that Pylyshyn championed. For example, one finds on the Web site for the Molecular and Cellular Cognition Society (under "About") the claim that

unlike Cognitive Neuroscience, which historically has focused on the connection between human brain systems and behavior, the field of Molecular and Cellular Cognition studies how molecular (i.e., receptor, kinase activation), intra-cellular (i.e., dendritic processes), and inter-cellular processes (i.e., synaptic plasticity; network representations such as place fields) modulate animal models of cognitive function. (www.molcellcog.org)

The picture at work in the actual practices of this field, of how neurobiology operates to address cognitive phenomena, will be new to many cognitive scientists. To fill in this picture I turn to work at the "single-cell" level that addresses an aspect of cognition that Pylyshyn has been particularly interested in: visual attention.

Psychologists have long known of three ways that behavior improves with directed visual attention to specific objects, locations, or features. Improvements include (1) increased response speed (e.g., subjects respond faster to queries or tasks involving attended stimuli), (2) increased response accuracy (e.g., subjects correctly identify an attended object or feature more often), and (3) improved detection thresholds (e.g., subjects recognize attended stimuli at lower luminance or salience). But what are the mechanisms by which attention affects behavior? Work from John Maunsell's lab at Harvard Medical School (previously at the Baylor College of Medicine) addresses this question at the level of cell physiology. Maunsell works with alert, behaving primates outfitted for single-cell electrophysiological recording while they perform various cognitive tasks. According to Maunsell, one of his principal research projects "examines how attention affects the responses of individual neurons in cerebral cortex" (http://maunsell.med.harvard.edu/projects.html). He observes behaving primates (rhesus monkeys) performing tasks that require visual attention while outfitted for single-neuron recording throughout the visual streams.

One specific question Maunsell has addressed concerns the neurophysiological effects that attention to a given visual neuron's receptive field has on its action potential frequency. (A visual neuron's receptive field is the location in space that particular kinds of stimuli evoke increased action potential frequency in it.) One possible effect is that visual attention increases a neuron's action potential frequency to all stimuli that it responds to. For example, if we were measuring the frequency of action potentials in a visual neuron activated by stimulus orientation in a particular region of visual space—a neuron that was most responsive (in terms of number of action potentials per second) to one particular orientation, a bit less responsive to similar orientations, and unresponsive (above its baseline action potential rate to no stimuli) to opposite orientations—selective visual attention to that region might "turn up the gain" of that neuron's responses to all orientation stimuli. (See figure 5.1A, B.) This effect is called *multiplicative scaling.* It would account for improvements in behavioral performance with visual attention because the neural activity coding for the stimulus in the attended area would be more robust, leading to increased activity in "downstream" neurons in particular sensory-to-motor circuits and ultimately in motor response.

Another possibility is that selective visual attention sharpens the tuning curves of neurons whose activity is keyed to its focus. Perhaps neurons respond more vigorously (more action potentials per time unit) to preferred stimuli (e.g., some particular orientation and closely related ones), but their activity is dampened to orientations that activate them less. (See figure 5.1C.) This result would account for improved behavioral performance with attention because it would increase the signal-to-noise ratio in neurons most activated by the specific visual stimulus, also leading to increased activity in "downstream" neurons in the sensory-to-motor circuits tied specifically to features of the attended stimulus.

To investigate which (if either) of these effects obtain, Maunsell and his colleagues developed an experimental technique that isolates the effects of selective visual attention on individual neuron activity. Suppose that the monkey is fixating on the central spot in the first frame of figure 5.2 and that we are recording action potentials from a neuron whose receptive field is the dashed circle to the right. Call that location the "Attend In" region, because if we force the monkey to direct its visual attention to that region, he will be "attending into" the recorded neuron's visual field. (In figure 5.2, taken from McAdams and Maunsell 1999, this location is dubbed "Attended.") Call the location opposite the "Attend In" location vis-à-vis the central fixation point (where the shaded stimulus appears in later

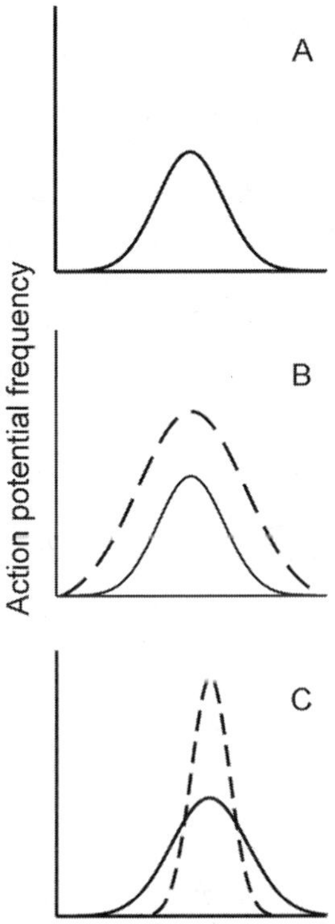

Figure 5.1

(A) Schematic illustration of a sensory neuron with a Gaussian response. The x-axis represents particular features of sensory stimuli the neuron responds to (e.g., degree of stimulus orientation for orientation-selective neurons); y-axis represents frequency of action potentials generated by the stimuli (higher y-values represent higher action potential frequencies). (B) Schematic illustration of "multiplicative scaling" of action potential frequency resulting from attention to the neuron's receptive field (scale exaggerated for illustration). (C) Schematic illustration of "sharpening of tuning curve" resulting from attention directed to the neuron's receptive field (scale exaggerated for illustration). Figure constructed by Marica Bernstein.

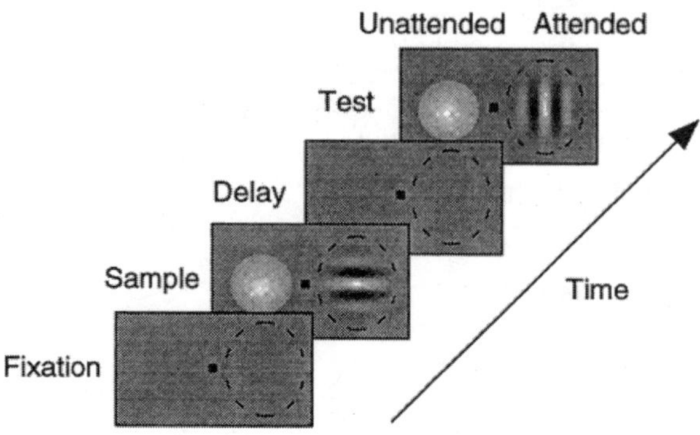

Figure 5.2
Schematic illustration of Maunsell and colleagues' delayed match-to-sample task. See text for timing details. "Attended" here denotes the "Attend In" region discussed in the text; "Unattended" here denotes the "Attend Out" region. In the case presented here, if the monkey had been cued to attend to the "Unattended" ("Attend Out") region, he would have to hold onto the lever for the full one second after test stimuli appeared (to indicate sample-test stimuli match in that region). If the monkey had been cued to attend to the "Attended" ("Attend In") region, he would have to release the lever within 500 milliseconds after test stimuli appeared (to indicate sample-to-test stimuli non-match in that region). From McAdams and Maunsell 1999, figure 1, p. 432. Reprinted with permission. (Copyright 1999 by the Society for Neuroscience)

frames of figure 5.2) the "Attend Out" location, because if we force the monkey to attend there, he will be "attending out" of the recorded neuron's visual field. (In figure 5.2 this location is dubbed "Unattended.") Now we can present stimuli to the recorded neuron's visual field and measure the differences in action potential frequency that a given stimulus elicits under "Attend In" versus "Attend Out" conditions.

How does Maunsell ensure that his monkeys "Attend In" or "Attend Out"? While single-cell electrophysiological recordings take place, the monkey performs a nonmatching-to-sample task that requires selective visual attention to a specific location for successful completion. The monkey fixates his vision on a visible dot on a computer screen. (The monkey's fixation can be located precisely because a scleral coil has been implanted in his eye.) "Attend In" and "Attend Out" regions for the neuron being recorded from are established. On a given trial, the monkey is cued visually to attend to either the "Attend In" or "Attend Out" location and grasps a lever to indicate readiness. Stimuli then appear for

500 milliseconds in both "Attend In" ("Attended") and "Attend Out" ("Unattended") regions and are extinguished (while the monkey maintains fixation on the central spot). These visual stimuli are followed by a 500-millisecond delay period (during which only the fixation spot is visible), after which another pair of stimuli appears in the two regions for up to one second. During the final stimulus presentation, the monkey must indicate whether the first and second stimuli *in the cued region* ("Attend In" or "Attend Out") were identical (a "match") by continuing to grasp the lever during the entire stimulus presentation, or different (a "nonmatch") by releasing the lever within 500 milliseconds of the onset of the second stimuli pair. (See figure 5.2.) Trials in which the monkey fails on this matching task are scrubbed and electrophysiological data are not included in the statistical analyses. Electrophysiological data gathered from successful "Attend In" trials to a particular stimulus provide data for action potential frequency when the monkey is attending to the neuron's receptive field; data gathered from successful "Attend Out" trials to that same stimulus (presented in the "Attend In" region) provide data for action potential frequency when the monkey is attending to the other region of visual space.

Maunsell and his colleagues employ a variety of sophisticated statistical techniques to analyze their experimental data, but most are based on a simple subtraction method for calculating the effects of selective visual attention on single neuron activity. For a given neuron n and a particular stimulus parameter s (e.g., a specific orientation), they first average the action potential frequency of n's responses to s on trials where the monkey successfully completed "Attend In" matching tests. They then do the same thing for n's action potential frequency to s (still presented in n's visual field) where the monkey successfully completed "Attend Out" matching tests. They subtract the second value from the first to get the specific effects of selective visual attention on n to s. They do this analysis for a variety of s's that prompt activity in n (e.g., different orientation stimuli), and then for a large number of n's.

Maunsell and others have now used variations of this technique for studying the effects of attention on single neuron activity with a variety of visual stimuli and neural regions. (All of the studies listed below were performed using rhesus macaque monkeys.) Treue and Maunsell (1999) used motion direction stimuli, recording in middle temporal (MT) and medial superior temporal (MST) cortex. Treue and Martinez-Trujillo (1999) used motion stimuli and a feature-based detection task, recording in MT. Recanzone and Wurtz (2000) used motion stimuli and a pursuit integration task, recording in MT and MST. Cook and Maunsell (2002) used motion stimuli in a change detection task, recording in MT and ventral intraparietal (VIP) cortex. Cook and Maunsell (2004) used motion stimuli and an integration-of-stimulus task, recording in MT. Here I'll describe in some detail the use of this technique and the results from a study by McAdams and Maunsell (1999), using combinations of color and orientation stimuli and recording from single neurons in macaque areas V4 and V1.

V4 contains orientation-selective neurons with Gaussian receptive fields. Each has a preferred orientation—stimuli oriented to that degree elicit highest action potential frequency. Its response falls off slightly to stimuli with different but similar orientations, and more so as orientations get less similar to its preferred degree. Finally, for stimuli with very different orientations, its responses remain around its baseline frequency. With stimulus orientation on the x-axis and action potential frequency on the y-axis, these neurons' response profiles fit a Gaussian curve.[1] (See figure 5.3 below.) In McAdams and Maunsell's (1999) selective attention matching task, orientation stimuli (Gabors) always appeared in the "Attend In" region and color stimuli (colored Gaussians) always appeared in the "Attend Out" region.

Figure 5.3 presents results from a single V4 orientation-selective neuron with a Gaussian receptive field. Notice that it demonstrates clear evidence of multiplicative scaling with selective attention to its visual field. For every stimulus orientation that elicits a response, the monkey's successfully performing an "Attend In" trial involving that stimulus orientation increased the neuron's action potential frequency, compared to its frequency when the monkey successfully performed an "Attend Out" trial while that stimulus orientation appeared in the Attend In location. However, attention had no significant effect on the width of this neuron's tuning curve, and thus produced no sharpening.

As figure 5.4 demonstrates, the evidence for multiplicative scaling is even stronger in the population-tuning curve data. Stimulus orientation was first normalized for all neurons by assigning a relative orientation of 0° to each neuron's most preferred orientation, and the value of the divergence from that orientation (e.g., 30°, –60°, and so on) to the other orientations. Response profiles under "Attend In" and "Attend Out" conditions of many neurons could then be averaged. Figure 5.4 presents the averaged, normalized population tuning-curve data for 197 orientation-selective V4 neurons under "Attend In" and "Attend Out" modes.[2] As with the individual V4 neuron profile just presented, these neurons' averaged response amplitudes increased for all stimulus orientations with explicit attention to their

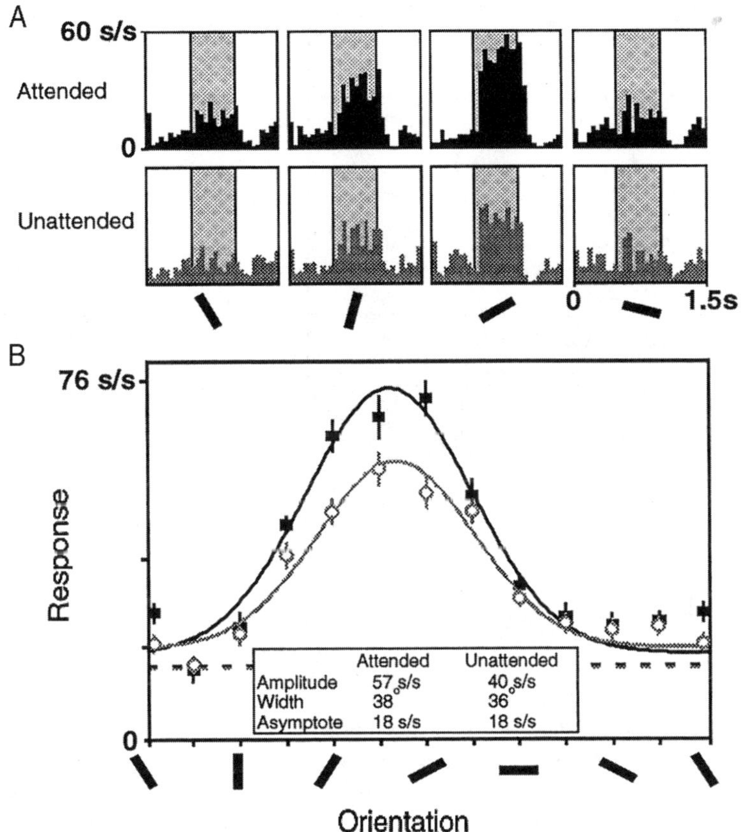

Figure 5.3

Data from one V4 neuron in the McAdams and Maunsell (1999) study showing multiplicative scaling in the "Attend In" mode (black squares, denoted "Attended") relative to the "Attend Out" mode (open circles, "Unattended"). Tuning curves were constructed for each mode by fitting responses to a Gaussian. This neuron showed a significant increase in action potential frequency in the "Attend In" mode relative to the "Attend Out" mode to each orientation stimulus, but no significant changes in preferred orientation or width. From McAdams and Maunsell 1999, figure 2, p. 434. Reprinted with permission. (Copyright 1999 by the Society for Neuroscience)

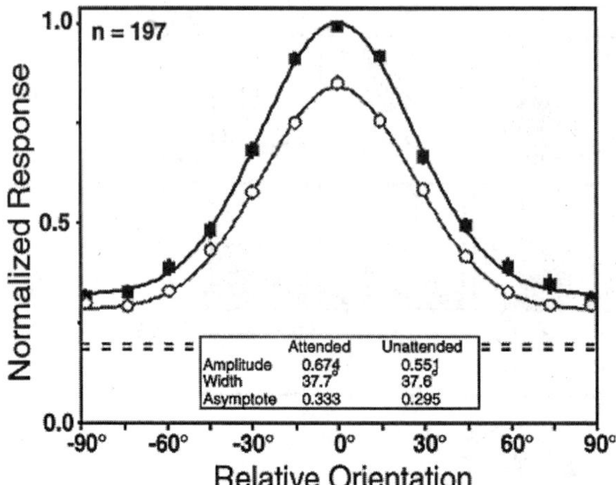

	Attended	Unattended
Amplitude	0.674	0.553
Width	37.7	37.6
Asymptote	0.333	0.295

Figure 5.4

Population-tuning curves for the V4 neurons in the McAdams and Maunsell study whose responses could be fit to Gaussians, tuned in both "Attend In" (black squares, denoted "Attended") and "Attend Out" (open circles, "Unattended") modes. See text for discussion. From McAdams and Maunsell 1999, figure 4, p. 435. Reprinted with permission. (Copyright 1999 by the Society for Neuroscience)

receptive fields, without any significant change to response profile widths. This same effect was found in normalized population tuning-curve data for all 262 orientation-selective V4 neurons recorded from, including 65 with individual tuning curves that could not be fit to a Gaussian (data not presented).

Multiplicative scaling of action potential frequency is a widespread effect of selective visual attention to a neuron's receptive field. McAdams and Maunsell (1999) found this effect as far back in the visual processing hierarchy as V1 (primary visual cortex—data not presented). This demonstrates that visual attention to a neuron's receptive field has measurable effects on action potential frequency all the way back to the first stage of cortical visual processing.

Why are these results important for cognitive science? For one thing, they suggest that at the level of cell physiology, the mechanisms of selective visual attention are commonplace. Multiplicative scaling of neuron action potential frequency is also elicited, for example, by simply increasing stimulus salience and contrast (by making the external visual stimulus brighter or more unlike surrounding stimuli). McAdams and Maunsell take note of this fact when discussing their results:

The phenomenological similarity between the effects of attention and the effects of stimulus manipulation raises the possibility that attention involves neural mechanisms that are similar to those used in processing ascending signals from the retinas, and that cortical neurons treat retinal and attentional inputs equivalently. (1999, 439)

Selective attention seems to be just one of the brain's common cellular tricks for "turning up the gain" of particular neurons, yielding predictable effects all the way down processing circuits to muscle tissue. There is nothing special or unique about this mechanism, despite its "cognitive" label and phenomenological vividness.[3]

A second reason why these results are important for cognitive science stems from their reductionistic potential, especially when they are combined with results from another vision lab. Charles Gilbert's group used an adenovirus vector to insert a gene expressing green fluorescent protein (GFP) into V1 neurons. When this gene is expressed in infected neurons, GFP fills their somas, dendrites, and axons. Since the protein is absent from extracellular space, this tracer (synthesized by the neuron itself) shows both a clear demarcation of the injection site and a count of labeled cells. This technique has several advantages over more widely used extracellularly injected tracers, including better labeling of axonal processes and improved localization of labeled cell soma. Gilbert and his colleagues used this technique to compare the relationship of V1 intrinsic and V2-to-V1 feedback connections to the functional architecture of V1. (See Stettler et al. 2002 for a description of this anatomical method and some striking images of labeled neurons.)

To chart the functional architecture of a given visual region, Gilbert and his colleagues have long used "optical imaging" in behaving primates (see Das and Gilbert 1995). A small portion of visual cortex is exposed and the brain surface is illuminated with red light (605–700 nm). Active cortical regions absorb more light than less active regions (thought to be due in part to local changes in oxygenated versus deoxygenated hemoglobin that accompany activity—similar to the signal tracked more globally in functional magnetic resonance imaging [fMRI]). Using a sensitive video camera and averaging over numerous trials, experimenters can visualize these differences and use them to map cortical patterns of activation in response to different visual stimuli. Gilbert and his colleagues then imposed the images of detailed axonal patterns gathered from the GFP labeling study on the optical imaging orientation maps to infer the orientation specificity of V1 horizontal connections and feedback projections from V2 to V1 neurons. Within 500 microns of a neuron's soma (measured using precise data about the viral injection site), V1 neurons do not make preferential synaptic connections with other neurons sharing similar preferred orientation stimuli; the cells project instead to neurons with a variety of preferred stimulus orientations. But these neurons' "long-range horizontal connections"—their axonal projections to other V1 neurons at distances greater than 500 microns—are significantly greater to other neurons with the same preferred orientation stimulus. In Gilbert's study nearly one-quarter of V1 neurons' long-range projections went to other V1 neurons with the same preferred orientation. (For quantitative data, see Stettler et al. 2002, figure 7, 743.)

Gilbert and his colleagues recognize the potential of their discoveries as a mechanism for selective visual attention. Discussing the results sketched above, they write: "While the character of its distribution suggests that V2 to V1 feedback is not the primary mechanism underlying contextual interactions in V1, it might yet influence such interactions, perhaps by mediating the top-down influences of attention and perceptual task" (Stettler et al. 2002, 747). In another paper they suggest: "One possible mechanism underlying the attention effects is a gating or modulation of the synaptic effects of long-range horizontal connections by feedback connections from higher cortical areas" (Gilbert et al. 2000, 1224). Notice that multiplicative scaling would result from the increased activation in all neurons induced by activated long-range horizontal connections and top-down feedback, in conjunction with the feedforward activation from retinal stimulation.

Whether similar patterns of horizontal axonal projections activated by feedback projections from higher visual centers exist further up in the visual processing streams (e.g., in V4) remains an open question. But it is important to see the potential reduction lurking in these cell-physiological and anatomical results. They suggest that selective visual attention *is* multiplicative scaling of action potential frequency in specific visual neurons, driven by endogenously activated feedback axonal projections from specific neurons in regions further up the visual processing hierarchies. These feedback projections activate long-range horizontal connections between similarly tuned visual neurons earlier in the hierarchy, leading ultimately to increased downstream activations in specific pathways.

Does this work meet the standards on "accomplished reductions" implicit in current molecular and cellular cognition? (For an initial attempt to articulate the "reductionism-in-practice" in molecular and cellular cognition, see Bickle 2003, 2006, and Silva and Bickle in press.) Not yet. Instead of characterizing psychoneural reductionism in terms of chains of intertheoretic or "functional" reductions, reduction-in-actual-neuroscientificpractice stresses *intervening* into the causal mechanisms at increasingly lower levels of biological organization and then *tracking* significant behavioral effects of that intervention in well-accepted experimental protocols for the cognitive phenomenon being investigated. When successful (and properly controlled!), the lower-level mechanisms intervened into—along with the anatomical pathways those intervened-into neurons are part of—are taken to directly explain the behavioral data.

At least four conditions must be met before one can justifiably claim to have found a lower-level causal mechanism for—to have reduced—a cognitive phenomena.[4] These are conditions on sufficient evidence for connecting a hypothesized cellular or molecular mechanism with a cognitive phenomenon (like selective visual attention), implicit in the experimental practices of molecular and cellular cognition.

1. Observation   Various temporal relationships between mechanism and phenomenon must be established (for example, the hypothesized cellular or intracellular molecular mechanisms must precede their behavioral effects in time, and the temporal dimensions of the mechanisms must correlate correctly with system activity in light of the known anatomical circuits).
2. Negative alteration   Intervening directly to decrease activity of the mechanisms must reliably decrease the behaviors that are taken as experimental measures for the cognitive phenomenon.
3. Positive alteration   Intervening directly to increase activity of the mechanisms must reliably increase the behaviors that are taken as experimental measures for the cognitive phenomenon.
4. Integration   The hypothesis that the proposed mechanism yields the behavioral consequences taken as experimental measures of the cognitive phenomenon must be connected up with as much experimental data as is available about both the hypothesized mechanism and the phenomenon.[5]

Notice that conditions (1) and (4) require higher-level scientific investigations. To establish the required temporal criteria on proposed mechanisms and system activity, and to establish the theoretical plausibility of the proposed mechanisms for the cognitive phenomenon in question, we need precise knowledge of what the system does (under controlled experimental conditions). This means both having precise data of system behavior (as grist for our lower-level mechanistic explanations) and good behavioral measures for the cognitive phenomenon at issue. These are jobs for cognitive scientists and experimental psychologists, not electrophysiologists or molecular geneticists. We also need to know where to insert our cellular and molecular interventions. The "decomposition and localization" investigations of cognitive neuroscientists are crucial for this knowledge. (For a useful discussion of this strategy, see Bechtel and Richardson 1992.) We also need to know what sorts of neuronal activity to intervene into. Action potential frequency? Action potential dynamics? Field potentials? Something else entirely? The work of neurocomputational modelers and simulators will be important here. Molecular and cellular cognition needs a lot of higher-level cognitive science and neuroscience to accomplish its potential reductions.

Yet in the final analysis, it is conditions (2) and (3) that clinch the empirical case for proposed lower-level mechanisms. Meeting those conditions establishes that potential mechanisms are actually doing the job. When conditions (2) and (3) are established, our best causal-mechanistic story then resides at the lowest level of effective interventions.[6] This lesson is implicit in the ways that molecular and cellular cognitivists discuss their results and develop their experimental strategies (Bickle 2003, 2006; Silva and Bickle in press). From the perspective of molecular and cellular cognition, when all four conditions are met, higher-level explanations *lose* their status as causal-mechanistic—although they retain their status as providing structural support for at least two of the four "legs" that the experimental case rests upon for our best causal-mechanistic explanation of the cognitive phenomenon.

With just these brief remarks in place, we can see what still needs to be accomplished to achieve a reduction of selective visual attention to the cell-physiological mechanism suggested by Maunsell's and Gilbert's work. Cognitive-scientific work generating behavioral data and measures of visual attention, along with Maunsell's and Gilbert's physiological work on multiplicative scaling in visual neurons under conditions of selective attention, already provides strong structure for "legs" (conditions) (1) and (4). What is now needed are successful interventions in behaving animals into the hypothesized cell-physiological mechanisms, and tracking of the resulting behaviors using accepted experimental measures of visual attention. Cortical microstimulation might be a useful strategy. Bill Newsome's laboratory has used this technique with alert, behaving primates to induce activity in tiny clusters (typically 250–500 cubic microns) of similarly tuned visual neurons in medial temporal lobe (area MT), tracking startling effects on visual motion and stereoscopic depth detection tasks (Salzman et al. 1992; D'Angelis, Cumming, and Newsome 1998). Other labs have employed it to induce cellular activity with significant behavioral effects on other visual, somatosensory, and working memory discrimination tasks, and on multiple joint motor behaviors. (For a useful review with an extensive annotated bibliography, see Romo and Salinas 2003.) Maunsell himself has begun using it as an experimental tool. Could cortical microstimulation with existing microelectrodes induce multiplicative scaling that matches the effects on both cellular physiology and behavior that explicit attention produces? Could experimenters cue the monkey to "Attend Out," microstimulate the appropriate neurons to match the physiological effects of "Attending In," and induce the "Attend In" behavioral response?

Existing microstimulation technology might not (yet) be up to the specific interventions required to get the behavioral effects just suggested. (The continued failures using this procedure to intervene into color processing in the primate visual system is instructive. Sometimes the functional anatomy of the system does not cooperate with our existing electrode technology, impressive though that technology is.) But as we learn more about the subcellular details of the long-range horizontal and feedback connections—the specific neurotransmitters and receptor subtypes involved, for example—perhaps genetic interventions that manipulate these molecules and proteins will yield new results meeting conditions (2) and (3). It is difficult to argue against the successes that such approaches have had in other areas of molecular and cellular cognition.

I opened this essay by contending that we now confront *competing* accounts of cognitive-scientific *practice*, of how best to search for the causal mechanisms of cognition. I've illustrated one example of how molecular and cellular cognition, broadly construed, addresses the cognitive phenomenon of visual attention—a phenomenon toward which Zenon Pylyshyn has been a major scientific contributor. And I've pointed out some general reductionistic features of the alternative approach. These features stand in sharp contrast to the three-level picture in standard cognitive science that Pylyshyn has eloquently expressed and defended for more than two decades. Both approaches ascribe a role for the scientific successes of the other. Molecular and cellular cognition neither ignores nor disparages discoveries made by higher-level investigators. It locates them under the particular conditions that they help to fulfill, each condition individually necessary to establish an empirical case for lower-level mechanisms for a higher-level phenomenon. But this approach also emphasizes two reductionistic necessary conditions and, when these two conditions are met, revokes the status of causal-mechanistic explanation given to higher-level explanations. This contrasts sharply *and irreconcilably* with the "implementationalist" view ascribed to cellular and molecular neuroscience in the practices of Pylyshyn-style cognitive science (and in cognitive neuroscience, a recent supplement to that branch). Pylyshyn correctly articulated the practices of one approach to cognitive science, the "only game in town" twenty years ago. Those practices blossomed and continue to the present day. But there's a new kid on the block in the cognitive sciences, with a very different set of practices and a different account of what has been accomplished and where to go next. We now confront two images of how to pursue a science of cognition, images that ascribe contradictory status to investigations pursued by the other. My bets rest on molecular and cellular cognition.

Notes

1. A Gaussian curve is a symmetrical curve representing the normal ("bell-shaped") distribution.

2. Data were collected from two male rhesus macaques.

3. In Bickle 2003 (chapter 4), I urge that this experimentally justified fact has serious consequences for fans of consciousness. Some "consciophiles" are now willing to bite the bullet and admit that consciousness is some kind of neural activity, but they still hold out for its realization being *a special, perhaps unique type* of neural mechanism. By showing that the causal effects of selective visual attention—a phenomenologically robust species of conscious experience—on individual neuron firing rates are the same as so mundane an effect as simply increasing external stimulus salience, Maunsell's results seem to strip consciousness of even the status of being a unique and special neural mechanism. However, since Pylyshyn has never shown any special affinity for consciophilia, I won't press this point here.

4. The conditions stated here are the preliminary results of a collaborative metascientific investigation by neurobiologist Alcino Silva and philosopher of neuroscience John Bickle. Silva has long advocated versions of these conditions, although not explicitly in print. Many details must still be elaborated, and detailed case studies must still be described, to defend these conditions as independently necessary and collectively sufficient.

5. Condition (4) obviously needs extensive explication. One task it accomplishes is to reject silly objections to the account such as "removing oxygen from the animal's environment significantly alters its behavior in this memory task. Is oxygen consumption thereby a mechanism of memory?" or "Does memory thereby reduce to oxygen consumption?" Clearly, the empirical background against which serious experimental studies are performed already rule out such mechanisms or reductions. Yet condition (4) must do far more than this. Basically, it provides the empirical reasons why we investigate whether a particular lower-level mechanism is the crucial step in the process generating the behavioral data.

6. I offer this thesis as a metascientific claim based on an analysis of the experimental practices and their interpretations offered by molecular and cellular cognitivists. I do not offer it as a metaphysical thesis. For a start toward articulating the distinction between metascience and metaphysics and a defense of pursuing metascience exclusively, see Bickle 2003, chapter 1.

References

Bechtel, W., and R. C. Richardson (1992). *Discovering Complexity*. Princeton: Princeton University Press.

Bickle, J. (2003). *Philosophy and Neuroscience: A Ruthlessly Reductive Account*. Dordrecht: Kluwer.

Bickle, J. (2006). Reducing mind to molecular pathways: Explicating the reductionism implicit in "molecular and cellular cognition." *Synthese* 152: 411–434.

Cook, E. P., and J. H. R. Maunsell (2002). Attentional modulation of behavioral performance and neuronal responses in middle temporal and ventral intraparietal areas of macaque monkey. *Journal of Neuroscience* 22: 1994–2004.

Cook, E. P., and J. H. R. Maunsell (2004). Attentional modulation of motion integration of individual neurons in the middle temporal visual area. *Journal of Neuroscience* 24: 7964–7977.

Das, A., and C. D. Gilbert (1995). Long-range horizontal connections and their role in cortical reorganization revealed by optical recording of cat primary visual cortex. *Nature* 375: 780–784.

DeAngelis, G. C., B. G. Cumming, and W. T. Newsome (1998). Cortical area MT and the perception of stereoscopic depth. *Nature* 394: 677–680.

Gilbert, C. D., M. Ito, M. K. Kapadia, and G. Westheimer (2000). Interactions between attention, context and learning in primary visual cortex. *Vision Research* 40: 1217–1226.

McAdams, C. J., and J. H. S. Maunsell (1999). Effects of attention on orientation-tuning functions of single neurons in macaque cortical area V4. *Journal of Neuroscience* 19: 431–441.

Pylyshyn, Z. W. (1984). *Computation and Cognition*. Cambridge, Mass.: MIT Press.

Recanzone, G. H., and R. K. Wurtz (2000). Effects of attention on MT and MST neuron activity during pursuit initiation. *Journal of Neurophysiology* 83: 777–790.

Romo, R., and E. Salinas (2003). Flutter discrimination: Neural codes, perception, memory and decision making. *Nature Reviews Neuroscience* 4: 203–216.

Salzman, C. D., C. M. Murasugi, K. H. Britten, and W. T. Newsome (1992). Micro-stimulation in visual area MT: effects on direction discrimination performance. *Journal of Neuroscience* 12: 2331–2355.

Silva, A. J., and J. Bickle (in press). Understanding the strategies for the search for cognitive mechanisms. In *Oxford Handbook of Philosophy and Neuroscience*, ed. J. Bickle. Oxford: Oxford University Press.

Stettler, D. D., A. Das, J. Bennett, and C. D. Gilbert (2002). Lateral connectivity and contextual interactions in macaque primary visual cortex. *Neuron* 36: 739–750.

Treue, S., and J. C. Martines Trujillo (1999). Feature based attention influences motion processing gain in macaque visual cortex. *Nature* 399: 575–579.

Treue, S., and J. H. R. Maunsell (1999). Effects of attention on the processing of motion in macaque middle temporal and medial superior temporal visual cortical areas. *Journal of Neuroscience* 19: 7591–7602.

6 Inhibition of Return Is Cognitively Penetrable

Richard D. Wright and Lisa N. Jefferies

To what extent do expectations affect what we see? For example, if a moving object disappears from view and seems to be behind an occluding object, do our beliefs about its presence or absence affect visual analysis of the scene? In 1980, Pylyshyn proposed that questions like this can be tested using a criterion called *cognitive penetrability*. In particular, a process is said to cognitively penetrable if it can be influenced by factors such as a perceiver's beliefs, expectations, and general knowledge of the world (see also Pylyshyn 1984, 130–145). If that same process is not influenced by such higher-order changes, however, it is not cognitively penetrable. Pylyshyn originally developed the notion of cognitive penetrability in the 1980s as a means of distinguishing between two different explanations of mental imagery. In the years that followed, he also examined the extent to which visual processing is cognitively penetrable. The results are summarized in his seminal 2003 book, *Seeing and Visualizing*. The current consensus is that at least some types of visual processing are not and cannot be cognitively penetrable (see, e.g., the many open peer comments accompanying Pylyshyn's 1999 article). A visual search effect called *inhibition of return* (IOR), for example, has traditionally been viewed by attention researchers as cognitively impenetrable. In this chapter, we make the case that this assumption is incorrect and that, in fact, IOR is cognitively penetrable.

A description of the IOR effect was first published in 1984 by Posner and Cohen. IOR is indicated by an inhibition of target detection (and in some cases target discrimination) response times. It is typically studied in the laboratory by presenting location cues prior to the onset of the target to be detected. There are many excellent descriptions in the attention literature of the location cueing paradigm developed by Posner and colleagues (see, e.g., the description in Wright and Richard 2003, 925–927; Wright and Ward 2008). In general, a location cue of some type is presented for a brief period of time before the onset of a target to be detected. When the cue accurately indicates the impending target's location, detection response times are improved relative to an experimental trial on which no cue was presented (see, e.g., Posner, Snyder, and Davidson 1980). This is the response-time "benefit" of cueing. On the other hand, when the cue does not accurately indicate the target's location, detection response times are inhibited relative to an experimental trial on which no cue was presented (ibid.). This is the response-time "cost" of cueing. Responses are typically made by manually pressing a button. Throughout each experimental trial, observers are required to direct their eyes toward a small cross in the center of the stimulus display. With eye movements controlled for in this way, it is assumed that any effects of location cueing on target detection is due to covert shifts of attention to cued locations.

One variant of the typical location cueing paradigm is to present two location cues in succession prior to the onset of the target. Doing so led to the discovery of IOR (Posner and Cohen 1984). In particular, the effect occurred when one cue was presented at a peripheral location and was followed, 200 ms later, by the presentation of a second cue at the center of the stimulus display. Then, a short time later (200 ms), the target appeared at either the first cued location, the second cued location, or the uncued location (see figure 6.1). The time required to detect the target and press a response button was significantly slower for targets presented at the first cued location than for those presented at the second cued location, and even the uncued location (ibid.). This is the IOR effect. Its magnitude is typically measured by comparing response times to targets appearing at the first of two previously cued locations with response times to targets appearing at uncued locations. It is a robust effect that has been replicated using many different paradigms.

Inhibition of Return and Spatial Indexing

Posner and colleagues proposed that the IOR effect is associated with a mechanism that biases visual search toward novel items or locations (e.g., Clohessy et al. 1991; Harman et al. 1994; Posner and Cohen 1984; Posner et al. 1985). Put simply, when someone examines a visual scene containing many items, it may be the case that a mechanism is available to mark items that have already been inspected in order to keep them separate from the remaining items. Doing so would reduce the frequency of rechecking previously inspected items and, instead, guide analysis more efficiently toward uninspected items. Using Posner and colleagues' terms, processing is

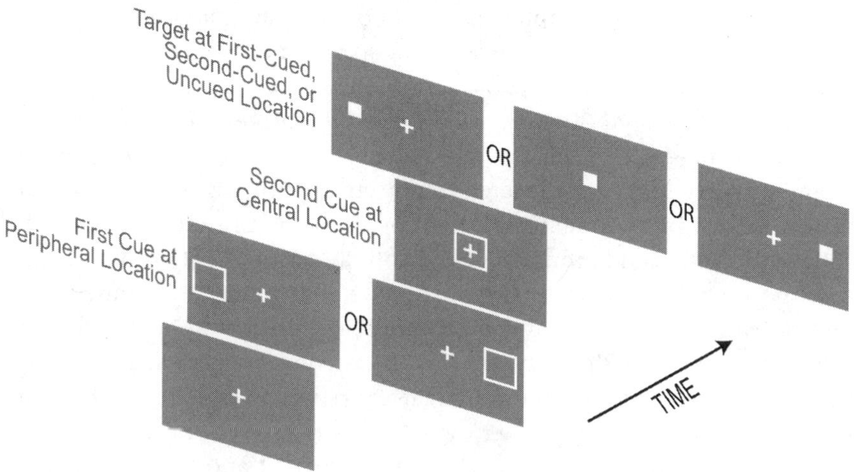

Figure 6.1

Example of the stimulus display used in a typical IOR experiment. In this figure, the first cue (an outline box) is presented at a peripheral location, the second cue is presented at the central location, and then the target appears at either of these locations or at the uncued location. The IOR effect is indicated by slower response times to targets appearing at the first cued location than at the uncued location.

"inhibited from returning" to previously inspected objects and/or locations.

There is a growing body of evidence indicating that a mechanism that marks items during visual analysis could involve an operation called *spatial indexing*. Perhaps most compelling are the results of a study of our ability to visually track multiple moving objects (Pylyshyn and Storm 1988). When performing this task, observers are typically shown a number of identical objects (e.g., 10) and asked to keep track of a subset of them. As the objects move randomly and independently, observers can usually track at least four or five with nearly the same efficiency as tracking one. Pylyshyn and Storm (1988) claimed that tracking appears to involve some form of processing that dynamically maintains the location of as many as four or five objects simultaneously. Multiple-object tracking (MOT) is a robust phenomenon and has been replicated several times (e.g., Alvarez et al. 2005; Jovicich et al. 2001; Liu et al. 2005; Oksama and Hyönä 2004; Pylyshyn 2004; Scholl and Pylyshyn 1999; Sears and Pylyshyn 2000; Viswanathan and Mingolla 2002).

As outlined in other chapters in this book, Pylyshyn accounted for the MOT result in terms of the allocation of spatial indexes he called FINSTs (an acronym for Fingers of INSTantiation; for a detailed description of this account, see Pylyshyn 2003, 223–232). For the purpose of this chapter, it will suffice to say that he proposed that a limited number of indexes (approximately four) are allocated to objects, and that this allocation can be maintained independently of attention as the objects move. Yantis and colleagues (Yantis and Johnson 1990; Yantis and Jones 1991) also proposed that the number of indexes (which they refer to as tags) is limited (i.e., four or five). When performing a MOT task, the primary role of indexes is simply to stay "glued" to the objects. If, at some point, observers must verify that a particular object is a target, Pylyshyn proposed that they can respond on the basis of whether the object is indexed or not.

In the early 1990s, the connection between IOR and spatial indexing became apparent when it was discovered that IOR can be object based (Tipper, Driver, and Weaver 1991; Tipper et al. 1994). Object-based processing, in simple terms, operates on a representation of an object rather than a representation of spatial locations (see, e.g., Duncan 1984; Egly, Driver, and Rafal 1994; Moore, Yantis, and Vaughan 1998). The latter is said to be space based or location based. The discovery of object-based IOR was made when an experiment was conducted like that described previously, but in which the first of the two cues was moved after it was presented (Tipper, Driver, and Weaver 1991). It was found that IOR occurred at the new location of the cue after its movement was complete. In other words, IOR "traveled" with the cue object as it moved, and the effect was, therefore, object based. In other experiments, it was found that IOR also continued to occur at the original location of the cue, even after it had been moved to a new location (Tipper et al. 1994). That is, IOR's occurrence at the original cue location was independent of the presence of a cue object and was, therefore, location based (see also Jordan and Tipper 1998; Leek, Reppa, and Tipper 2003).

One property of the spatial index mechanism proposed by Pylyshyn is that it can remain allocated to an object, even if that object should change locations during the course of an experimental trial. When the IOR continued to be associated with the first of the two cues, even when it changed locations, this could have been due to dynamic spatial indexing. That is, when the cue moved and the inhibition associated with that object "traveled" along with it, the mechanism responsible for this may be the same spatial indexing thought to mediate the visual tracking of multiple moving objects. Moreover, if the object-based IOR in Tipper et al.'s (1994) experiment was mediated by one of a small pool of spatial indexes that can remain dynamically assigned to moving objects, the location-based IOR that also occurred at the same time at the cue's original location could have been due to a second index being allocated to that location.

Pylyshyn and Storm's (1988) hypothesis that there appear to be four or five spatial indexes is consistent with the discovery that IOR can occur at more than one location at the same time. The occurrence of object-based and location-based IOR on the same trial is one instance in which the effect is associated with two locations simultaneously (see, e.g., Tipper et al. 1994). Posner and Cohen (1984) also found evidence for the IOR effect occurring at two locations simultaneously. Inspired by Posner and Cohen's (1984) finding that IOR can occur at multiple locations, we conducted a replication experiment in the early 1990s to determine whether or not the effect could be obtained at as many as four locations simultaneously (Wright and Richard 1996). One to four cues were presented at the same time, followed by a centrally located cue, and then the target. Like Posner and Cohen, we found that IOR occurred at multiple locations simultaneously (in this case, as many as four) with roughly equal magnitude (i.e., the inhibitory effect of cueing on target-detection response times was roughly the same on multiple-cue trials as on single-cue trials). Also, like Posner and Cohen (1984, 539), we concluded that multiple-location IOR may not be explainable purely in terms of attentional processing. We suggested instead that cued locations may be initially encoded by spatial indexes, and that this marking operation contributed to the multiple-location IOR effect that was obtained. Multiple-location IOR has since been found by other researchers (e.g., Danziger, Kingstone, and Snyder 1998; Dodd, Castel, and Pratt 2003; Paul and Tipper 2003; Snyder and Kingstone 2001; Tipper, Weaver, and Watson 1996).

Posner and Cohen (1984) speculated that some form of marking operation associated with IOR helps us to search the environment more efficiently. This idea was tested with a cleverly designed visual search experiment that involved a secondary task requiring detection of a probe stimulus (Klein 1988). Immediately after the presentation of a search display containing distractor items and a target item, the display was removed from view and occasionally a probe dot was presented somewhere in the field of view. The main finding was that observers were slower to detect the onset of this probe dot when it appeared at a location previously occupied by a nontarget item (i.e., a distractor) than when it appeared at a location that was not previously occupied by a nontarget item. It was concluded that the slower responses to probe dots at distractor locations relative to empty locations indicated that the former were "tagged" during the serial search that preceded the secondary task, and that, despite the removal of the search stimuli, these tags remained allocated to locations previously occupied by distractors when the probe detection task was performed. Klein (1988) referred to this as *inhibitory tagging*. This is consistent with the claim that the IOR effect is mediated by a marking operation.

Inspired by the finding that object-based IOR occurs when location cues move (e.g., Tipper et al. 1994), other researchers demonstrated that the inhibitory tagging effect also occurs when visual search sets are composed of moving items (Ogawa, Takeda, and Yagi 2002). This suggests that spatial indexes with the capacity to remain dynamically bound to moving objects mediate serial search for a target among distractors.

In summary, Pylyshyn's (1989, 1998, 2003) spatial index hypothesis holds that there is a limited pool of four to five indexes that can remain dynamically bound to objects as they move. If these spatial indexes are involved in the processing that mediates the IOR effect, then two predictions yielded by the hypothesis are: (1) IOR should be object based (i.e., dynamically bound to moving objects) and (2) it should occur at as many as four or five locations. Both of these predictions have been confirmed.

Inhibition of Return Is not Reflexive

It is tacitly assumed by many researchers that IOR is a mandatory, reflexive consequence of location cueing. One reason for the popularity of this assumption is that the IOR effect is robust and almost always occurs when location cues and targets are presented in the typical temporal sequence. There are numerous indications and a growing body of evidence, however, that IOR is not reflexive (see, e.g., Schendel, Robertson, and Treisman 2001). In this section, we make our case that IOR is not a mandatory consequence of location cueing and, moreover, is cognitively penetrable.

One indication that IOR is not reflexive is a finding that manipulation of location cue validity (i.e., the probability that a target will appear at the cued location) affects the occurrence of IOR (Wright and Richard 2000). In particular, when the first of the two successive cues had a low validity (10% probability of target appearing at cued location), IOR did not occur. Similarly, when the first of the two successive cues had a high validity (80% probability of target appearing at cued location), IOR also did not occur—although these cues did have a facilitative effect on detection times for targets appearing at their locations. IOR occurred only when cue validity was uninformative (i.e., target was equally likely to appear at a cued or uncued location). The dependence of IOR's occurrence on cue validity indicates that inhibitory effects of location cueing are not reflexive and are, to some extent, cognitively mediated.

This finding suggests that IOR may not occur at a cued location unless there is a reasonable degree of uncertainty about the target appearing there. This is consistent with the idea that IOR is associated with procedures that make serial visual analysis more efficient. In particular, when observers realize, over trials, that targets are very likely to appear at high-validity cue locations, then IOR might be less likely to occur when targets appear there because this would impair search efficiency. In contrast, when observers realize, over trials, that targets are very unlikely to appear at low-validity cue locations, then IOR will be also less likely to occur when targets appear there. In the case of low-validity cue locations, however, perhaps the attenuation of the IOR effect occurs because there would be little need to invoke a visual routine that spatially indexes and maintains these locations. That is, the visual system is unlikely to use limited processing resources (indexes) for the purpose of keeping track of locations where the target very probably will not occur. Search would be more efficient if, instead, these locations were eliminated entirely from the search set or at least not given special (inhibitory) treatment (cf. Treisman 1998; Treisman and Sato 1990).

Attenuation of the IOR effect with increases in target-location predictability was also found in a study involving targets presented at the same locations on successive trials (Maylor and Hockey 1987). That is, inhibition decreased slightly when a target was presented at the same location as the previous two targets, and continued to decrease with further increases in the number of target location repetitions (see also Posner et al. 1984; cf. Taylor and Donnelly 2002). The experimenters claimed that this target-location repetition effect on IOR is attributable to observers' subjective expectancies about target location, and that, in order for IOR to occur, the locations of successive events may need to be random (Maylor and Hockey 1987, 53). In other words, the operations associated with IOR appear to be, to some extent, under the observer's control.

These findings imply that there is a strategic aspect to the occurrence of IOR. This is consistent with the fact that serial visual search is, by nature, strategic and improves with practice (cf. Lee and Quessy 2003). A number of studies have shown, for example, that this improvement is continual over the first 10 years of life. In one experiment, serial search by 10-year-olds was as efficient (in terms of the time it took to decide whether a target was present or absent) as that of young adults, but serial search by 5-year-olds was significantly less efficient (Enns and Cameron 1987). Perhaps search efficiency improves with age because strategic reallocation of a limited number of indexes to different search set items becomes more refined with practice.

The results of other experiments indicated that the magnitude of IOR decreased as subjects got more practice performing the task (e.g., Lupiáñez et al. 2001). More recently, experimenters reported an IOR practice effect that is related to the direction in which people first learned to read text—a skill that is acquired only with extensive practice (Spalek and Hammad 2005). That is, the IOR effect appeared to show a left-to-right bias in people who normally read from left to right (e.g., English) but a right-to-left bias in people who normally read from right to left (e.g., Arabic). These indications that the IOR effect is associated with learned rather than reflexive processes are consistent with Wright and Richard's (1998) claim that IOR is similar, in some ways, to an automatized routine that appears to be reflexive, but can be influenced in a goal-driven manner (cf. Cavanagh 2004).

A more direct test of the goal-driven nature of IOR would be to create a scenario in which the occurrence of the effect would depend on the observer's beliefs about objects in the visual scene. This would indicate whether or not IOR is cognitively penetrable. One intriguing finding is that object-based IOR occurs in response to a moving object, even if that object is not visible when it is cued, when it terminates its motion sequence, or both (Yi, Kim, and Chun 2003). This result was replicated in an experiment involving objects that moved across the stimulus display and disappeared behind an illusory occluding surface, a Kanisza square as seen in figure 6.2 (Jefferies, Wright, and Di Lollo 2005).

In this experiment, two groups of observers were presented with displays in which an object appeared in one quadrant of the screen and subsequently moved to the other side of the screen. On half of the trials, the object moved such that it disappeared behind an occluding square at the end of its motion path (occluder trials); on the remaining trials, it moved without intersecting the occluder and therefore remained visible (context trials). The purpose of the context trials was to develop and manipulate the observer's expectation about the general behavior of the object— specifically, whether it continued to exist or ceased to exist at the end of its motion path. The context trials differed for the two groups of observers. For one group, the object always remained present for the 400 ms cue-target onset asynchrony (CTOA) before the onset of the target, leading to the expectation that the object would similarly continue to exist when

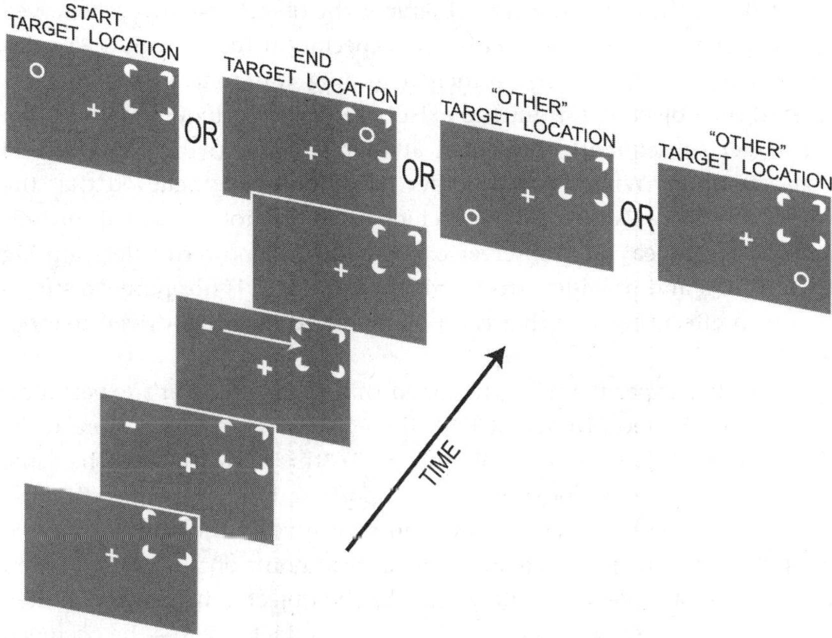

Figure 6.2

An example of the stimulus display in Jefferies et al.'s (2005) study. Trials began with the movement of a cue across the display toward an illusory square as though it was moving behind it. This was followed by the onset of a target at the original location of the cue, the perceived destination of the cue, or one of two movement-independent locations above or below the cue movement vector. When subjects believed that a moving cue was occluded by a square composed of illusory contours, IOR occurred when a target was presented there. When subjects believed the moving cue had simply disappeared, IOR did not occur. This shows that IOR is cognitively penetrable.

occluded. For the other group of observers, the object was always removed for the 400 ms CTOA, resulting in the expectation that it would also disappear when occluded. The principal finding was that when observers believed the object to continue to exist behind the occluder, IOR occurred to a target subsequently presented at that location (which supports Yi et al.'s finding). When, on the other hand, observers believed that the object ceased to exist behind the occluder, IOR did not occur. This dichotomy makes it clear that observer expectation is a powerful determining factor in IOR and provides strong evidence that IOR is not purely a stimulus-driven effect, but one that can be influenced in a goal-driven manner as well.

In a further experiment, we reasoned that if the observer's expectation is critical to the occurrence of IOR, then if that expectation were to be disconfirmed, IOR should be eliminated. To this end, we used the same procedure as described above with one important difference: The occluding square slid to an empty screen location shortly after the object appeared to move behind it (see figure 6.3). The critical point of this manipulation was that the observer could now see that the object, which was expected to continue to exist behind the occluder, was in fact absent—the occluded location was blank except for the four inducing disks. As expected, this led to an elimination of IOR at the occluded location. Also as expected, IOR persisted in the context trials, in which the object was always visible and hence no expectation was required. It could be argued, however, that the observers perceived the object as being "stuck" behind the occluder, and traveling with the occluder to its new location. This seems not to be the case, though, since IOR did not occur to targets presented at the occluder's (and presumably the object's) new location.

Occlusion has a similar effect on performance of the MOT task. In one experiment, multiple-object tracking was unaffected by the presence of occluders if objects disappeared and reappeared in a way that was consistent with occlusion of a persisting object (Scholl and Pylyshyn 1999). If the objects disappeared behind occluders and then reappeared in a way that was inconsistent with observers' beliefs about a continuously present object, however, multiple-object tracking was disrupted. The experimenters concluded that spatial indexes can remain dynamically bound to moving objects, even when these objects appear to momentarily pass behind an occluding surface and then reappear again. Allocation of indexes to moving objects will be terminated, however, if the observer does not interpret the disappearance and reappearance as consistent with a persisting object (see Mitroff, Scholl, and Wynn 2004). The same conclusion

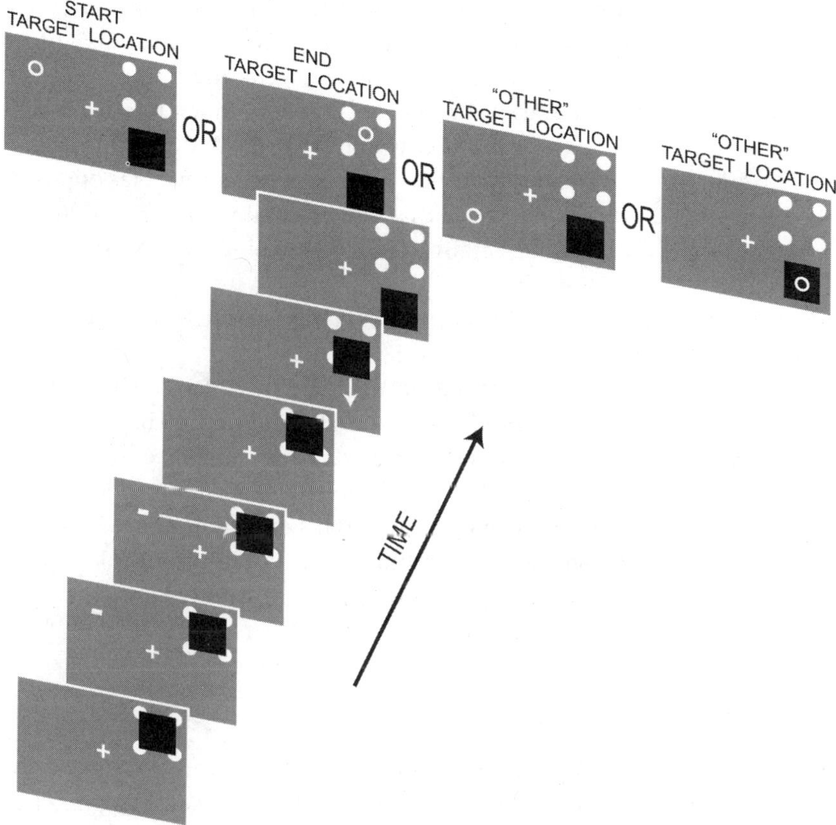

Figure 6.3

In a second experiment in Jefferies et al.'s (2005) study, the cue moved across the display and seemed to disappear behind a red square. Then the square moved to a different location to reveal that the cue was not actually present. As a result, subjects did not believe the moving cue was still present but occluded, and IOR did not occur.

could be made about the experiments outlined in figures 6.2 and 6.3 (Jefferies, Wright, and Di Lollo 2005). That is, when the observers' interpretation is that the cue has moved behind an occluding figure bounded by illusory contours (but does not reappear), a spatial index might remain bound to the cue and, as a result, IOR would occur at that location. Conversely, when their interpretation is that cue has been removed (as opposed to being occluded), the spatial index that was originally bound to it during the movement would also be removed and IOR would not occur at that location.

In summary, there is a growing body of evidence that IOR is not reflexive and is cognitively penetrable.

It appears that the marking operation involved in IOR is mediated by a limited number of dynamic spatial indexes. Depending on the observer's expectations, an index could be removed from a moving object that disappears from view, or could remain assigned to the expected location of an object that is believed to be occluded. Pylyshyn's proposals about dynamic spatial indexing and cognitive penetrability play an important role in this explanation. And, more generally, his ideas have contributed greatly to our understanding of phenomena like IOR.

Acknowledgments

This project was supported by Discovery Grant 133551 and a postgraduate scholarship awarded to the authors by the Natural Science and Engineering Research Council of Canada.

References

Alvarez, G. A., T. S. Horowitz, H. C. Arsenio, J. S. DiMase, and J. M. Wolfe (2005). Do multielement visual tracking and visual search draw continuously on the same visual attention resources? *Journal of Experimental Psychology: Human Perception and Performance* 31: 643–667.

Cavanagh, P. (2004). Attention routines and the architecture of selection. In *Cognitive Neuroscience of Attention*, ed. M. I. Posner, 13–28. New York: Guilford Press.

Clohessy, A. B., M. I. Posner, M. K. Rothbart, and S. P. Vecera (1991). The development of inhibition of return in early infancy. *Journal of Cognitive Neuroscience* 3/4: 345–350.

Danziger, S., A. Kingstone, and J. J. Snyder (1998). Inhibition of return to successively stimulated locations in a sequential visual search paradigm. *Journal of Experimental Psychology: Human Perception and Performance* 24: 1467–1475.

Dodd, M. D., A. D. Castel, and J. Pratt (2003). Inhibition of return with rapid serial shifts of attention: Implications for memory and visual search. *Perception and Psychophysics* 65: 1126–1135.

Duncan, J. (1984). Selective attention and the organization of visual information. *Journal of Experimental Psychology: General* 113: 501–517.

Egly, R., J. Driver, and R. D. Rafal (1994). Shifting visual attention between objects and locations: Evidence from normal and parietal lesion subjects. *Journal of Experimental Psychology: General* 123: 161–177.

Enns, J. T., and S. Cameron (1987). Selective attention in young children: The relations between visual search, filtering, and priming. *Journal of Experimental Child Psychology* 44: 38–63.

Harman, C., M. I. Posner, M. K. Rothbart, and L. Thomas-Thrapp (1994). Development of orienting to objects and locations in human infants. *Canadian Journal of Experimental Psychology* 48: 301–318.

Jefferies, L. N., R. D. Wright, and V. Di Lollo (2005). Inhibition of return to an occluded object depends on expectation. *Journal of Experimental Psychology: Human Perception and Performance* 31: 1224–1233.

Jordan, H., and S. P. Tipper (1998). Object-based inhibition of return in static displays. *Psychonomic Bulletin and Review* 5: 504–509.

Jovicich, J., R. J. Peters, C. Koch, J. Braun, L. Chang, and T. Ernst (2001). Brain areas specific for attentional load in a motion-tracking task. *Journal of Cognitive Neuroscience* 13: 1048–1058.

Klein, R. (1988). Inhibitory tagging system facilitates visual search. *Nature* 334: 430–431.

Lee, D., and S. Quessy (2003). Visual search is facilitated by scene and sequence familiarity in rhesus monkeys. *Vision Research* 43: 1455–1463.

Leek, E. C., I. Reppa, and S. P. Tipper (2003). Inhibition of return for objects and locations in static displays. *Perception and Psychophysics* 65: 388–395.

Liu, G., E. L. Austen, K. S. Booth, B. D. Fisher, R. Argue, M. I. Rempel, and J. T. Enns (2005). Multiple-object tracking is based on scene, not retinal, coordinates. *Journal of Experimental Psychology: Human Perception and Performance* 31: 235–247.

Lupiáñez, J., B. Weaver, S. P. Tipper, and E. Madrid (2001). The effects of practice on cueing in detection and discrimination tasks. *Psicológica* 22: 1–23.

Maylor, E. A., and R. Hockey (1987). Effects of repetition on the facilitory and inhibitory components of orienting in visual space. *Neuropsychologia* 25: 41–54.

Mitroff, S. R., B. J. Scholl, and K. Wynn (2004). Divide and conquer: How object files adapt when a persisting object splits into two. *Psychological Science* 15: 420–425.

Moore, C. M., S. Yantis, and B. Vaughan (1998). Object-based visual selection: Evidence from perceptual completion. *Psychological Science* 9: 104–110.

Ogawa, H., Y. Takeda, and A. Yagi (2002). Inhibitory tagging on randomly moving objects. *Psychological Science* 13: 125–129.

Oksama, L., and J. Hyönä (2004). Is multiple object tracking carried out automatically by an early vision mechanism independent of higher-order cognition? An individual difference approach. *Visual Cognition* 11: 631–671.

Paul, M. A., and S. P. Tipper (2003). Object-based representations facilitate memory for inhibitory processes. *Experiment Brain Research* 148: 283–289.

Posner, M. I., and Y. Cohen (1984). Components of visual attention. In *Attention and Performance*, vol. 10., ed. H. Bouma and D. G. Bouwhuis, 531–555. Hillsdale, N.J.: Lawrence Erlbaum.

Posner, M. I., Y. Cohen, L. S. Choate, R. Hockey, and E. A. Maylor (1984). Sustained concentration: Passive filtering or active orienting? In *Preparatory States and Processes*, ed. S. Kornblum and J. Requin, 49–65. Hillsdale, N.J.: Lawrence Erlbaum.

Posner, M. I., R. D. Rafal, L. Choate, and J. Vaughan (1985). Inhibition of return: Neural basis and function. *Cognitive Neuropsychology* 2: 211–218.

Posner, M. I., C. R. R. Snyder, and B. J. Davidson (1980). Attention and the detection of signals. *Journal of Experimental Psychology: General* 109: 160–174.

Pylyshyn, Z. (1980). Cognitive representation and the process-architecture distinction. *Behavioral and Brain Sciences* 3: 154–169.

Pylyshyn, Z. (1984). *Computation and Cognition*. Cambridge, Mass.: MIT Press.

Pylyshyn, Z. (1989). The role of location indexes in spatial perception: A sketch of the FINST spatial-index model. *Cognition* 32: 65–97.

Pylyshyn, Z. (1998). Visual indexes in spatial vision and imagery. In *Visual Attention*, ed. R. D. Wright, 215–231. New York: Oxford University Press.

Pylyshyn, Z. (1999). Is vision continuous with cognition? The case for cognitive impenetrability of visual perception. *Behavioral and Brain Sciences* 22: 341–423.

Pylyshyn, Z. (2003). *Seeing and Visualizing: It's Not What You Think*. Cambridge, Mass.: MIT Press.

Pylyshyn, Z. (2004). Some puzzling findings in multiple object tracking: I. Tracking without keeping track of object identities. *Visual Cognition* 11: 801–822.

Pylyshyn, Z., and R. W. Storm (1988). Tracking multiple independent targets: Evidence for a parallel tracking mechanism. *Spatial Vision* 3: 179–197.

Schendel, K. L., L. C. Roberson, and A. Treisman (2001). Objects and their locations in exogenous cueing. *Perception and Psychophysics* 63: 577–594.

Scholl, B. J., and Z. Pylyshyn (1999). Tracking multiple items through occlusion: Clues to visual objecthood. *Cognitive Psychology* 38: 259–290.

Sears, C. R., and Z. Pylyshyn (2000). Multiple object tracking and attentional processing. *Canadian Journal of Experimental Psychology* 54: 1–14.

Snyder, J., and A. Kingstone (2001). Inhibition of return at multiple locations in visual search: When you see it and when you don't. *Quarterly Journal of Experimental Psychology* 54: 1221–1237.

Spalek, T. M., and S. Hammad (2005). The left-to-right bias in inhibition of return is due to the direction of reading. *Psychological Science* 16: 15–18.

Taylor, T. L., and M. P. W. Donnelly (2002) Inhibition of return for target discriminations: The effect of repeating discriminated and irrelevant stimulus dimensions. *Perception and Psychophysics* 64: 292–317.

Tipper, S. P., J. Driver, and B. Weaver (1991). Object-centred inhibition of return of visual attention. *Quarterly Journal of Experiment Psychology* 43(A): 289–298.

Tipper, S. P., B. Weaver, L. M. Jerreat, and A. L. Burak (1994). Object-based and environment-based inhibition of return of visual attention. *Journal of Experimental Psychology: Human Perception and Performance* 20: 478–499.

Tipper, S. P., B. Weaver, and F. L. Watson (1996). Inhibition of return to successively cued spatial locations: Commentary on Pratt and Abrams (1995). *Journal of Experimental Psychology: Human Perception and Performance* 22: 1289–1293.

Treisman, A. (1998). The perception of features and objects. In *Visual Attention*, ed. R.D. Wright, 26–54. New York: Oxford University Press.

Treisman, A., and S. Sato (1990). Conjunction search revisited. *Journal of Experimental Psychology: Human Perception and Performance* 16: 459–478.

Viswanathan, L., and E. Mingolla (2002). Dynamics of attention in depth: Evidence from multi-element tracking. *Perception* 31: 1415–1437.

Wright, R. D., and C. M. Richard (1996). Inhibition-of-return at multiple locations in visual space. *Canadian Journal of Experimental Psychology* 50: 324–327.

Wright, R. D., and C. M. Richard (1998). Inhibition of return is not reflexive. In *Visual Attention*, ed. R. D. Wright, 330–347. New York: Oxford University Press.

Wright, R. D., and C. M. Richard (2000). Location cue validity affects inhibition of return of visual processing. *Vision Research* 40: 2351–2358.

Wright, R. D., and C. M. Richard (2003). Sensory mediation of stimulus-driven attentional capture in multiple-cue displays. *Perception and Psychophysics* 65: 925–938.

Wright, R. D., and L. M. Ward. (2008). *Orienting of Attention*. New York: Oxford University Press.

Yantis, S., and D. N. Johnson (1990). Mechanisms of attentional priority. *Journal of Experimental Psychology: Human Perception and Performance* 16: 812–825.

Yantis, S., and E. Jones (1991). Mechanisms of attentional selection: Temporally modulated priority tags. *Perception and Psychophysics* 50: 166–178.

Yi, D. J., M. S. Kim, and M. M. Chun (2003). Inhibition of return to occluded objects. *Perception and Psychophysics* 65: 1222–1230.

II Foundations

7 Computation and Cognition—and Connectionism

Michael R. W. Dawson

Pylyshyn and the Foundations of Cognitive Science

Computation and Cognition (Pylyshyn 1984) is a seminal publication in the annals of cognitive science. First and foremost it provides a definitive statement of the representational theory of mind. It is a manifesto for classical cognitive science, in which cognition is computation: the manipulation of formal symbols. It has become one of the most cited monographs in cognitive science because it delivers an extremely cogent account of why a cognitive vocabulary is required to capture explanatory generalizations in the study of cognition.

Second, *Computation and Cognition* was one of the pioneering appeals for using the trilevel hypothesis within cognitive science (Marr 1982). According to the trilevel hypothesis, a complete account of a cognitive phenomenon requires the use of the three qualitatively different vocabularies, each of which captures generalizations at different levels of analysis, and each of which answers a qualitatively different research question (Dawson 1998). At the computational level, one asks "what information processing problem is being solved by a cognitive agent?," and usually answers this question with a formal proof. At the algorithmic level, one is concerned with "what sequence of information processing steps are being used to solve the information processing problem?," and usually answers this question by making empirical observations of the agent's behavior. At the implementational level, one inquires "what physical states are required to carry out particular information processing steps?," and usually answers this question by appealing to the methods of neuroscience.

Third, *Computation and Cognition* is a blueprint for a "comparative cognitive science." Pylyshyn details how one might validate the relationship between a model and the agent being modeled. He points out that if two systems are equivalent according to the Turing test, all that this means is that they are computing the same input–output function. In other words, they are only equivalent at the computational level. Pylyshyn calls this weak equivalence, and argues that this kind of equivalence is not sufficient for cognitive science. Instead, he makes the case that cognitive science is required to establish the strong equivalence of its models. For a model to be strongly equivalent to an agent, the two must be solving the same problem in the same way. Strong equivalence occurs when the model is using the same algorithm as the agent, and is also using the same primitive information-processing operations. This can only occur when the model and the agent are equivalent in terms of the computational level, the algorithmic level, and in terms of their functional architecture. The functional architecture is the set of information-processing primitives that bridges the algorithmic and implementational levels (Cummins 1983).

I was privileged to be a graduate student in Pylyshyn's lab at the time that *Computation and Cognition* was being released. This experience had a profound influence on my thinking. However, after leaving his lab my interests turned to connectionism—an area far from Pylyshyn's heart, and which he has argued to be far from cognitive science (Fodor and Pylyshyn 1988). The purpose of this essay is to argue that I haven't really strayed too far from his teachings. In particular, my own work on connectionism pays attention to the central idea that its contributions to cognitive science depend crucially on the context of the trilevel hypothesis (Dawson 1998, 2004). The point is to show that if one considers connectionism from the perspective of the trilevel hypothesis, then one discovers that connectionism and classical cognitive science have many crucial similarities.

This chapter proceeds as follows: First, it briefly introduces the conflict between classical and connectionist approaches to cognition. It then proceeds to examine connectionism in the context of the trilevel hypothesis in an attempt to show that this conflict might be more contrived than real. Second, it considers the in-principle power of connectionist networks. Third, it provides two case studies of connectionism at the algorithmic level, one involving music perception, the other involving mushroom identification. Both of these case studies involve interpreting the internal structure of networks in an attempt to discover how networks solve particular problems. It is observed that these interpretations can be used to inform classical theory—and sometimes they show that classical and connectionist theories are identical. Finally, a brief treatment of connectionism and the implementational level is provided.

Cognition versus Connectionism

Even a casual glance at the history of the study of mentality reveals a constantly repeating pattern in which a new school of thought is born in reaction to the existing views. Cognitive science is no exception. The representational theory of mind was born in the late 1950s as a reaction to psychological behaviorism. In the 1980s, connectionism arose as a reaction to the representational theory of mind, or at least as a reaction to the particular version of that theory which construed thought as the rule-governed manipulation of symbols.

The connectionist revolution in cognitive science was largely due to three related factors (McClelland, Rumelhart, and Hinton 1986). First, researchers who were drawn to connectionism shared the view that the representational theory of mind had stalled. Second, researchers were discovering new learning algorithms that permitted the training of powerful multilayer networks. Third, connectionists argued that the biologically inspired architecture of their networks, in which parallel processing and distributed representations provided a flexible similarity-based type of information processing, would solve the problems that they felt had stalled classical cognitive science.

The representational theory of mind was viewed by connectionists as being too disparate from what we know of the brain (Clark 1989). Classical theories were seen as being overly motivated by formal logic and by digital computers, resulting in a marked separation between symbols and rules, and in an assumption of serial processing. Connectionists argued that these assumptions necessarily led to slow, brittle models that had little chance of being translated into neural architecture (Feldman and Ballard 1982). Furthermore, they felt that the reason computers were poor at accomplishing tasks that were natural to humans (vision, locomotion, etc.) was because the brain was involved in a radically different kind of information processing than the kind that was being espoused by classical cognitive science (Churchland and Sejnowski 1992).

The connectionist alternative was to blur the symbol–rule distinction. They proposed information processing that was accomplished by a network of simple, interconnected processing units. These units were analogous to neurons, and the connections between them were analogous to synapses. Rather than being programmed, connectionist networks were taught from examples. The information processing conducted by a network in order to mediate a particular input–output function was described as being subsymbolic, and as being more akin to statistical mechanics than to logic. In the mid- to late 1980s a number of connectionist models of complex cognitive functions appeared in the literature, and were offered as radical alternatives to extant classical theory. Connectionism was described by some as a paradigm shift for psychology (Schneider 1987).

Connectionism and the Computational Level

Given his position as a pioneer of the representational theory of mind, it is not surprising that Pylyshyn was also at the forefront of classical cognitive science's reaction against the rise of connectionism. For example, the following Pylyshyn quote appeared in *Scientific American:* "Voodoo. People are fascinated by the prospect of getting intelligence by mysterious Frankenstein-like means—by voodoo. And there have been few attempts to do this as successful as neural nets" (Stix 1994).

Pylyshyn's negative reaction is completely consistent with the historical tradition that led to his own research program. The representational theory of mind was itself a reaction against psychological behaviorism and against associationism. At the birth of cognitive science, researchers argued that associationist theories simply did not have the power to capture psychological regularities. Most notably, computational proofs that associationism was incapable of dealing with natural human language appeared in the literature (Bever, Fodor, and Garrett 1968; Chomsky 1959). Connectionism shares many of the foundational features of associationism. Given this, it was natural that Pylyshyn focused his criticism against connectionism by attacking its computational power. "Connectionism appears to have fatal limitations. The problem with Connectionist models is that all the reasons for thinking that they might be true are reasons for thinking that they couldn't be psychology" (Fodor and Pylyshyn 1988).

The general move made by classical cognitive science against associationism was to argue that associationist architectures didn't have the computational power of a universal Turing machine. For instance, Bever, Fodor, and Garrett (1968) argued that associationist models were equivalent in power to finite state automata, and as such were formally incapable of dealing with a recursive structure of natural language. The spirit of this argument against associationism can be found in Fodor and Pylyshyn's famous 1988 critique of connectionism, in which they argued that computational limits on artificial neural networks prevented them from being componential and systematic.

However, a good deal of formal analysis has shown that connectionist networks do indeed have sufficient computational power to be of interest to cognitive science. In particular, some researchers have shown that networks are indeed systematic, reacting against the criticism of Fodor and Pylyshyn (Hadley 1994a,b, 1997; Hadley and Hayward 1997). More generally, many results have established the vast "in principle" computational power of artificial neural networks. First, networks with no more than two layers of hidden units are capable of arbitrary pattern classification (Lippmann 1989). Second, networks with a single layer of hidden units are capable of being universal function approximators (Cotter 1990; Cybenko 1989; Funahashi 1989; Hartman, Keeler, and Kowalski 1989; Hornik, Stinchcombe, and White 1989). Third, it was long ago proved that the tape head of a universal Turing machine could be constructed from a McCulloch-Pitts network (McCulloch and Pitts 1943). More recently, researchers have shown how modern neural networks can attain UTM equivalence without the need for infinite external memory (Siegelmann 1999; Siegelmann and Sontag 1991, 1995).

As a whole, results like these have established the computational power of connectionist networks. Classical cognitive science cannot use computational arguments to dismiss their connectionist rivals. If the two approaches are indeed qualitatively different, then these differences must appear at other levels of analysis. In the next section, we turn to an examination of connectionism at the algorithmic level.

Connectionism and the Algorithmic Level

Connectionism and Bonini's Paradox

In cognitive science, computer simulations are supposed to offer rigorous accounts of cognitive phenomena. Unfortunately, things are not quite that simple. Lewandowsky (1993) has noted that computer simulation methods are not without disadvantages, including what has been called Bonini's paradox. If a computer simulation falls into this trap, then this means that it is no easier to understand than the phenomenon that the simulation was supposed to illuminate.

Connectionist researchers freely admit that in many cases it is extremely difficult to determine how their networks accomplish the tasks that they have been taught. "If the purpose of simulation modeling is to clarify existing theoretical constructs, connectionism looks like exactly the wrong way to go. Connectionist models do not clarify theoretical ideas, they obscure them" (Seidenberg 1993, 229). This has raised serious doubts about the ability of connectionists to provide fruitful theories about cognitive processing. McCloskey (1991, 387) suggested that "connectionist networks should not be viewed as theories of human cognitive functions, or as simulations of theories, or even as demonstrations of specific theoretical points." In a nutshell, this dismissal of connectionism relies on the position that parallel distributed processing (PDP) networks are generally ι:ninterpretable (see also Dawson and Shamanski 1994).

If connectionists want to contribute to cognitive science by providing algorithmic accounts of cognitive processing, then some way must be found to avoid Bonini's paradox. Fortunately, several different approaches to interpreting the algorithmic structure of PDP networks have been described in the literature (Dawson 2004). Two of these methods are briefly described in the next sections. The first, the interpretation of a network trained to classify musical chords, demonstrates the utility of examining the connection weights in a trained network (Yaremchuk and Dawson 2005). The second, the interpretation of a network trained to classify mushrooms, illustrates analysis of networks by studying regularities in hidden unit activities (Dawson et al. 2000).

Case Study 1: Classifying Musical Chords

One of the motivators of the connectionist revolution was the desire to explore information processing that did not rely upon a strict segregation between symbols and rules. It was argued that networks were demonstrating "classical regularities," but were doing so by being "subsymbolic" or by not relying upon "explicit rules." In other words, connectionism was explicitly abandoning the formal logical roots upon which classical cognitive signs had been founded.

In some research domains, the alleged informal nature of artificial neural networks was viewed as an advantage. For instance, music is an area in which there is a great deal of formal understanding. However, many researchers believe that a formal account of musical structure is not capable of capturing the complete nature of music. For this reason many researchers have turned to connectionist models in an attempt to explore less formal properties of music. Artificial neural networks have been used to study a dizzying variety of musical tasks such as perception of pitch, perception of tonal structure, perception of musical sequences, and perception of rhythm; and networks have also been used as models of composition (see the many examples in Griffith and Todd 1999; Todd and Loy 1991).

However, it can also be argued that connectionist networks have a great deal in common with their classical cousins. Indeed, it is quite likely that one can achieve formal insights into music by studying the internal structure of a network trained to accomplish a musical task. This section explores this possibility by describing the interpretation of a network trained to classify different types of musical chords.

Defining the chord classification problem Imagine a small piano keyboard consisting of only 24 keys, black and white. The first 12 keys of this mini-piano represent the following notes: A, A#, B, C, C#, D, D#, E, F, F#, G, and G#. In this pattern, every note paired with the # symbol corresponds to a black key on the keyboard, and all of the other notes correspond to white keys. Moving from the left to right in this pattern, each note is a semitone higher than the note on its left. The thirteenth key on this keyboard plays another A that is an octave higher than the A that started the keyboard. From this thirteenth key to the last (twenty-fourth) key on the piano, the pattern of notes is repeated. So, while there are 24 different keys on this keyboard, they are only associated with 12 different note names, and each note is repeated an octave higher than its first instance. Each of these 12 different notes can serve as the starting note, or root, of a major scale. For instance, we could have a scale in the key of A-major that starts on the root A, a scale in the key of A#-major that starts on the root A#, and so on, up to the key of G#-major.

For any root note that we choose, there exists a basic harmonic structure. Harmony is the combination two or more notes into a compound in which all of the notes are played at the same time. For example, let us consider the major tetrachord. For the root note C, this chord is the set of notes C, E, and G that are the first, third, and fifth notes in the C-major scale, followed by a second C that is an octave higher than the root note that begins the chord. The ordered set of notes C, E, G, and C is called the *root position* of the chord.

However, this chord can be constructed in other ways too. For instance, we could start with the lowest E on the keyboard, and play the notes E, G, C, and E, where the last E is an octave higher than the first. In this version of the chord, the same notes are being played, but they are arranged in a different order. This order is called the *first inversion* of C-major. We could also start with the lowest G that we can find on the keyboard, and play the notes G, C, E, and G. This is called the *second inversion* of the chord.

One can take the major tetrachord built upon a root note, make a minor modification to one of its notes, and create a very different sounding type of chord. For instance, if one takes the second note of a major tetrachord in root position and lowers this note by a semitone, the result is a minor tetrachord. The C-major chord (C, E, G, and C) can be converted into the

C-minor chord (C, D#, G, and C) by lowering the E by a semitone to the note D#. As was the case for the major chords, we can write minor chords in first and second inversions as well. Similarly, if one takes the fourth note of a major tetrachord in root position and lowers this note by two semitones, the result is a dominant tetrachord. So, the dominant tetrachord that is built on the root note C is the four-note pattern C, E, G, and A#. In addition to the root version, one can create first, second, and third inversions of dominant tetrachords. Finally, if the second, third, and fourth notes of a dominant tetrachord (in root position) are all lowered by a semitone, the result is a diminished tetrachord. The diminished tetrachord built on the root of C is the set of notes C, D#, F#, and A. Diminished chords can also appear in first, second, and third inversion forms.

Regardless of the root note upon which a chord is based, and regardless of its inversion, each of the four different kinds of chords described above has a distinct quality or "sound." As a result, with a fair amount of training it is possible for a human musician to hear a chord and classify it as belonging to one of these four major chord categories. We were interested in training a network to also have this ability, and to then interpret its internal structure and discover how it accomplishes this task.

Training a network to classify chords   In Western dodecaphonic music, there are 12 different possible root notes. For each, one can build a major, a minor, a dominant, and a diminished tetrachord in root position. One can then convert these chords into their possible inversions. As a result, any root note is associated with 14 different chords, leading to a total set of 168 different chords. Our goal was to train a network to correctly classify each of these into one of the four different chord categories.

Our network used 24 input units, each of which corresponded to one of the keys of a "mini-piano keyboard." Any of the chords in the training set could be presented to the network by turning four of these input units on and by turning the remaining input units off. The network also used four output units, one for each chord type. Finally, the network had four hidden units, which was determined to be the minimum number that would permit the network to solve this particular problem. The network was "fully connected" between layers. That is, every input unit had a connection to every hidden unit, and every hidden unit had a connection to every output unit. There were no direct connections between input units and output units.

The hidden units and the output units in the network were *value units*. Value units are similar to traditional processing units in multilayer perceptrons, but instead of using a logistic activation function, they use a Gaussian equation (Dawson and Schopflocher 1992). We chose value units because many studies have shown that they have advantages over traditional processing units. In particular, we have found that value unit networks are often easier to interpret. The network was trained using a variation of the generalized delta rule for networks of value units, and generated a correct response for every stimulus after 5,230 sweeps of training (i.e., after 5,230 presentations of the full set of chords).

Interpreting the trained network   How does this network correctly classify this set of chords? The first step in interpreting the network's structure was to examine the connection weights from the input units to the hidden units. This was because the connection weight between an input unit and a hidden unit could be considered the hidden unit's "name" for that note.

Two of the hidden units exhibited a repeating pattern of four different weights assigned to the 24 input "piano keys." As a result, both of these units assigned the same connection weight value to three different notes in the Western dodecaphonic scale, dividing this scale into four groups of three notes each. The first group of notes was (A, F, C#), the second was (D, F#, A#), the third was (G, D#, B), and the final group was (C, G#, E). Because the three notes in a group were given the same connection weight, to a hidden unit the three notes were functionally equivalent. That is, these two hidden units would be unable to distinguish an A from an F or from a C#. By assigning different weights to each group of three notes, the hidden units were treating each group as being functionally distinct from the others. In other words, rather than assigning 12 different names analogous to the dodecaphonic system, these two hidden units were in essence only using four different note names. This was a radically different set of equivalence classes of input unit notes than we had expected.

Importantly, the equivalence classes that were being used by these hidden units have a definite formal structure. Each group of three notes can be arranged in a circle of major thirds. That is, if one takes any note in the group, one of the other notes is exactly a major third (i.e., four semitones) above it, while the other is exactly a major third below it. For example, if one moves a major third up from the note A, one reaches the note F. Moving another major third up, one reaches the note C#. Moving yet another major third up, one returns to an A. This hidden unit encoding of inputs reveals a structure that is rarely commented upon in formal treatments of music, but nevertheless can easily be characterized by extending these treatments.

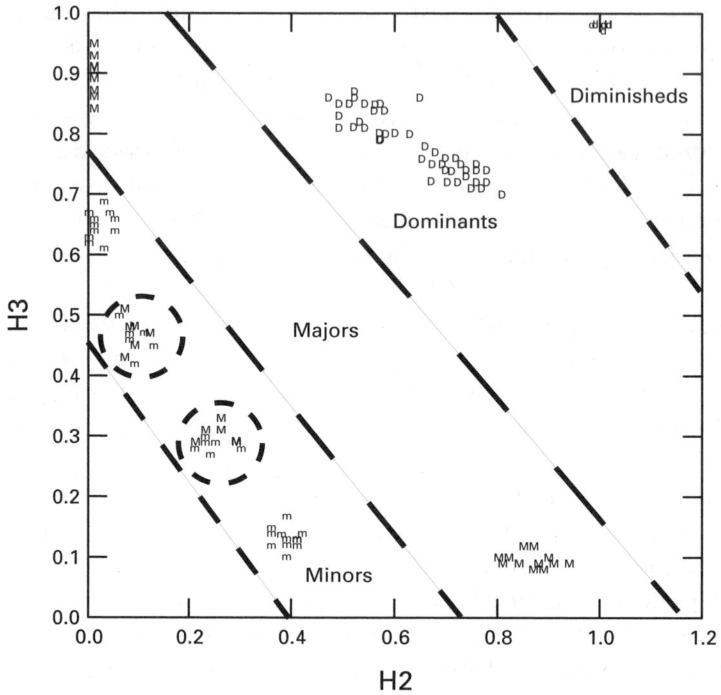

Figure 7.1

A two-dimensional map of the different chord stimuli using the activity that they produce in two of the hidden units as coordinates. All of the chords are separated into chord types except for the second-inversion major chords, which fall into the regions of minor chords that are referenced with the dashed circles.

Why do these units adopt this circle of major thirds representation of input notes? One answer to this question can be provided by graphing the position of each input chord as a point in a two-dimensional space, where the coordinates of this point are provided by the activation values produced by the chord in these two hidden units. When this is done, it can be seen that these two hidden units arrange all of the chords in distinct "diagonal layers" across this space (see, e.g., Dawson 2004, 231). These diagonal layers are important in the value unit architecture, because an output value unit makes two parallel cuts through this kind of hidden unit space to separate one class of patterns from others. In fact, these two hidden units can by themselves create a space in which output units can correctly segregate 92.9 percent of all of the different chords. The only exception is one distinct subset of chords: the 12 input patterns that are second-inversion majors. These chords are mistakenly grouped with the minor chords.

The remaining two hidden units add two dimensions to the space, and work to pull the second-inversion major chords away from the minors. These two hidden units assign the different input notes into two different classes. The first is the set of notes (A, B, C#, D#, F, G), and the second is the set (A#, C, D, E, F#, G#). Each of these sets of notes can be arranged as a circle of major seconds, in which adjacent notes in the circle are exactly a major second (i.e., two semitones) apart. Again, this is a very atypical organization of notes, but one that can be quite easily described using formal musical terms. For instance, the two circles of major seconds are strongly related to French composer Olivier Messiaen's first mode of limited transposition (Messiaen 1956).

Implications of Case Study 1 for Cognitive Science

The purpose of presenting the chord network is to illustrate a few themes that are relevant to considering the role of connectionism within cognitive science, particularly at the algorithmic level of analysis.

First, the extent to which the chord network reveals interesting properties depends upon the degree to which the network's internal structure can be interpreted. This is in contrast to an early era of research, "gee whiz connectionism." Gee whiz connectionism is characterized by training a network to accomplish some task, usually one that is characteristic of classical cognitive science. The claim is then made that the result is a nonclassical model of a classical phenomenon, usually without any evidence to support the model's nonclassical nature. However, this research tradition makes no sense if one notes, as was done earlier, that artificial neural networks possess extreme computational power. If this power is assumed, then one must expect to be able to train a network to perform any task of interest. The mere ability to create a network is not really of interest. What is instead of interest comes from peering inside the creation, particularly if one wants to make constructive comparisons between classical and connectionist cognitive science.

Second, when one interprets the internal structure of a trained network, one begins to tell a representational story. In other words, the interpretation of the network is the process of assigning representational contents to internal network components and to network states. Furthermore, one appeals to these representational contents in an attempt to account for the behavior of the network. For instance, predictions of the behavior of the chord network involve considering the implications of encodings based on circles of major thirds and major seconds. The interpretation of a connectionist network might make novel contributions to cognitive science by revealing new kinds of representations that have not been considered by classical cognitive science (Hinton, McClelland, and Rumelhart 1986). However, the account is still going to be representational and therefore will still be quite at home in the kind of cognitive science espoused by Pylyshyn.

Third, the interpretation of a connectionist network reveals more about the domain in which the network has been trained than it does about the network per se. For example, the chord network is interesting precisely because it reveals alternative properties of music that might be important for chord perception or recognition. The network is providing insight into music, not into itself. For this particular network, this is even more interesting because it shows that artificial neural networks can provide information about formal properties of domains and are not restricted to capturing informal or subsymbolic regularities. Again, this is to be expected if the kind of algorithmic account of a connectionist network is not dramatically different from a similar account of a classical model.

The interpretation of the chord network illustrates a particular methodology in which classical and connectionist algorithms can be related. However, this network is not the best example of establishing strong relationships between these two approaches. For such an example let us turn to a second case study.

Case Study 2: Classifying Mushrooms

In the philosophy of science, if two apparently different theories are in fact identical, then one should be able to translate one theory into the other. This is called *intertheoretic reduction* (Churchland 1985, 1988; Hooker 1979, 1981). The widely accepted view that classical and connectionist cognitive science are fundamentally different (Schneider 1987) amounts to the claim that intertheoretic reduction between a symbolic model and a connectionist network is impossible. Below, we examine this directly by asking whether we can translate a classical theory into a network using standard training techniques.

Classical algorithms for classifying mushrooms A benchmark problem in the machine learning literature is the classification of mushrooms as being either edible or poisonous (Schlimmer 1987). This problem consists of 8,124 different mushrooms, each defined as a set of 21 different features.

The task of interest is to use the descriptive features to correctly classify each mushroom in the training set.

A classical algorithm that accomplishes this task can be obtained using standard machine learning techniques (Dawson et al. 2000). For example, a variation of the ID3 algorithm (Quinlan 1986) was used to induce a decision tree for the mushroom problem. The decision tree that was generated is a sequence of five rules, given in table 7.1, that correctly classify all the mushrooms. This decision tree is a classical algorithm because its rules are explicit, local, and digital (Haugeland 1985), and must be executed in a particular serial order.

Another way to demonstrate the classical nature of this decision tree is to translate it into an alternative classical algorithm. In particular, the

Table 7.1

A five-step decision tree for classifying the mushrooms. Decision points in this tree where mushrooms are classified (e.g., Rule 1 Edible) are given in bold.

Step 1	*What is the mushroom's odor?* If it is almond or anise then it is edible. **(Rule 1 Edible)** If it is creosote or fishy or foul or musty or pungent or spicy then it is poisonous. **(Rule 1 Poisonous)** If it has no odor then proceed to Step 2.
Step 2	*Obtain the spore print of the mushroom.* If the spore print is black or brown or buff or chocolate or orange or yellow then it is edible. **(Rule 2 Edible)** If the spore print is green or purple then it is poisonous. **(Rule 2 Poisonous)** If the spore print is white then proceed to Step 3.
Step 3	*Examine the gill size of the mushroom.* If the gill size is broad, then it is edible. **(Rule 3 Edible)** If the gill size is narrow, then proceed to Step 4.
Step 4	*Examine the stalk surface above the mushroom's ring.* If the surface is fibrous then it is edible. **(Rule 4 Edible)** If the surface is silky or scaly then it is poisonous. **(Rule 4 Poisonous)** If the surface is smooth the proceed to Step 5.
Step 5	*Examine the mushroom for bruises.* If it has no bruises then it is edible. **(Rule 5 Edible)** If it has bruises then it is poisonous. **(Rule 5 Poisonous)**

decision tree in table 7.1 can be translated into an equivalent set of production rules (Dawson et al. 2000). Each production describes the properties of mushrooms that must be true at each decision point in the decision tree. These properties define a production's condition; its consequent action is asserting that the mushroom is either edible or poisonous. For instance, at the "Rule 1 Edible" decision point in table 7.1, one could create the production rule "If the odor is anise or almond, then the mushroom is edible." Similar productions can be created for later decision points in the algorithm; these productions will involve a longer list of mushroom features. The complete set of productions that were created for the decision tree algorithm is provided in table 7.2.

Table 7.2
The translation of the decision tree in table 7.1 into an equivalent set of nine production rules. The mapping from these rules to network states is provided in the "Network Cluster" column, which is described in more detail later in the chapter.

Decision Point From Table 7.1	Network Cluster	Equivalent Production
Rule 1 Edible	2 or 3	P1: if (odor = anise) ∨ (odor = almond) → edible
Rule 1 Poisonous	1	P2: if (odor ≠ anise) ∧ (odor ≠ almond) ∧ (odor ≠ none) → not edible
Rule 2 Edible	9	P3: if (odor = none) ∧ (spore print color ≠ green) ∧ (spore print color ≠ purple) ∧ (spore print color ≠ white) → edible
Rule 2 Poisonous	6	P4: if (odor = none) ∧ ((spore print color = green) ∨ (spore print color = purple)) → not edible
Rule 3 Edible	4	P5: if (odor = none) ∧ (spore print color = white) ∧ (gill size = broad) → edible
Rule 4 Edible	7 or 11	P6: if (odor = none) ∧ (spore print color = white) ∧ (gill size = narrow) ∧ (stalk surface above ring = fibrous) → edible
Rule 4 Poisonous	5	P7: if (odor = none) ∧ (spore print color = white) ∧ (gill size = narrow) ∧ ((stalk surface above ring = silky) ∨ (stalk surface above ring = scaly)) → not edible
Rule 5 Edible	8 or 12	P8: if (odor = none) ∧ (spore print color = white) ∧ (gill size = narrow) ∧ (stalk surface above ring = smooth) ∧ (bruises = no) → edible
Rule 5 Poisonous	10	P9: if (odor = none) ∧ (spore print color = white) ∧ (gill size = narrow) ∧ (stalk surface above ring = smooth) ∧ (bruises = yes) → not edible

Connectionist networks for classifying mushrooms   One can also train connectionist networks to solve the mushroom problem. Dawson et al. (2000) trained one network that used a single output unit to represent the mushroom classification, four hidden units, and 21 input units to represent mushroom features. Each input unit corresponded to a single feature; different activation values for an input unit were used to encode different feature values. The output unit and the hidden units were all value units. When trained with Dawson and Schopflocher's (1992) learning rule, the network generated a correct response to each of the 8,142 mushrooms in the training set after 1,852 sweeps.

Dawson et al. (2000) conducted a variety of analyses to determine how this network accomplished this task. Rather than examining connection weights (as was done above in case study 1), they recorded the activities of each of the four hidden units to all of the patterns in the training set. In other words, they converted the 21-dimensional input unit representation of each mushroom into a 4-dimensional hidden unit representation. Then, k-means cluster analysis was performed on the 8,124 vectors of hidden unit activities that represented each mushroom. It was found that 13 was the minimum number of different clusters required such that each member of each cluster resulted in the network generating the same output response. The features that characterized the mushrooms that fell into each cluster were then examined in order to find definite features associated with each cluster. A definite feature is one that is shared by every stimulus in a cluster, and can be identified using descriptive statistics (Berkeley et al. 1995; Dawson 2005).

Dawson et al. (2000) found a number of definite features that characterized each cluster. An examination of these features indicated that the artificial neural network was exploiting very different regularities than those that were revealed in the decision tree above. The features that were being exploited by the hidden units were: (cap color = cinnamon), (gill color = white), (stalk color above ring = white), (ring type = evanescent), (habitat = meadows), and (habitat = woods). A discriminant function that employs a linear weighting of these features can correctly classify all of the mushrooms. However, it is clearly not the same as the decision tree algorithm, because it exploits different features, and exploits them in a different manner.

Of greater interest was a second network trained by Dawson et al. (2000) to perform this classification task. This network was trained with a different definition of its output states in an attempt to translate the decision tree into the network's internal states.

A pattern classification system is normally only informed about what the correct label for a pattern should be. For instance, in the mushroom problem, the system would normally only be taught to generate the label "edible" or the label "poisonous." But, it is often the case that more information than this is actually available. Specifically, there often exists prior information about why an input pattern belongs to one class or another. Thus, one could add this information to the pattern classification problem by teaching the system not only to generate a label of interest (e.g., "edible," "poisonous") but also to generate a reason for assigning this label (e.g., "passed Rule 1," "failed Rule 4").

Elaborating a classification task along such lines has been called *injection of hints* or *extra output learning* (Abu-Mostafa 1990; Suddarth and Kergosien 1990). We hypothesized that extra output learning could be used to insert the decision tree described above into a network. Our prediction was that after training was complete an examination of the network's internal structure would reveal an internal representation of the classical algorithm. If this were the case, then we would have used standard training practices to translate the classical algorithm into a PDP network.

The network that was trained used the same input representation as the previous network, and required five hidden units to accomplish the more demanding classification task. The primary difference between it and the preceding network was its output unit configuration. Ten different output value units were used. One output unit encoded the edible/poisonous classification, and the other nine output units were used to inject the hints that were available from the decision tree of table 7.1. That is, table 7.1 lists nine different points in the decision tree at which a definite classification of a mushroom is possible ("Rule 1 edible," "Rule 1 poisonous," "Rule 2 edible," etc.). The second network was trained to indicate a mushroom classification with its first output unit, and to also use one of the remaining output units to indicate at which point in the decision tree this classification could be made. When trained on this more difficult version of the problem (because it requires the network to generate many different sub-classifications of stimuli), the network achieved convergence after 8,699 epochs of training.

Clearly, the question of interest was whether the internal structure of this network would reveal that the classical algorithm had been translated into network form. Dawson et al. (2000) represented each mushroom as the vector of five hidden unit activation values that it produced when presented to the network. They then performed a k-means clustering of this data. They found that when the hidden unit activities were assigned to 12 different clusters, each mushroom in the cluster produced the same network output, indicating that this was the appropriate number of clusters to use to describe this network. They then proceeded to determine a set of distinct mushroom features that were associated with each cluster.

The sets of definite features associated with each cluster can be thought of as conditions, represented internally by the network (as a vector of hidden unit activities), that result in the network producing a particular response (in particular, the edible/poisonous judgment represented by the first output unit). For example, when a network is presented with a mushroom that belongs to Cluster 2, its hidden units will adopt a particular vector of activities. This vector of activities represents the fact about the mushroom that its odor is either almond or anise. Either of these properties is in turn sufficient to support the claim that the mushroom is edible.

Importantly, this way of considering a hidden unit vector as representing condition features that are prerequisites to network responses permitted Dawson et al. (2000) to examine the relationships between the clusters and the set of productions given earlier. They discovered that there exists a unique mapping from internal network states (i.e., vectors of hidden unit activities) to the productions that define a classical algorithm. That is, each distinct class of hidden unit activities (i.e., each cluster) corresponds to one, and only one, of the productions listed in the table (a complete listing is provided in table 7.2). In other words, when one describes the network as generating a response because its hidden units are in one state of activity, one can translate this into the claim that the network is executing a particular production. This shows that the extra output learning translated the classical algorithm into a network model.

Implications of Case Study 2 for Cognitive Science

Why is it interesting that one can use a standard connectionist learning algorithm to translate a classical algorithm into a network model? One implication of this finding concerns the issue of reducing one theory to another. One modern version of reductionism, the *new wave* (Bickle 1996; Endicott 1998), has its origins in the work of Hooker and Churchland that was cited earlier. The main innovation of the new wave is that one does not reduce the secondary theory directly to the primary theory. Instead, one takes the primary theory and constructs from it a structure that is analogous to the secondary theory, but which is created in the vocabulary of the primary theory. Theory reduction involves constructing a mapping between the secondary theory and its image constructed from the primary theory. "The older theory, accordingly, is never deduced; it is just the target of a relevantly adequate mimicry" (Churchland 1985).

The analysis of the second network provides a new wave intertheoretic reduction between a classical algorithm and a PDP model. The goal of new wave reductionism is to demonstrate that one theory performs an "adequate mimicry" of another. This has clearly been accomplished by mapping different classes of hidden unit states to the execution of particular productions as shown in table 7.2. In turn, there is a direct mapping from any of the productions back to the decision tree algorithm. This provides extremely strong evidence that Dawson et al. (2000) were able to use extra output learning to provide an exact translation of a classical algorithm into the network of value units.

What are the implications of this finding for the relationship between classical and connectionist cognitive science? The main implication is that one cannot assume that classical models and connectionist networks are fundamentally different at the algorithmic level, because one type of model can be translated into the other. In other words, the main result of the second case study is to demonstrate that at the algorithmic level it is possible to have a classical model that is exactly equivalent to a PDP network.

Von Eckardt (1993) has suggested that if one considers "higher-level" representations in PDP models (i.e., patterns of activity distributed across processors, instead of the properties of individual processing units), then connectionist networks can be viewed as computers analogous to those brought to mind when one thinks of classical architectures. This is because when examined at this level, connectionist networks have the capacity to input and output represented information, to store represented information, and to manipulate represented information. This is the position that the subsymbolic properties of networks approximate the symbolic properties of classical architectures (Smolensky 1988).

The relationship between hidden unit activities and productions in the mushroom network is an example of the apparent equivalence between symbolic and subsymbolic accounts. This type of relationship also has implications for another debate that involves the algorithmic comparison between classical and connectionist architectures. Consider a recent attempt to incorporate situated action theories (including connectionism) into classical cognitive science (Vera and Simon 1993). Vera and Simon argue that any situation-action pairing can be represented either as a single production in a production system, or (for complicated situations) as a set of productions. "Productions provide an essentially neutral language for describing the linkages between information and action at any desired (sufficiently high) level of aggregation" (Vera and Simon 1993, 42). They go on to describe such systems as ALVINN (a neural network that is part of the navigational component of an autonomous vehicle [Pomerleau 1991]) as being equivalent to a classical set of productions.

However, such translations of nonclassical models into classical systems have been strongly challenged. For example, Vera and Simon's (1993) definition of "symbol" has been deemed too liberal by connectionist researchers Touretzky and Pomerleau, who argue that ALVINN's hidden unit "patterns are not arbitrarily shaped symbols, and they are not combinatorial. Its hidden unit feature detectors are tuned filters" (Touretzky and Pomerleau 1994, 348). Greeno and Moore take the middle road in their analysis of ALVINN, suggesting that "some of the processes are symbolic and some are not" (1993, 54). Disagreements about what counts as a symbol are clearly at the heart of the debate that Vera and Simon initiated (Vera and Simon 1994).

In our view, one reason Vera and Simon's (1993) interpretation of connectionist networks (and other systems) as being production systems is not completely satisfactory is that it is not specific enough. For instance, Vera and Simon did not have direct access to ALVINN, and therefore were not capable of explicitly analyzing all of its internal states. Furthermore, they did not generate a specific set of productions that were equivalent to ALVINN. As a result, they were not in a position to provide a detailed translation of ALVINN into a production system (i.e., statements of the form "ALVINN State x is equivalent to Production y").

The interpretation of the second mushroom network provides a much stronger example for the general position that Vera and Simon (1993) propose. This is because it provides (a) a detailed analysis of the internal states of a PDP network and (b) a precise mapping from these states to a set of equivalent productions. As a result, we can confidently make claims of the type "Network State x is equivalent to Production y." Of course, this one result cannot by itself validate Vera and Simon's argument. If there is progress to be made in the discussion that they started, then detailed analyses of the type described above will be required. For instance, can any classical theory be translated into a network? This is one type of algorithmic-level issue that requires a great deal of additional research.

Connectionism and the Implementational Level

The final level of analysis to consider for a comparison between classical and connectionist models is the level of implementation. This level has been the source of a great deal of controversy in the debate between these two approaches to cognitive science. On the one hand, many proponents of connectionism have argued that PDP models are more biologically plausible than classical systems (Clark 1993, 1997; Dreyfus and Dreyfus 1988; McClelland, Rumelhart, and Hinton 1986). On the other hand, classical supporters have claimed that if connectionist models are to be taken as biological accounts, then they are not part of cognitive science because they do not appeal to a cognitive vocabulary (Broadbent 1985; Fodor and Pylyshyn 1988; Pylyshyn 1991).

However, there are many reasons to delay a comparison between the two approaches at the implementational level. First, many researchers have pointed out that many properties of connectionist networks are not biologically plausible (Crick and Asanuma 1986; Douglas and Martin 1991; Smolensky 1988). Second, many analyses of connectionism indicate (at the very least) that it is unclear whether networks are to be understood as implementational theories or as cognitive theories (Broadbent 1985; Dawson 1998). Third, it has been shown that novel cognitive (as opposed to implementational) theories can be extracted from connectionist networks (Dawson, Medler, and Berkeley 1997). In short, the implementational story about connectionism is in exactly the same state as the same story about classical models: vague and incomplete, and requiring further study.

Computation and Cognition and Connectionism

Connectionist models are properly viewed using the trilevel hypothesis that characterizes modern cognitive science. Connectionist research has often emphasized its contributions at the implementational level. However, we saw earlier that there is a history of computational-level analyses of artificial neural networks that has established that they have the same computational power as classical models. At the algorithmic level, we have seen that one can generate an account of the methods used by a connectionist network to solve a problem if one abandons "gee whiz" connectionism and conducts an analysis of a network's internal organization. In short, the current essay has argued that connectionist networks can lead to interesting results when studied at all three levels.

A close cousin of "gee whiz" connectionism is "look, ma, it's different" connectionism. Those who adopt this latter position make strong claims about the differences between connectionist models and other types of models, but usually don't accompany these claims with the necessary supporting evidence. Interestingly, the view is held by both connectionists and nonconnectionists. The former group uses this position when they rely on intuitions that connectionist networks are qualitatively different from classical models in their attempt to display connectionism as an alternative modeling approach. The latter group uses this position when they argue that connectionism is so radically different from the classical approach that it can't be taken seriously as a component of cognitive psychology or of cognitive science.

The perspective that has been briefly illustrated in this essay is that "look, ma, it's different" connectionism isn't really a viable approach. If one takes the time to conduct a thoughtful and objective comparison of connectionism to classical cognitive science, then one finds that these two approaches are far more similar than a casual glance at the extant literature would suggest. Importantly, a particularly useful framework for guiding this comparison can be found in Pylyshyn's (1984) *Computation and Cognition*. The issues that have been introduced in this chapter demonstrate that when Pylsyhyn's version of the trilevel hypothesis is applied to connectionism, one can find equivalences between it and classical cognitive science at both the computational and the algorithmic levels. Further insights into the relationships between these two approaches need to be obtained by examining other potential equivalences (i.e., architectural and implementational) that are fundamental to Pylyshyn's view of cognitive science as a whole.

Acknowledgment

The research reported in this chapter was supported by grants from NSERC and from SSHRC.

References

Abu-Mostafa, Y. S. (1990). Learning from hints in neural networks. *Journal of Complexity* 6: 192–198.

Berkeley, I. S. N., M. R. W. Dawson, D. A. Medler, D. P. Schopflocher, and L. Hornsby (1995). Density plots of hidden value unit activations reveal interpretable bands. *Connection Science* 7: 167–186.

Bever, T. G., J. A. Fodor, and M. Garrett (1968). A formal limitation of associationism. In *Verbal Behavior And General Behavior Theory*, ed. T. R. Dixon and D. L. Horton, 582–585. Englewood Cliffs, N.J.: Prentice-Hall.

Bickle, J. (1996). New wave psychophysical reductionism and the methodological caveats. *Philosophy and Phenomenological Research* 56: 57–78.

Broadbent, D. (1985). A question of levels: Comment on McClelland and Rumelhart. *Journal of Experimental Psychology: General* 114: 189–192.

Chomsky, N. (1959). A review of B. F. Skinner's *Verbal Behavior*. *Language* 35: 26–58.

Churchland, P. M. (1985). Reduction, qualia, and the direct introspection of brain states. *Journal of Philosophy* 82: 8–28.

Churchland, P. M. (1988). *Matter and Consciousness*, revised edition. Cambridge, Mass.: MIT press.

Churchland, P. S., and T. J. Sejnowski (1992). *The Computational Brain*. Cambridge, Mass.: MIT Press.

Clark, A. (1989). *Microcognition*. Cambridge, Mass.: MIT Press.

Clark, A. (1993). *Associative Engines*. Cambridge, Mass.: MIT Press.

Clark, A. (1997). *Being There: Putting Brain, Body, and World Together Again*. Cambridge, Mass.: MIT Press.

Cotter, N. E. (1990). The Stone-Weierstrass theorem and its application to neural networks. *IEEE Transactions On Neural Networks* 1: 290–295.

Crick, F., and C. Asanuma (1986). Certain aspects of the anatomy and physiology of the cerebral cortex. In *Parallel Distributed Processing*, vol. 2, ed. J. McClelland and D. E. Rumelhart, 333–371. Cambridge, Mass.: MIT Press.

Cummins, R. (1983). *The Nature of Psychological Explanation*. Cambridge, Mass.: MIT Press.

Cybenko, G. (1989). Approximation by superpositions of a sigmoidal function. *Mathematics of Control, Signals, and Systems* 2: 303–314.

Dawson, M. R. W. (1998). *Understanding Cognitive Science*. Oxford: Blackwell.

Dawson, M. R. W. (2004). *Minds and Machines : Connectionism and Psychological Modeling*. Malden, Mass.: Blackwell.

Dawson, M. R. W. (2005). *Connectionism*. Malden, Mass.: Blackwell.

Dawson, M. R. W., D. A. Medler, and I. S. N. Berkeley (1997). PDP networks can provide models that are not mere implementations of classical theories. *Philosophical Psychology* 10: 25–40.

Dawson, M. R. W., D. A. Medler, D. B. McCaughan, L. Willson, and M. Carbonaro (2000). Using extra output learning to insert a symbolic theory into a connectionist network. *Minds and Machines* 10: 171–201.

Dawson, M. R. W., and D. P. Schopflocher (1992). Modifying the generalized delta rule to train networks of nonmonotonic processors for pattern classification. *Connection Science* 4: 19–31.

Dawson, M. R. W., and K. S. Shamanski (1994). Connectionism, confusion, and cognitive science. *Journal of Intelligent Systems* 4: 215–262.

Douglas, R. J., and K. A. C. Martin (1991). Opening the grey box. *Trends in Neuroscience* 14: 286–293.

Dreyfus, H. L., and S. E. Dreyfus (1988). Making a mind versus modeling the brain: Artificial intelligence back at the branchpoint. In *The Artificial Intelligence Debate*, ed. S. Graubard. Cambridge, Mass.: MIT Press.

Endicott, R. P. (1998). Collapse of the new wave. *Journal of Philosophy* 95: 53–72.

Feldman, J. A., and D. H. Ballard (1982). Connectionist models and their properties. *Cognitive Science* 6: 205–254.

Fodor, J. A., and Z. W. Pylyshyn (1988). Connectionism and cognitive architecture. *Cognition* 28: 3–71.

Funahashi, K. (1989). On the approximate realization of continuous mappings by neural networks. *Neural Networks* 2: 183–192.

Greeno, J. G., and J. L. Moore (1993). Situativity and symbols: Response to Vera and Simon. *Cognitive Science* 17: 49–59.

Griffith, N., and P. M. Todd (1999). *Musical Networks: Parallel Distributed Perception and Performance*. Cambridge, Mass.: MIT Press.

Hadley, R. F. (1994a). Systematicity in connectionist language learning. *Minds and Machines* 3: 183–200.

Hadley, R. F. (1994b). Systematicity revisited: Reply to Christiansen and Chater and Niclasson and van Gelder. *Mind and Language* 9: 431–444.

Hadley, R. F. (1997). Cognition, systematicity, and nomic necessity. *Mind and Language* 12: 137–153.

Hadley, R. F., and M. B. Hayward (1997). Strong semantic systematicity from Hebbian connectionist learning. *Minds and Machines* 7: 1–37.

Hartman, E., J. D. Keeler, and J. M. Kowalski (1989). Layered neural networks with Gaussian hidden units as universal approximation. *Neural Computation* 2: 210–215.

Haugeland, J. (1985). *Artificial Intelligence: The Very Idea*. Cambridge, Mass.: MIT Press.

Hinton, G. E., J. McClelland, and D. Rumelhart (1986). Distributed representations. In *Parallel Distributed Processing*, vol. 1, ed. D. Rumelhart and J. McClelland, 77–109. Cambridge, Mass.: MIT Press.

Hooker, C. A. (1979). Critical notice: R. M. Yoshida's *Reduction in the Physical Sciences*. *Dialogue* 18: 81–99.

Hooker, C. A. (1981). Towards a general theory of reduction. *Dialogue* 20: 38–59, 201–236, 496–529.

Hornik, M., M. Stinchcombe, and H. White (1989). Multilayer feedforward networks are universal approximators. *Neural Networks* 2: 359–366.

Lewandowsky, S. (1993). The rewards and hazards of computer simulations. *Psychological Science* 4: 236–243.

Lippmann, R. P. (1989). Pattern classification using neural networks. *IEEE Communications Magazine* (November): 47–64.

Marr, D. (1982). *Vision*. San Francisco: W. H. Freeman.

McClelland, J. L., D. E. Rumelhart, and G. E. Hinton (1986). The appeal of parallel distributed processing. In *Parallel Distributed Processing*, vol. 1, ed. D. Rumelhart and J. McClelland, 3–44. Cambridge, Mass.: MIT Press.

McCloskey, M. (1991). Networks and theories: The place of connectionism in cognitive science. *Psychological Science* 2: 387–395.

McCulloch, W. S., and W. Pitts (1943). A logical calculus of the ideas immanent in nervous activity. *Bulletin of Mathematical Biophysics* 5: 115–133.

Messiaen, O. (1956). *The Technique of My Musical Language*. Paris: A. Leduc.

Pomerleau, D. A. (1991). Efficient training of artificial neural networks for autonomous navigation. *Neural Computation* 3: 88–97.

Pylyshyn, Z. W. (1984). *Computation and Cognition*. Cambridge, Mass.: MIT Press.

Pylyshyn, Z. W. (1991). The role of cognitive architectures in theories of cognition. In *Architectures For Intelligence*, ed. K. VanLehn, 189–223. Hillsdale, N.J.: Lawrence Erlbaum.

Quinlan, J. R. (1986). Induction of decision trees. *Machine Learning* 1: 81–106.

Schlimmer, J. S. (1987). Concept acquisition through representational adjustment. Unpublished doctoral dissertation, University of California Irvine, Irvine, California.

Schneider, W. (1987). Connectionism: Is it a paradigm shift for psychology? *Behavior Research Methods, Instruments, and Computers* 19: 73–83.

Seidenberg, M. (1993). Connectionist models and cognitive theory. *Psychological Science* 4: 228–235.

Siegelmann, H. T. (1999). *Neural Networks and Analog Computation: Beyond the Turing Limit*. Boston, Mass.: Birkhauser.

Siegelmann, H. T., and E. D. Sontag (1991). Turing computability with neural nets. *Applied Mathematics Letters* 4: 77–80.

Siegelmann, H. T., and E. D. Sontag (1995). On the computational power of neural nets. *Journal of Computer and System Sciences* 50: 132–150.

Smolensky, P. (1988). On the proper treatment of connectionism. *Behavioral and Brain Sciences* 11: 1–74.

Stix, G. (1994). Bad apple picker: Can a neural network help find problem cops? *Scientific American* 271: 44–46.

Suddarth, S. C., and Y. L. Kergosien (1990). Rule-injection hints as a means of improving network performance and learning time. In *Neural Networks, Lecture Notes In Computer Science*, vol. 412, ed. L. B. Almeida and C. J. Wellekens, 120–129. Berlin: Springer-Verlag.

Todd, P. M., and D. G. Loy (1991). *Music and Connectionism*. Cambridge, Mass.: MIT Press.

Touretzky, D. S., and D. A. Pomerleau (1994). Reconstructing physical symbol systems. *Cognitive Science* 18: 345–353.

Vera, A. H., and H. A. Simon (1993). Situated action: A symbolic interpretation. *Cognitive Science* 17: 7–48.

Vera, A. H., and H. A. Simon (1994). Reply to Touretzky and Pomerlau: Reconstructing physical symbol systems. *Cognitive Science* 18: 355–360.

Von Eckardt, B. (1993). *What Is Cognitive Science?* Cambridge, Mass.: MIT Press.

Yaremchuk, V., and M. R. W. Dawson (2005). Chord classifications by artificial neural networks revisited: Internal representations of circles of major thirds and minor thirds. *Artificial Neural Networks: Biological Inspirations—ICANN 2005*, Pt. 1, Proceedings, 3696, 605–610.

8 Intermodular Explanation in Cognitive Science: An Example from Phonology

Charles Reiss

1 Linguist(ic)s and Cognitive Science

Linguistics is often said to be one of the most advanced or mature branches of cognitive science; however, it is actually not very common to find among working linguists an interest in fundamental issues concerning the place of their field in cognitive science. A letter from a former student who enrolled in a prestigious linguistics Ph.D. program is telling:

> Most of the people here are simply not interested in linguistics as a cognitive science. In fact, I think the idea is generally considered worthless or stupid, although, everyone maintains, in a weird way, that questions about how language relates to people are OK, but really these questions are too grand, too philosophical, perhaps for people like Chomsky to think about, but not serious practicing linguists.

Comments from colleagues concerning my own interest in cognitive science are consistent with this impression:

> "Why are you guys always talking about the mind? Me, I'm interested in sound patterns."

> "What you do isn't really phonology—it's more like philosophy of phonology."

The first of these comments came from an MIT-trained phonologist, the second from a phonologist now at one of the top generative linguistics programs in the world.

Given this intellectual context, I would like to ask the following question:

(1) Can serious consideration of foundational issues in cognitive science have any bearing on the work of practicing linguists?

My answer will be "yes," obviously enough. Relating foundational issues and empirical work is one of the explicit goals of Pylyshyn's *Computation and Cognition* (C&C), and I will discuss how the book has been a source of inspiration in this regard for my own work on phonology.

2   What Is UG about?

The following quotation, selected basically at random from discussions on the Internet, represents a standard view of the goal of generative linguistics, universal grammar (UG)—a theory of the human language faculty:

the only theory of a language is a grammar of that language, and unless you believe in the Joos view that "languages can differ in innumerable ways" you must believe that an individual grammar must be based on a theory of grammar which must account for all and only the grammars of all and only the possible languages in the world. (Vickie Fromkin, writing informally, at http://linguistlist.org/issues/2/2–94. html)

While this idea is a commonplace in the linguistics literature, it is surprising how little thought has gone into making explicit the notion "possible language" (but see Newmeyer 2005 for a recent important contribution). We will explore this notion with reference to the following warning from C&C:

(2)   Potentially unobservable regularities

[T]he appropriate type of explanation depends on more than just the nature of the observed regularities; it depends on the regularities that are possible in certain situations not observed (and which may never be observed, for one reason or another). (C&C, 206)

To understand this quotation, it is useful to contrast it with Chomsky's and Pylyshyn's admonishments in support of competence models:

(3)   Importance of competence theories

In my opinion, many psychologists have a curious definition of their discipline. A definition that is destructive, suicidal. A dead end. They want to confine themselves solely to the study of performance—behavior—yet, as I've said, it makes no sense to construct a discipline that studies the manner in which a system is acquired or utilized, but refuses to consider the nature of this system. (Chomsky 1977, 49)

[I]f we confine ourselves to the scientific and intellectual goals of understanding psychological phenomena [as opposed to predicting observed behavior] one could certainly make a good case for the claim that there is a need to direct our attention away from superficial "data fitting" models toward deeper structural theories. (Pylyshyn 1973, 48)

The two quotations under (3) stress the importance of competence theories, but the quotation under (2) makes the point that not every regularity is to be attributed to competence. The quotation is part of a discussion of the following thought experiment (C&C, 205ff.): Consider a black box that outputs signals of spikes and plateaus. When a two-spike pattern and a one-spike pattern are adjacent, it is typically the case that the former precedes the latter, as on the left side in figure 8.1. However, we occasionally see the order switched, but only when the two- and one-spike patterns are preceded by the double plateau-spike pattern, shown on the right side of figure 8.1. Pylyshyn asks what we can conclude from such observations about the computational capacities of the system in the box. His answer, perhaps surprisingly, is that we can conclude almost nothing. This, he explains, is because "we would not find the explanation of the box's behavior in its internal structure, nor would we find it in any properties intrinsic to the box or its contents." Pylyshyn's claim is based on what he designed his imaginary black box to be doing. The spikes and plateaus in figure 8.1 correspond to the dots and dashes of Morse code, and the observed regularities reflect the English spelling rule "*i* before *e*, except after *c*." In other words, the system is processing English text. If we fed it German text, with

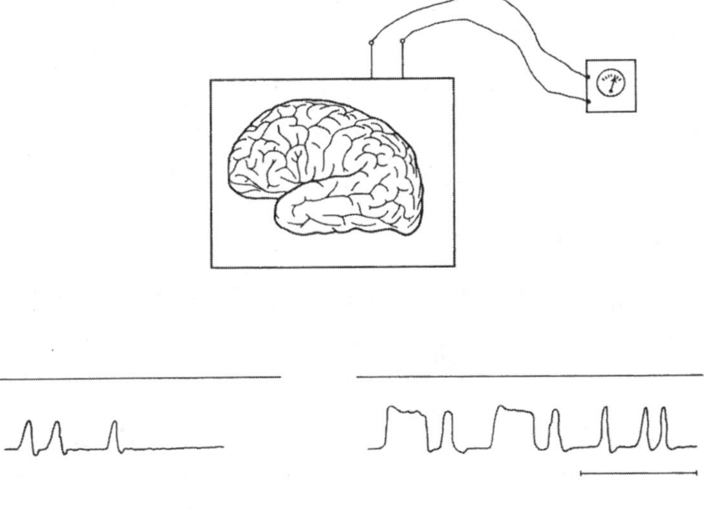

Figure 8.1
How do we figure out the computational capacity of the system inside the box? (Reproduced from Pylyshyn 1984 by permission of MIT Press.)

ie and *ei* clusters freely occurring in overlapping distribution, we would no longer observe the same output patterns.

Pylyshyn explains:

The example of the Morse-code box illustrates . . . that two fundamentally different types of explanation are available for explaining a system's behavior. The first type appeals to the intrinsic properties of the system. . . . The second type of explanation appeals, roughly, to extrinsic properties . . . of real or imagined worlds to which the system bears a certain relation (called representing, or, more generally, semantics). The example illustrates the point that the appropriate type of explanation depends on more than just the nature of the observed regularities; it depends on the regularities that are possible in certain situations not observed (and which may never be observed, for one reason or another). (C&C, 205ff.)

In linguistic terms, the explanation for the patterns we see in the data (either patterns we see or patterns in what we don't see, systematic gaps) may reflect not intrinsic properties of the language faculty, but instead properties of the kinds of information the language faculty has access to. In the remainder of this chapter, I explore the implications of the Morse-code example.

2.1   An Argument against Constraints

Pylyshyn does not draw this conclusion, but it seems to me that the Morse-code thought experiment leads naturally to the conclusion that constraints, statements about what cannot occur, are not the appropriate devices for characterizing cognitive systems. Observed regularities can tell us that a black box is capable of regular patterns of behavior, but there are infinitely many patterns that any given system cannot output. My standard example (see Hale and Reiss 2008 for discussion and references) is the linguistic "constraint" NoBanana, which states that no linguistic representation may contain a banana (an actual banana, not the word and not the representation of a banana). Obviously, there are infinitely many such constraints, and actual linguistic constraints proposed in the literature are necessarily interpreted in the context of an implicit universe of discourse that determines what is worth considering as a domain over which constraints apply. The Morse-code box suggests to me that cognitive modeling should involve characterizing correctly this universe of discourse in each domain.

I suspect that most readers will not find this argument compelling, on the grounds that constraints can typically be recast as, say, procedural rules. However, consider the following valid chain of reasoning implicit in much nativist linguistic literature.

(4)   Nativism and constraints

• Assumption: Grammars contain constraints (on syntactic movement, on tree structure, etc.).
• Constraints (statements about what cannot occur) cannot be learned on the basis of positive evidence, since the prohibited structure could show up in the next piece of data encountered.
• Constraints can only be learned on the basis of negative evidence (explicit information that a structure is ungrammatical).
• Children do not seem to receive negative evidence systematically, and they seem to ignore it when they do receive it (see Marcus 1993 for discussion).
• ∴ Constraints must be innate.

This chain of reasoning has led to proposals of innate grammatical knowledge of a level of specificity that strains belief—in the realm of phonology, for example, the amount of phonetic detail included in linguistic constraints such as those needed to model, say, the voicing patterns of obstruents in Japanese compounds, precludes a model of UG that is abstract enough to apply to both spoken and signed languages.

The chain of reasoning under (4) is valid, but the conclusion is not necessarily true. We can reject the conclusion if we reject the assumption that grammars contain constraints. Thus, the Morse-code box supports the attempt to apply in linguistics what I believe is standard scientific practice—posit just the minimum of theoretical apparatus necessary to account for observed phenomena. To make a simple analogy, physicists posit some minimum number of fundamental particles and some principles of combination to account for more complex structures. They do not additionally need to posit constraints against what cannot occur, as laws of nature. Any such constraint is understood as a derivative notion that follows from the characterization of what does occur.

One application of these ideas in my own work (Reiss 2003) is to replace a purported universal constraint commonly invoked in the literature (when convenient, often ignored when not) called the *obligatory contour principle* (OCP). Such a constraint is invoked, for example, to block the application of a general rule that deletes vowels in a certain context just in case the deletion would bring together identical consonants. Invoking such a constraint appears to illustrate a fairly general phenomenon in the linguistics literature of assuming "that the child's problem is that of learning how to constrain an over-hasty generalization" (Pullum and Scholz 2002). Instead, I propose formulating the deletion rule correctly—it applies only between nonidentical consonants. Such computation of nonidentity requires the power of existential quantification—it is necessary to find one arbitrary feature for which the two consonants disagree in value, say, [+voiced] versus [–voiced]. It then turns out that a fairly complex system of indexation is needed for phonological representations that, in turn, can lead us to reject the standard, less powerful feature geometry model. So, rejection of constraints like the OCP leads to careful consideration of what sort of computational power must be attributed to the phonology; this in turn leads to a revision in the theory of phonological representations. The Morse-code box example thus inspires discovery of what the phonological faculty's intrinsic properties are, which I take to be the goal of phonological theory.

2.2   Theory and Data in Linguistics

We can clarify the difficulty of determining what the computational resources of a cognitive system are by asking what universal grammar (UG) should be a theory of, and considering the relationship between this theory and available data. Should UG account for all and only the attested languages? Obviously, we do not want our theory of possible languages to just reflect the decisions of graduate admissions committees and the fate of empires, two factors that have played a major role in determining which languages have been studied by linguists. So, the scope of UG must be greater than just the set of attested languages.

Proposing that UG should be general enough to account for any statable language is an error in the other direction. For example, we can describe a language that lengthens vowels in prime-numbered syllables, but there is no reason to think that the human language faculty can represent categories like "prime number."[1] To equate the study of UG with formal language theory would reduce linguistic theory to a branch of mathematics, with no relation to the human language faculty as a natural object.

We know that there are extinct languages, and languages that have not yet come into being, and these are attestable in principle,[2] so a tempting intermediate hypothesis between the set of attested languages and the set of all describable languages is that UG should be understood as a theory of all attestable languages.

However, given Pylyshyn's point that "the appropriate type of explanation depends on more than just the nature of the observed regularities; it depends on the regularities that are possible in certain situations not observed (and which may never be observed, for one reason or another)," even this intermediate hypothesis turns out to be too narrow.

Why should we have to account for classes of languages that can never be observed? Consider that grammars are embedded in humans and that they are partially learned. It follows from this that the human transducers (input and output systems), the language acquisition inference systems, and performance systems place a limit on the set of attestable languages beyond the (upper) limits determined by S_0, the initial state of the language faculty.

In figure 8.2, we can see, as discussed above, that the set of attested languages, corresponding to the small dark circle, is a subset of the attestable languages, shown as the hatchmarked region. Obviously, this latter set is a subset of the statable languages, the box that defines the universal set in our diagram. However, there are two remaining regions defined in the diagram that need to be explained. Note that the set of attestable languages corresponds to the intersection of two sets, the set of humanly computable languages, the large gray circle, and the white circle, labeled as "processable/transducible/acquirable."

To be attestable, a language must be acquirable on the basis of evidence presented to a learner; an attestable language must also not overload the processing capacity of a human; and finally, an attestable language must be able to be presented to the language faculty via the perceptual and articulatory transduction systems. If a language failed to meet any of these criteria, it would not be attestable, even if it made use only of the representational and computational primitives of the human language faculty—

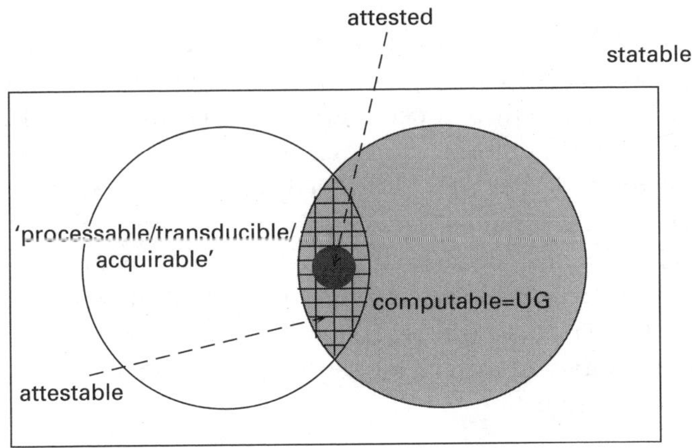

Figure 8.2
What is UG about?

that is, even if it were a member of the set represented by the large light gray circle.[3]

An example of an unprocessable language, one falling outside of the white circle, would be one in which all words contained at least 98 syllables—word recognition memory buffers would presumably not be able to handle such input. An example of an untransducible language would be one presented in a signal outside of the range of human hearing. We would not want to explain the fact that such a language is unattested or unattestable by appealing to properties of the language faculty *qua* computational system.

Languages that fail to fall inside the white circle may or may not fall inside the large gray circle. Those that do fall within the gray circle would fall in the part that is not hatchmarked.

2.3   Computable but Unacquirable

Our third argument utilizes a particular phonological theory, but should be accessible even to those unfamiliar with the details. The theory of stress computation developed by Halle and Idsardi (HI) is mathematically explicit, elegant and has a wide empirical coverage. We know that all theories in all domains are ultimately incomplete or otherwise flawed, but let's suppose that the HI model is the best theory of stress we have. In this model, syllables are projected onto a metrical grid, generating in the first instance grid lines of asterisks, each of which corresponds to a syllable. For example, in the simplest case a four-syllable word will project a grid line of this form: * * * *.

A further step in computing stress is the insertion of boundary markers, "(" and ")", which group asterisks into feet. One type of boundary insertion rule is the *edge-marking rule*, which is determined on a language-specific basis within a range defined by three parameters. The rule may insert a left or right parenthesis to the left or right of the left- or rightmost asterisk. We can thus characterize the essential elements of a given system by specifying a triplet of values for an "Edge:" parameter, each value ranging over L, R. "Edge:RLR" thus is to be read "insert a right parenthesis (the first R) to the left (the L) of the rightmost (the second R) asterisk." There are thus eight possible combinations of parameter settings, with eight distinct effects on a string of asterisks:

(5)   HI Edge-marking rules

1. Edge:RRR   * * * *)   Insert R paren to R of R-most *
2. Edge:RLR   * * *) *   Insert R paren to L of R-most *

3. Edge:RRL   *) * * *   Insert R paren to R of L-most *
4. ?Edge:RLL   ) * * * *   etc.
5. Edge:LLL   (* * * *
6. Edge:LRL   * (* * *
7. Edge:LLR   * * * (*
8. ?Edge:LRR   * * * * (

Note that edge-marking rules (4) and (8) are marked with a question mark. This denotes the fact that no conceivable data could indicate to the linguist that a language has such a version of the edge-marking rule for word stress. For the same reason, no child equipped with an HI-type stress computation module in its phonology would ever find evidence to set the edge-marking rule as either (4) or (8). Inserting parentheses in those ways has no effect on the grouping of asterisks and thus can play no role in stress computation.

Should the language faculty contain explicit statements that (4) and (8) are not possible edge-marking rules? Clearly not, since such statements serve no purpose. A learner will never posit (4) or (8), whether or not the innate knowledge of stress computation contains, say, constraints like *RRL and *LLR. It follows from our position that if neuroscience advanced to the point where we could program specific grammars into human brains, then (4) and (8) would be computable by human language faculties. The absence of such rules from the set of attested and attestable (in the absence of neural programming) languages is a fact about how specific languages are learned. It is not a fact about the cognitive architecture of the language faculty.

Chomsky (1957) points out that there is no straightforward way to restrict a generative grammar to sentences of a predefined length. In other words, the assumption of a unbounded set of sentences including ones of arbitrary length actually makes it possible to construct a simpler model. The same considerations hold for the stress example just discussed. In the case of sentence length, we can appeal to performance factors and the nature of corpora to explain the absence of sentences over some defined length in a given corpus. Similarly, we can adduce learnability considerations to explain the absence of certain combinations of independent parameters of the HI stress model. These absences need not arise from restrictions encoded in mental grammars (instantiated in individuals), or even in our models, since the empirical data that would lead us to posit such a restriction never arises. Such restrictions thus can never be empirically relevant to either the learner or the scientist. The absence of certain combinations is accidental from a grammatical perspective.

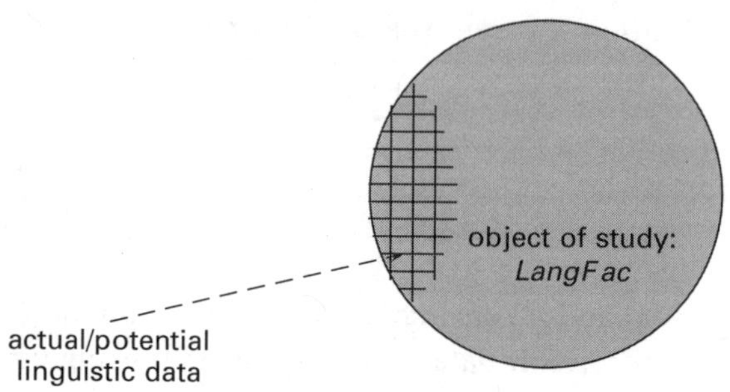

Figure 8.3
Evidence and object of study.

So, according to our discussion, the purview of linguistic theory should be the set of humanly computable languages, the large gray set, even though some such languages are unattestable—they "may never be observed, for one reason or another."

This situation, which I think may be relatively normal in science, can be best appreciated by extracting part of figure 8.2, as in figure 8.3. Our source of actual and potential data is restricted to the set of attestable languages, but we have to induce from this empirical data the nature of the larger set of potential languages. It is important to keep in mind the fact that inducing this larger set will probably be a matter of positing fewer properties for the language faculty—by being less specific, more general, we describe a larger set. To make this idea concrete, contrast a phonological UG that specifies just that there are rules that insert, delete, and change feature values with one that specifies all this, as well as stipulating that, in syllable codas, feature-changing rules affecting [voiced] always involve turning [+voiced] to [–voiced] and never [–voiced] to [+voiced]. The position I am pushing is that the first version of UG, the one that makes no mention of specific features in specific rules, is what we should aim for. This is not to say that UG does not specify a set of representational primitives—features. The claim is just that the specific attested combinations of representational and operational primitives found in particular languages, like "Change [+voiced] to [–voiced] in codas," are not encoded in UG. The building blocks for the phonology of attestable, and even some unattestable, languages must obviously be present in UG, but not the rules of particular languages.

The approach taken here is expressed in the following quotation concerning the ultimate goals of linguistic theory:

to abstract from the welter of descriptive complexity certain general principles governing computation that would allow the rules of a particular language to be given in very simple forms. (Chomsky 2000, 122)

Thus, the frequently attested pattern of coda devoicing (whatever the correct featural description of this may be), and the perhaps complete absence of coda voicing, as phonological processes in the languages of the world is thus, in my view, not to be accounted for by UG. This view is inspired by a long history of empirical work concerning the phonetics of sound change by John Ohala (see Hale and Reiss 2000a,b for discussion).

2.4   Temptation and Seduction

The substance-free approach advocated here directly contradicts the spirit of Optimality Theory, currently the dominant theory in phonology. This theory utilizes language-specific ranking of a set of universal, violable constraints to model individual grammars. Since there are $n!$ rankings of the n universal constraints, the range of possible variation in this theory of universal grammar is called the *factorial typology*. Prince and Smolensky (1993) see it as a central goal of grammatical theory to capture the facts of linguistic typology—the set of rankings determine (aside from accidental gaps) the set of observed linguistic phenomena. The constraints are so specific as to refer to a level of phonetic detail that includes even, say, combinations of feature values on vowels. The inclusion of such detail of phonetic substance, in opposition to more formalist theories, is seen as a major component of the enterprise:

(6)   Prince and Smolensky (1993, 198) on factorial typology

We urge a reassessment of [an] essentially formalist position. If phonology is separated from the principles of well-formedness (the "laws") that drive it, the resulting loss of constraint and theoretical depth will mark a major defeat for the enterprise.

In the linguistics literature, and especially the Optimality Theory literature, the putative complexity of particular speech sounds (or configurations of sounds), from either an acoustic or articulatory perspective, has led to the positing of the substantive constraints that Prince and Smolensky refer to as "laws" of well-formedness. These violable laws, more widely known as *markedness constraints* in the literature, define and delimit the range of possible languages by virtue of their interaction as determined by their language-specific ranking.

The universal constraint set includes, according to McCarthy and Prince (1995), a constraint whose effects are visible in Japanese, but not English, because of the ranking differences in the two languages, against the realization of [g] between vowels. McCarthy and Prince refer to this constraint as a "phonologization of Boyle's law" because the production of a voiced stop so far back in the mouth is supposedly a challenge from the perspective of aerodynamics, a challenge that only some grammars overcome by the appropriate constraint ranking.

However, arguments against building substance into the grammar have been around for a while. I recently discovered (thanks to Ash Asudeh) a paper by Ronald Kaplan (1987/1995, 346–347) that discusses these issues very cogently:

A formal theory may have a relatively smooth outline . . . [t]hen you start taking chunks out of it . . . because you claim that no human language or grammar has such and such a property. . . . It's a mistake to carry premature and unjustified substantive hypotheses into our computational and mathematical work, especially if it leads to mathematically complex, even if more restrictive, theories. . . . [W]e should be wary of the seduction of substance.

Complementing the ideas in C&C, Pylyshyn (2003, 8) unwittingly diagnoses the fundamental mistake that Prince and Smolensky and McCarthy and Prince are making. Pylyshyn is introducing a long treatise on vision, but the applicability to phonology is apparent: The phonologists haven't been "wary of the seduction of substance" (as Kaplan puts it), but rather have fallen prey to "the temptation to make the mistake of attributing to a mental representation the properties of what it represents." If one accepts this analysis, the potential benefit of improved communication among cognitive scientists becomes apparent—seduction and temptation are perhaps not so different across the field's subdomains.

The current trend of building so much substance into the theory of grammar is particularly striking given the fact that even as Optimality Theory was emerging, some sober voices, apparently more continent, and sensitive to the need for modular explanation, could be heard:

Presumably, a true understanding of why certain features tend to assimilate or dissimilate under certain adjacency conditions will rely on considerations of historical linguistics, acoustics, and articulation. The explanation for asymmetries . . . therefore probably lies outside of the domain of phonology proper. (Odden 1994)

A lesson from the Morse-code example is that part of the task of understanding phonology, or grammar more generally, requires that we understand the nature of the systems that pass information to or receive information from the grammar, either via direct interface or through the mediation of other systems.

The relevance of the competence–performance distinction, discussed above, is sometimes obscured by the fact that discussion of performance tends to focus on so-called performance errors, which include mispronunciations, failures to mark obligatory agreement, and the like. In fact, every utterance "in a language" reflects competence and performance. It is sometimes said that insistence on competence theory is not valid unless we provide a theory of performance, but I adopt the position that there should not be a single theory of performance since performance includes all the components of the "speech chain" that are not grammar (that is, not phonology, morphology, syntax, or semantics). By better understanding what is not phonology, by pursuing modular explanation, we can better understand what phonology is—we won't mistakenly attribute a property to the phonology that rightly belongs elsewhere.[4] It is to this problem that we now turn.

3   Inherited Limitations

In addition to the insight it provides into empirical phenomena, C&C also inspires the kind of speculation that can lead to future experimental work. In this section, I sketch some ideas for intermodular explanation that are inspired by the transduction–computation distinction of C&C.

Pylyshyn calls transduction the "bridge from the physical to the symbolic" and provides the following discussion:

This, then is the importance of a transducer. By mapping certain classes of physical states of the environment into computationally relevant states of a device [e.g., a human], the transducer performs a rather special conversion: converting computationally arbitrary physical events into computational events. A description of a transducer function shows how certain nonsymbolic physical events are mapped into certain symbolic systems. (C&C, 152)

Bregman (1990) is relying on the notion of transduction in the following quotation:

In using the word "representations," we are implying the existence of a two-part system: one part forms the representations and another uses them to do such things as calculate. . . . (Bregman 1990, 3)

For example, our visual systems have transducers with color and edge detectors that may detect properties of two noncontiguous parts of the retinal image, but there must also be a system of inference or calculation or computation that reaches the conclusion that the two regions correspond to parts of a single object that is partially occluded by objects corresponding to the intervening region.

Given such an understanding of a system such as vision or audition, it is clear that the limitations of the transducers are inherited by, or reflected in, the set of symbols over which the computational system can compute. For example, once a physical distinction is lost in transduction, it cannot be made available to the computational system; the set of equivalence classes that are output by any module sets an upper limit on the set of representational primitives available to the modules that receive that output as input.

Now, since the auditory system and speech processing systems must serve as input transducers to the phonological systems of spoken languages, it seems clear that the nature of the former must determine to some extent what is attestable as a phonological system. Audition and speech processing determine the semantics, in the sense of the quotation about the Morse-code example, of phonological representations.

In other work (see Hale and Reiss 2008) I have speculated that the streaming of speech (streams are the "objects" of audition in Bregman's auditory scene analysis model) by the audition and speech perception modules creates representations whose components within the stream can only be in one of two possible relationships: For two components of a stream, x and y, it is either the case that one precedes the other, or that one contains the other.[5] While the physical correlates of x and y may overlap in very complex ways, the proposal is that the transduction process filters this complexity into streams whose components are organized by these two relations, and thus, there are no interlocking elements in a stream.

(7)   Relationships of components x and y within a stream:

a. $[_x \quad]_x \qquad [_y \quad]_y$ Precedence: x precedes y—*Possible*
b. $[_x \quad [_y \quad]_y \quad [_z \quad]_z \quad]_x$ Containment: x contains y—*Possible*
c. $[_x \quad [_y \qquad]_x \quad]_y$ interlocked—*Not possible*

This speculative proposal concerning the nature of streams was inspired by work on a set of auditory illusions by Nakajima and colleagues (see Nakajima 1996, 2004; Nakajima et al. 2000). If the proposal is valid for audition generally, then it is reasonable to assume that the "immediate constituent" structure of auditory streams is inherited by the speech perception module, thus providing a lower-level explanation for the discrete nature of phonological representation, in spite of the continuous and overlapping nature of acoustic cues and articulatory gestures.

4 Conclusions

I have tried to illustrate the various ways in which a working linguist can draw inspiration from foundational work in cognitive science such as C&C. Both the discussion of the acquisition of stress rules and the discussion of the auditory transducers that feed phonology illustrate the principle suggested by the Morse-code box—the principle that an appropriate model of a cognitive module may overgenerate beyond what is observable even in principle, since such systems are embedded in complex structures possessing input and output and learning systems whose individual properties obscure those of the module under analysis. If these conclusions are valid, they vindicate the book's contention that doing good cognitive science requires doing some philosophy of cognitive science:

> I try to show that the kinds of theories cognitive scientists entertain are intimately related to the set of tacit assumptions they make about the very foundations of the field of cognitive science. In cognitive science the gap between metatheory and practice is extremely narrow. (C&C, xix)

In addition to any practical benefits it may lead to, grappling with foundational issues raised in C&C turns out to be a highly gratifying way of uncovering some intellectual coherence to the field of cognitive science.

Notes

1. Actually, the notion of prime number appears to have no relevance in any empirical field. This point leads to an issue that has arisen in numerous discussions of the proposal that phonology is pure computation and thus substance-free, as discussed by Hale and Reiss (2000a,b). It has been objected that our claim is uninteresting since it appears that we are proposing that the phonology is basically a universal Turing machine. This is not a valid conclusion: Our position is that phonology is all, that is, only, computation, not that all computations can be used by the phonological faculty of the mind.

2. Of course, in the context of mentalistic, I-linguistics, we have to recognize that only an infinitesimal number of attestable languages have been described in any detail.

3. The careful reader will notice that this diagram has to be interpreted as fairly informal, since the languages represented are sometimes conceptualized as grammars, sometimes as sets of sentences, or even utterances. I think the expository usefulness of the diagram outweighs this inconsistency.

4. As I finish writing this essay, I realize that Pylyshyn 1983 did exactly what I am trying to do in another linguistic domain—he showed that a theory of pronoun interpretation needed to be modular, with syntax providing only part of the explanation for speakers' judgments of acceptability and reference.

5. Coterminous, completely overlapping representations can be represented as ones that contain each other.

References

Bregman, A. (1990). *Auditory Scene Analysis*. Cambridge, Mass.: MIT Press.

Chomsky, N. (1957). *Syntactic Structures*. The Hague: Mouton.

Chomsky, N. (1971). *Problems of Knowledge and Freedom*. New York: Random House.

Chomsky, N. (1977). *Language and Responsibility*. New York. Pantheon.

Hale, M., and C. Reiss (2008). *The Phonological Enterprise*. Oxford University Press.

Hale, M., and C. Reiss (2000a). Substance abuse and dysfunctionalism: Current trends in phonology. *Linguistic Inquiry* 31: 157–169.

Hale, M., and C. Reiss (2000b). Phonology as cognition. In *Phonological Knowledge*, ed. N. Burton-Roberts, Philip Carr, and Gerry Docherty, 161–184. Oxford: Oxford University Press.

Halle, M., and W. Idsardi. (1995). Stress and metrical structure. In *Handbook of Phonological Theory*, ed. J. Goldsmith. Oxford: Blackwell.

Kaplan, Ronald (1987/1995). Three seductions of computational psycholinguistics. In *Formal Issues in Lexical-Functional Grammar*, ed. Mary Dalrymple, Ronald Kaplan, John Maxwell III, and Annie Zaenen. Palo Alto: CSLI Publications.

McCarthy, John J., and Alan S. Prince (1995). Faithfulness and reduplicative identity. In *University of Massachusetts Occasional Papers in Linguistics: UMOP 18*, edited by J. Beckman, S. Urbanczyk, and L. Walsh, 249–384. Amherst, Mass.: GLSA.

Marcus, Gary F. (1993). Negative evidence in language acquisition. *Cognition* 46: 53–85.

Nakajima, Y. (1996). A simple grammar for auditory organization: Streams, events, and subevents. *Approaches to Auditory Organization: XXVI International Congress of Psychology*. Montreal, Canada.

Nakajima, Y. (2004). *Demonstrations of Auditory Illusions and Tricks*, 2nd ed. http:// www.kyushu-id.ac.jp/ynhome/ENG/Demo/illusions2nd. html.

Nakajima, Y., T. Sasaki, K. Kanafuka, A. Miyamoto, G. Remijn, and G. ten Hoopen (2000). Illusory recouplings of onsets and terminations of glide tone components. *Perception and Psychophysics* 62: 1413–1425.

Newmeyer, F. (2005). *Possible and Probable Languages: A Generative Perspective on Linguistic Typology*. Oxford: Oxford University Press.

Odden, D. (1994). Adjacent parameters in phonology. *Language* 70 (2): 289–330.

Ohala, John J. (1990). The phonetics and phonology of aspects of assimilation. In *Papers in Laboratory Phonology I: Between the Grammar and Physics of Speech*, ed. J. Kingston and M. Beckman, 258–275. Cambridge: Cambridge University Press.

Prince, A., and P. Smolensky (1993). Optimality Theory: Constraint interaction in generative grammar. Technical Report RUCCS, Rutgers University, New Brunswick, N.J.

Pullum, G., and B. Scholz (2002). Empirical assessment of stimulus poverty arguments. *Linguistic Review* 19 (special issue, nos. 1–2: A Review of "The Poverty of Stimulus Argument," edited by N. Ritter): 9–50.

Pylyshyn, Z. W. (1973). The role of competence theories in cognitive psychology. *Journal of Psycholinguistic Research* 2: 21–50.

Pylyshyn, Z. W. (1983). Syntax as an autonomous component of language. In *Psychobiology of Language*, ed. M. Studdert-Kennedy. Cambridge, Mass.: MIT Press.

Pylyshyn, Z. W. (1984). *Computation and Cognition: Toward a Foundation for Cognitive Science*. Cambridge, Mass.: MIT Press.

Pylyshyn, Z. W. (2003). *Seeing and Visualizing: It's Not What You Think*. Cambridge, Mass.: MIT Press.

Reiss, C. (2003). Quantification in structural descriptions: Attested and unattested patterns. *Linguistic Review* 20: 305–338.

9 The Evolution of Cognition: The Case of Number

Claudia Uller

Lest the prospect of being a sibling of the computer appear as disturbing as the prospect of being the nephew or niece of the great ape once was, we should keep in mind that these are merely ways of classifying individuals for the purpose of discovering some of their operating principles.

—Z. W. Pylyshyn, *Computation and Cognition* (1984)

It is, therefore, highly probable that with mankind the intellectual faculties have been mainly and gradually perfected through natural selection; and this conclusion is sufficient for our purpose. Undoubtedly, it would be interesting to trace the development of each separate faculty from the state in which it exists in the lower animals to that in which it exists in man.

—C. Darwin, *The Descent of Man* (1871)

In *Computation and Cognition* (1984), Zenon Pylyshyn proposes an agenda for cognitive science, a new framework to study thinking things—humans, animals, machines—*cognizers* or *informavores*, that acquire, represent, and use information. According to his view, behavioral responses happen according to how information is mentally represented. This is the representational theory of mind, which takes primarily from philosophy and computer science, but also psychology, linguistics, and I would like to think, animal research at large, including but not restricted to biology, traditional ethology, cognitive ethology, behavioral ecology, animal behavior, and anthropology.

In *Computation and Cognition*, Pylyshyn contends that we should not fear the computer–mind analogy; it should rather be understood as a means to discover how different minds operate and to characterize their operating principles. In the following pages, I subscribe to the same idea without using the computer analogy (although making reference to representational states and the conceptual structure of minds) but by resorting to evolutionary ideas. I propose that the human mind has its roots in evolutionary history, and for this purpose, I draw my evidence from closely matched experimental protocols with different species that inform rather objectively about similarities and differences in cognition.

In adopting the analogy proposed by Pylyshyn, I take no position in the debate of whether cognitive traits correspond in structure due to common descent (*homogeny*) or whether the similarities in structure have been produced by the operation of random independent processes (*homoplasy*) (Lankester 1909). Because the field of comparative cognition is still in its infancy (the field I refer to makes use of methodologies adapted from those used in human-infant research, no pun intended) and because such homogeny-homoplasy considerations can seem, at times, rather fruitless and sterile, I will assume that empirical evidence brought to light with the use of closely matched tasks employed across species reveal quite a lot about the nature of human cognition and will eventually tease apart which cognitive abilities have or have not a common descent.

The continuity hypothesis, as I have outlined it above, may seem to violate Morgan's *Canon* (1894)—that one should not interpret a behavior as the expression of a higher psychological faculty if it can be interpreted as the outcome of some simpler capacity—because the evolution of cognition entails intrinsic linkages from order to order that disregard, in part, specific environmental attributes. I recall Pylyshyn's beautiful analogy, recaptured in Gallistel's (1990) seminal book, *The Organization of Learning*.

Much the same way as proposed by Darwin (1871/1981) in *The Descent of Man*, who suggested that it would be interesting to study each *intellectual faculty* separately so as to determine the development of each of them, Gallistel proposed isolating the particular *faculties* into *domains of specialization* for a computational-representational approach analysis. In this framework, animals represent information about the environment through simple computations performed by the nervous system. The representations are isomorphic to what is being represented. This relationship between the brain process that represents the environment and the environment itself allows the animal to adapt its behavior to the environment. Hence, Morgan's *Canon* considerations can be satisfied in the sense that there is no need to resort to higher cognitive functions—representations are computations that even a machine can perform.

According to Gallistel, animals as distant from humans as ants and bees, as well as other classes such as avians and, within mammals, nonhuman primates, are specialized *informavores* (in Pylyshyn's terminology) that represent time (temporal intervals, time of occurrence), space (navigation, dead reckoning, cognitive maps, geometric modules), number (rate, vector spaces), and so forth. It is, therefore, within this framework that we develop our idea that number embodies a domain of study across species, because it enables researchers not only to establish the computational resources of specializations particular to each class or species, but also to consider similarities or differences in cognition, given the nature of the computations.

But do we really think that animals have numerical concepts the same way that humans do? There are distinctions to be drawn when discussing the concept of number, and the question above may become irrelevant once we make some commitments. One aspect of the concept of number refers to its very nature and origins: The building blocks of cognition—one argues—have to include a precursory system for number because it is evolutionarily advantageous. A second aspect of the concept of number refers to the cultural construction that humans engage as part of their cognitive architecture. We are unique in developing algebra, trigonometry, sophisticated systems of logic, airplanes, and skyscrapers. No other species has been shown to produce mathematical artifacts even remotely close to the way humans can produce mathematics.

Thus, the concept of number stands at the core of unique human accomplishments in science, architecture, and engineering. The evolutionary origins of its existence in the human lineage, however, remain uncertain. It is possible that precursors of social interaction among humanoids would have required engaging in quantity discrimination in activities such as trade at least as early as 12000 BCE. Take the Natoufian, for example, a sedentary Epipalaeolithic culture living in the high mountains of Lebanon across the Sinai and the Syro-Arabian desert. Site excavations reveal evidence that the Natoufian produced shellfish and malachite beads that presumably could have been used as currency for the exchange of goods (Bar-Yosef and Valla 1991).

Might number have evolved prior to the evolution of modern primates? If so, then one must entertain the hypothesis that there must have been a purpose for the selection of number as a capacity that merited the attention of an evolutionary process. We would like to argue that number may have ranked high through the evolutionary process because animals have to eat, mate, and avoid death in order to be an evolutionary success.

Researchers suggest that *foraging* might be one function that could have triggered number as a selected domain for our cousin primates millions of years ago. Animals such as salamanders, pigeons, and rats in the laboratory and a variety of animals from ducks to monkeys in the wild seem to be hard-wired to detect at least more from less, and discriminate between two quantities (see Gallistel 1990 and Uller et al. 2003 for reviews of this literature). There is no direct evidence that number is evolutionarily relevant for *mate selection* and *mating*, but there is some indirect experimental evidence showing monkey preferences for relevant sexual information. Researchers have shown that male monkeys willingly sacrifice trading juice for pictures of female monkey perinea, while they will require fluid payment to view faces of low-ranking male monkeys (Deaner, Khera, and Platt 2005). As far as number being valuable for *survival*, researchers have suggested that number may be a relevant evolutionary domain in the establishment of coalitions and the detection of competition among groups of female lions in the Serengeti desert (McComb, Packer, and Pusey 1994). Thus, it is highly probable that number has been selected by evolutionary pressures to compose the core of intellectual faculties, as Darwin (1871/1981) speculated.

The representation of quantities and number has been at the core of psychological studies for several reasons. Number encompasses an attribute of sets of things, rather than an attribute of entities or a property of objecthood. In this regard, number has been taken as *abstract* because it cannot be detected as a property of an object in the way that shape or size can be detected. As a property *descriptor,* number possesses an intrinsic feature itself: One can count anything. We count physical objects, events, (auditory and haptic, for example) and "nonphysical" objects. I can count how many students are taking my Comparative Cognition course, how many times I went downstairs to the main office of the Psychology building, how many notes are at the entrance of Beethoven's *Fifth*, how many ideas I had today, and how many unicorns I dreamed of after watching *Blade Runner, The Director's Cut.*

The sensory modality of the numerical encounter will determine the kind of representation to be set up. One can see Canadian geese flying in the sky in parallel, at once, and decide fast and accurately that there are three in the flock. Because of the intrinsic nature of the visual experience, if one desires, one can also individuate and attend to each object (bird) in separate. The same is not true for auditory events. Once a sound or, say, notes in harmony are heard or listened to, they go out of existence. One can only access them by either having the sound or notes repeated or accessing them from memory. Professional musicians experience this phenomenon every time they practice on their own a piece that is being played with an ensemble.

Another dimension of number involves determining the units as we parse the world. For example, we may attend to wholes or parts of objects or events. When I go to the ballet, I can count the number of ballerinas and wonder about the harmony of the group. But I can reduce the group to parts and count how many legs I see, or focus on a single ballerina as she moves around and count her legs. I can count the scenes while reading *Macbeth*, or, while at the *Globe*, in London, count how many times Macbeth is involved in crime in that particular production, or the number of laces on Lady Macbeth's corset.

What constitutes a *unit* for an animal? Roughly speaking, in order to exist in the world, animals need to individuate objects and tell them apart. This is the process of categorization that has been consistently described in the literature on the metaphysics of concepts such as *objecthood* and their origins in human babies (e.g., Xu and Carey 1996) and animals (e.g., Herrnstein 1990). Simply put, the mind uses criteria to determine what constitutes an object, where an object ends and begins, whether an object is the same one or a distinct individual. In order to accomplish that, there are principles that guide reasoning about objects (continuity, cohesion, solidity) and there are criteria for object individuation—feature/kind and space-time. Human adults, for example, use shape as a primary feature to individuate objects, but other features such as color and size can enter into consideration. Parallel to these features, criteria on space and time (object $[x_i]$ seen at location L_1 at time T_1 the same or not the same as object $[x_{ii}]$ seen at location L_1 at time T_2) will also help determine individuation and identity. Most important, the concept of what constitutes an *object* clearly determines the selection of objects in a set. Therefore, the concept of *number* is intrinsically related to the concept of object (see Trick and Pylyshyn 1993 for an extensive discussion of this relationship). These considerations on the concept of *objecthood* in the study of the origins of the human mind have led to the proposal for the investigation of comparable topics in other species.

The New Age of Comparative Cognition Study

In 1994, while I met with my graduate advisor, it occurred to me that we might apply methods common to developmental cognitive science— namely, the same kinds of experiments I was employing with babies—to investigations of the same domain of knowledge (in this case, number) in other species, say, apes or monkeys. In this new field of inquiry, the usefulness of the methods would therefore lie in the fact that they investigate nonverbal creatures, namely, nonhuman primates and human babies. The main reason for this methodological expedition was that one can compare cognitive abilities of species in very closely matched tasks, that is, tasks that do not require any training. One of the reasons why this is relevant is because traditionally, in comparative psychology, comparisons of different species have required the use of language, in the case of human children, and training, in the case of, say, chimpanzees, making the comparison difficult to interpret.

The use of nonlinguistic methods drawn from the study of cognitive development in human children provides a unique opportunity to compare cognitive abilities among species, especially because these methods assess cognitive abilities that are spontaneously present in species (for a discussion of this point, see Uller 1996, 2003, and Hauser and Carey 1999). This is particularly useful because tracking ontogenetic and phylogenetic roots to cognitive abilities will help determine what minds have been originally and naturally endowed with, and what minds can spontaneously generate (for arguments regarding the relationship between ontogeny and phylogeny, see Gould 1977). In this vein, let us review first the evidence for ontogenetic development of the concept of number and focus specifically on infant research.

Number in Babies

A consistent and overwhelming amount of research into the domain of number has been done with human babies within the past twenty years. The measure mostly used in these studies resorts to the fact that babies stare at things, in general, objects, events, people. This has been dubbed the *looking time paradigm,* which can sometimes be coupled with a habituation phase that is devised to measure thresholds of familiarity to a particular object or event. In looking time tasks, infants are presented with a certain entity/object(s) during a familiarization or habituation phase. In the test phase, they are then shown a contrast between a consistent event and an inconsistent event. An infant's attention to these displays is measured by a computer program that counts the amount of time spent looking at the displays. Generally, the results show a significant difference between time spent looking at to an inconsistent event and that spent looking at a consistent event, namely, looking times are longer for the inconsistent than the consistent event. These results have been taken as evidence for infant and nonhuman primate capacity for small number representation, in the sense that the operations performed on these sets

(1 + 1 = 2 versus 1, 1 + 1 = 2 versus 3, 1 + 1 = 2 versus a big 1, 2 − 1 = 1 versus 2, etc.) happen behind a screen, and creatures have to be able to keep track and store in memory the representations of the entities in the set (hence, number).

Another measure used to assess numerical understanding in young human babies is haptic reaching/searching for objects. For example, in a classic search task, babies are shown three toy ducks being placed into a box, one at a time, through an opening on the top of the box. After a delay (or not), infants are then allowed to reach into the opening to search for the ducks. The experimenter, however, surreptitiously removes one of the ducks and the babies can only find two. The question is—will the babies reach again in the box to search for the duck that is missing, showing that they have represented *threeness* and the finding of *twoness* does not correspond to the representation stored in memory? This and other alternative reaching methods have been shown to be productive in the sense that (1) objects are placed out of sight with no continuous visual experience and (2) objects are serially presented. These constraints are relevant in the sense that they require representations to be set up and to be kept alive in memory.

With the use of visual preferential paradigms, young preverbal infants are able to discriminate between 2 and 3 visual objects, but not 3 and 4, or 4 and 6 in presentations of dots or familiar objects (Antell and Keating 1983; Strauss and Curtis 1981; Starkey and Cooper 1980; Treiber and Wilcox 1984). Infants are also able to discriminate 2 from 3, but not 4 from 6 dots in moving displays (van Loosbroek and Smitsman 1990). Five-month-olds can discriminate between two "collections" of dots in a 2-versus-4 condition (Wynn, Bloom, and Chiang 2002), namely, infants habituated to a collection of 2 dots dishabituate to a collection of 4 dots, and vice versa. Still with the use of visual parallel discrimination, in the domain of "larger numerosity," 6-month-old infants can discriminate 8 from 16, but are unable to discriminate 8 from 12 (Xu and Spelke 2000). Six-month-old babies can discriminate "intermediate" numerosities of 4 versus 8, but not 2 versus 4 (Xu 2003), and were found to discriminate large sets as in 16 versus 32, but not 16 versus 24 (Xu, Spelke, and Godard 2005).

Five-month-olds are able to "add" and "subtract" numerosities in 1 + 1= 2, 1, or 3 conditions (Wynn 1992; Koechlin, Dehaene, and Mehler 1997; Simon, Hespos, and Rochat 1995; Uller et al. 1999), but 7-month-olds cannot add 1 + 1 on the basis of number when surface area and contour length, for example, are controlled for (Clearfield and Mix 1999; Feigenson,

Carey, and Hauser 2002). Recently, researchers have extended the addition and subtraction experiments to larger numerosities, showing that, in video formatted tasks, infants can add and subtract $5 + 5$ and $10 - 5$ (McCrink and Wynn 2004). As for addition of events, infants have been shown to add jumps of a puppet in $1 + 1 = 2$ or 1 or 3 conditions (Sharon and Wynn 1998). They can also visually discriminate number of events, namely, 6-month-olds can discriminate 4 versus 8 jumps of a puppet, but not 2 versus 4, while 9-month old infants will discriminate 2 versus 4 and 4 versus 6 (Wood and Spelke 2005).

With the use of searching/reaching methods, research shows that 10- and 12-month-old infants choose the larger numerosity in discrimination choice conditions of 1 v 2 and 2 v 3, but not 2 v 4, 3 v 4, and 3 v 6. Using a search in a box test, 12-month-old babies search for the exact number of objects in a box when the number <4, namely, when they see 1 object and 1 object going into the box and they retrieve 2, but not 3, and when they see $1 + 1 + 1$ and they retrieve 3, but not 4 (Feigenson and Carey 2003; Uller, Gaudin, and Fradella in preparation), suggesting that in these conditions, babies' memory for tracking objects one by one breaks down at around 4.

Recently, infants' ordinal choices were measured in a spontaneous forced-choice task in which 10- and 12-month-old infants were shown two buckets containing different numbers of cookies, 1 versus 2, 2 versus 3, 3 versus 4, and 3 versus 6 (Feigenson, Carey, and Hauser 2002). Here, the mother sat with the baby 100 cm away from the buckets. The experimenter showed the infant the cookies being placed inside each of the containers. The baby was then released to go for the bucket of choice. Each baby was tested in one condition only, and received only one trial. The overall result was that both age groups successfully chose the bucket containing the larger numerosity when 1 versus 2 and 2 versus 3 were contrasted, but not in the 3 versus 4. The researchers concluded that, in order to succeed in the task, infants had to recognize the ordinal relationships between the two numerosities (1 versus 2, 2 versus 3), and they had to track spontaneously the number of cookies because there was no training involved and thus no opportunity for learning. Infants therefore established the ordinal relationship between the two numerosities, choosing the container that yielded "more."

In the auditory domain, very young babies can discriminate between 2 and 3 syllables (Bijeljac-Babic, Bertoncini, and Mehler 1991). Recently, Lipton and Spelke (2003) tested 6-month-olds on an auditory task consisting of natural sounds such as bells, whistles, buzzes, drums, and horns.

Six-month-olds are able to discriminate 8 from 16 sounds, but not 8 versus 12. Nine-month olds can distinguish 8 from 12, when stimuli were controlled for element duration, sequence duration, interelement interval, and amount of acoustic energy.

In all these studies, only one modality at a time is being assessed, be it visual objects or auditory events. In the late '80s, Starkey, Spelke, and Gelman (1990) asked the question of whether the young human baby would have an *abstract* representation of number. The idea was to show whether 6-month-old babies would identify number across modalities, that is, when shown a picture of three familiar objects and when hearing a set of three beats, would babies match the two? This is an incredibly important question because it requires that one have a conceptual system that would enable the matching of a visual set of three objects with three auditory events. While the visual presentation would be static and parallel (all three items at once), aural events are not parallel in this case, and they are not static either. The drumbeats here were ephemeral (went out of existence) and serially presented.

Although replication of this study has remained an issue (cf. Moore et al. 1987), a recently published experiment with 6-month-olds using a looking time task shows support for the original study by Starkey et al. (1990). In a series of elegant and original results, Kobayashi, Hiraki, and Hasegawa (2005) showed that 6-month-olds can add across modalities. Babies see 1 object + 1 or 2 tone/s in the operation phase of the experiment. Then, in the test phase of the experiment, the babies are assessed on their *abstract* representation of the objects and the aural events by watching outcomes that correspond or violate number, irrespective of whether they are visually or auditorily presented. Their responses show that they are sensitive to number in an abstract way because the outcomes of the operation they had previously seen were shown in a way that differed from the original presentation in terms of modality. Therefore, they must have represented the concepts, say, *twoness* and *oneness*, regardless of modality to be able to (at a very minimum) discriminate the two.

We have been discussing empirical results with babies based on measures that, until 1994, were not used with any other population. Can these measures be used with nonhuman animals?

Number in Nonhuman Primates

Work in traditional comparative psychology has shown numerical cognition in animals at large with the use of training methods. For example, chimpanzees can select the larger and smaller numerosity in small sets (Beran 2001; Boysen and Berntson 1989, 1995; Rumbaugh, Savage-Rumbaugh, and Hegel 1987), can order numerosities up to 11 (Matsuzawa, Itakura, and Tomonaga 1991), can be taught to count and assign symbolic tags for numerosities from 1 to 9 (Boysen 1993; Boysen and Berntson 1989; Matsuzawa 1985). Sheba (Boysen 1993), for example, has reached a stage where she seemingly has a limited symbolic number system: she knows symbols (say, "***" or "3") that correspond to each numerosity ("three-ness"). Much the same way children learn that the word "one" corresponds to oneness, Sheba has learned that the Arabic symbol "1" corresponds to oneness, "2" to twoness, up to six. Sheba can also understand the ordinal relationships between numerosities. Most of the studies with chimpanzees require thousands of trials and years of training for the animals to learn not only the details of the task but also what is required of them.

With the use of the methods employed with human babies, in which cognition is assessed spontaneously without the need of training or language, researchers for the past ten years have investigated the origins of cognitive abilities using closely matched tasks across species.

Monkeys can determine the number of objects in a set that bear distinct properties (Uller, Carey, Hauser, and Xu 1997), add and subtract small numbers of objects (Hauser, McNeilage, and Ware 1996; Uller, Hauser, and Carey 2001), and can cross-modally match the number of voices they hear to the number of faces they see (Jordan et al. 2005). Cotton-top tamarins can add numbers of objects in sets (Flombaum, Junge, and Hauser 2005; Uller, Hauser, and Carey 2001) and discriminate small numbers auditorily (Hauser et al. 2002). Lemurs can add small numbers of objects in a set (Santos, Barnes, and Mahajan 2005) and discriminate between small sets (Jordan, Jaffe, and Brannon 2005; Cullen and Uller, under review). These studies make use of the *looking time* and *search/reach* methods described before, paradigms widespread in infant cognition as powerful tools to evaluate infants' expectations about outcomes of events.

One classic example of a looking time task with nonhuman primates is a groundbreaking experiment done with rhesus macaques in Cayo Santiago, Puerto Rico (Uller, Carey, and Hauser 1997; see also Uller 1996 for a more detailed account of the experiment). Uller, Carey, and Hauser (1997) used a 1 + 1 = 2 or 1 looking time task to assess the monkeys' object individuation abilities. Each monkey saw a familiarization trial in which they saw a carrot and a squash emerging from behind a screen (the screen was actually a side of a box roughly the size of a shoe box). This was meant to get a measure of their visual attention to the two objects—squash and carrot—that presumably existed inside the box. After they were familiarized with this outcome, they then saw the 1 + 1 operation. The experimenter placed the carrot inside the box, and then the squash, and then the screen was removed to reveal both objects or just one (in the one-object trial, the experimenter surreptitiously places the object into a pouch attached to the back of the screen and thus "invisible" to the subject being tested). The results showed that the monkeys looked longer in the test trials that violated their expectancies (one object) than in the trials that showed what they had expected (two objects), thus suggesting that they had expected to see two objects in the 1 + 1 task. The same methodology has been successfully used with a variety of primate species.

In the search/reach domain, Hauser, Carey, and Hauser (2000) asked whether nonhuman primates could discriminate between two numerosities paired in parallel. Rhesus monkeys were shown two buckets into which slices of apples were lowered. As in the infant case, this experiment involved no training and mimicked a natural foraging problem. The contrasts included 1 versus 2, 2 versus 3, 3 versus 4, 3 versus 5, 4 versus 5, 4 versus 6, 4 versus 8, and 3 versus 8. Each monkey was tested in one condition only, and received only one trial. The monkeys chose the container with the greater number in 1 versus 2, 2 versus 3, 3 versus 4, and 3 versus 5 slices, but not in 4 versus 5, 4 versus 6, 4 versus 8, and 3 versus 8 cases. The researchers concluded that the results show a spontaneous numerical ability for small numerosities that closely match the ones attained by human babies. They speculated that the failure to discriminate larger numerosities was in disagreement with a larger numerosity understanding shown in training experiments with rats, pigeons, and chimps. They also speculated that the small range of numerosities present spontaneously in ontogenetic and phylogenetic development seem to coincide with "number" encoded in the structure of natural languages.

The results with nonhuman primates and human babies, and the speculations derived from them, yield interesting predictions. If monkeys and human babies have a limited system for spontaneous representation of number, or at least a limited capacity to "go for more," then it may be part of the primate lineage only, in which case we would not expect it to occur in other species. Another prediction is that this limited system seems to correspond to the system encoded in natural languages. If only primates have this ability as an evolutionary trait, then other species would not have the same ability. The investigation of similar abilities in other classes could be fruitful insofar as it would shed light onto the validity of such predictions.

Although different numerical abilities have been reported in nonhuman primates at large, research on numerical abilities in other species remains at a cautious stage. My students and I are currently in the midst of developing this research. Here I will report on three series of studies with pigs (mammals), salamanders (amphibians), and crabs (arthropods) in which we chose to test for their capacity to "go for more." This is a relevant question for many reasons. For one, it taps into their understanding of *ordinality*, whether they have intuitions about whether 2 > 1, 3 > 2, 4 > 3, 6 > 4. Together with the ability to represent operational relations between small numerosities (1 + 1 = 2, 2 − 1 = 1, etc.), creatures should also have an ability to understand the "order" in which these numerosities are organized, because one could argue that showing an ability to discern visually that 1 + 1 = 2 and not 1 does not necessarily entail an understanding of number, but perhaps that the numbers are just visually dissimilar. Another reason is that, in ethology, theories of optimal foraging (MacArthur and Pianka 1966; Pyke, Pulliam, and Charnov 1977; Stephens and Krebs 1986) predict that animals "go for more." That is, animals evolve foraging strategies that maximize their net energy gain when foraging (i.e., the energetic profit exceeds the energetic loss during foraging).

The Piglet Experiments

As we have seen so far, human numerical representations may have originated early in the primate lineage, and can be found in different species of apes, monkeys and prosimians. Do these same abilities exist in other species of land mammals? The work reported here (Bull and Uller, under review) tested a species of land mammal, domestic pigs (*Sus scrofa*), in a closely matched experimental paradigm used with infants and monkeys, namely, their capacity to "go for more." This class of land mammal belongs to the order *Artiodactyla*, which, according to fossil records, emerged during the middle to late Eocene, about 48 million years ago (Bosma et al. 2004; Giuffra et al. 2000). Domestic pigs have been successfully trained to memorize sites of more food over less, thus showing that the animals may employ optimal foraging strategies (Held et al. 2004), but to date, there has been no work on pigs' spontaneous ability to "go for more."

Sixteen young Saddleback and old Gloucester male and female piglets were used in these tasks. They were kept outdoors in fenced woodland within the arboretum at Marks Hall Country Estate, Coggeshall, Essex, England. The piglets spontaneously fed on brambles growing in the arboretum, and were fed hard feed nuts twice daily by their keepers. Feed was given to the piglets by spreading it across the ground, in particular among the brambles, to encourage the piglets to eat the brambles. This method of feeding mimicked natural foraging for food. They lived in a one-acre fenced in area with an enclosure suitable for testing. They were completely familiarized with the enclosure because it is in fact a weighing station into which they were habituated to go weekly for body and weight measurements. This weighing station was only placed into a gap in the fence while the weekly weighing and experimental testing was taking place, allowing the piglets the freedom to move in and out of the holding pen whenever they chose to do so. This was to ensure that the holding pen was not a stressful environment for the piglets. The stimuli used for testing were "food balls" made from pig feed nuts mixed with warm water into a paste and molded into ball shapes each weighing 200 g.

Each piglet was individually lured into the weighing station, its sex and weight were recorded, and then it was tested. Once the piglet was in the weighing cage, it was shown two identical opaque empty buckets. The buckets were then placed onto circular trays on the ground between the keeper and the animal equidistant from each. The buckets were far enough apart that the piglets could not reach both at once. The keeper then showed the piglet each ball of food he would be dropping into each bucket to allow the animal to see and scent the stimuli. Immediately after each ball was shown to the piglet, it was lowered in quick succession into one of the buckets, sequentially. Only animals that watched all of the stimuli being lowered into the buckets were allowed to make a choice.

Coding of choice started when the front door of the weighing station was opened to allow the piglet to move out of the weighing station into the pen containing the buckets. After lowering all of the stimuli into the buckets, the keeper walked around to behind the right-hand board and opened the door of the weighing cage via a catch at the top, pulling the door back flush with the board. We measured whether the piglets would spontaneously select the bucket containing two food balls or the bucket containing three foodballs. Choice was coded as (1) body motion toward the bucket, (2) snout touching or head-banging the bucket, and/or (3) putting their snout into the bucket. The piglets' choice for one numerosity (x) over the other (y) was coded online and through video records.

Data from the 16 piglets that touched, head-butted, or put their snout into either the two-ball bucket or the three-ball bucket were recorded: 12 piglets chose the bucket containing three food balls and 4 piglets chose the bucket containing two food balls, which is a significant result. The piglets reliably chose three food balls over two. This result is original as there are no other empirical data available to date showing the ability to go for more over less in a species of nonprimate land mammal. The positive result of this experiment indicates that the method yields interpretable data with this species of mammal, and that further studies should be developed to shed light on the discrimination abilities in pigs.

The Salamander Experiments

The first salamander experiments were reported by Uller et al. (2003). Here we briefly offer a review of the original studies and present some further results of control experiments currently being developed with collaborators from the Institute of Brain Research at the University of Bremen.

The original experiments (Uller et al. 2003) were done with red-backed salamanders (*Plethodon cinereus*). Plethodontid fossil records indicate that this species exists since the Lower Miocene, 28 million years ago (Duellman and Trueb 1986). Red-backed salamanders employ an optimal foraging strategy in that they forage indiscriminately between two sizes of flies (*Drosophila*) when both are low in numbers but specialize in the larger flies when the numbers of prey increase (Jaeger, Barnard, and Joseph 1982). This ability to change foraging tactics suggests that a salamander can assess the number of prey items within its visual field.

We used adult male and female red-backed salamanders. These animals were collected in the forest near Mountain Lake Biological Station, Giles County, Virginia, and brought to the University of Louisiana. The animals were housed individually in Petri dishes until the date of testing, after which they were returned to the forest in Virginia.

The salamanders had to undergo a gradual procedure of familiarization to the conditions of testing over three days (for further details, see Uller et al. 2003). After this period elapsed, two empty laboratory tubes were introduced into their housing dishes. After these were covered with pheromones, and the animals were comfortable in their dishes, live fruit flies were introduced into the tubes for the numerical contrasts. The experiments included the contrasts 2 versus 3, 1 versus 2, 4 versus 6, and 3 versus 4.

Five minutes prior to testing, we removed the empty plastic tubes from the dish and replaced them with two identical tubes containing either x (e.g., 2) or y (e.g., 3) fruit flies. The two tubes were placed 20 cm apart and equidistant from the salamander's path of approach from the tunnel. The ends of the tubes were sealed to prevent the flies from escaping and to prevent chemical cues from the flies from emanating into the enclosure.

The flies could nonetheless freely move within each tube. For half the animals, (*x*) number of fruit flies were placed into the left tube, and for the other half, (*x*) number of fruit flies were placed into the right tube.

We coded choice as snout touching the selected tube or snapping at the selected tube. There were over 300 animals available for testing at the time. For each experiment, we used 30 animals that successfully made a choice, and discarded the animals that did not move. Each animal was tested only once—namely, if an animal was selected for the 1 versus 2 experiment, it was not used in any other experiment. Experiments 2 versus 3 and 4 versus 6 were replicated three times. Here we report only the results of the main experiment in each numerosity contrast.

We started our series of experiments with the discrimination of 2 versus 3 flies for a number of reasons. The primate literature has substantial evidence that monkeys discriminate between 2 and 3 apple pieces (Hauser, Carey, and Hauser 2000), and 12-month-old human babies discriminate between 2 and 3 cookies (Feigenson, Carey, and Hauser 2002). Second, as there is no prior evidence for this kind of discrimination in a species of amphibian, our intuitions about this ability in salamanders were poor. Third, as this is a completely new methodology, and it was unknown whether this experiment would yield interpretable data, numerosities that have yielded success in other species were taken as the appropriate candidates to start this investigation. In the 2 versus 3 contrast, twenty salamanders touched the 3-fly tube and 10 touched the 2-fly tube, which yields a significant result. The salamanders reliably chose 3 over 2. This result is original, as there are no scientific records to date showing numerical discrimination in a species of amphibian. We replicated this experiment with different populations of red-backed salamanders three times, and the results held nicely (Uller et al. 2003).

The next step was to see whether this same ratio holds when numbers are increased. Evidence from rhesus monkeys (Hauser, Carey, and Hauser 2000) and human infants (Feigenson, Carey, and Hauser 2002) indicates that these animals also show the same discrimination ability, but not beyond a set of 4 items. Rhesus monkeys, for example, cannot discriminate between 4 and 6 apple slices because they do not select the larger numerosity when given the choice, and human infants do not succeed in selecting 4 versus 6 cookies when given two jars to choose from. These results seem to imply that there is a limit on the number of items that monkeys and young babies can discriminate, namely, not beyond 4.

We thus contrasted the numerosities 4 and 6 as tested in experiments with nonhuman primates and human infants. A group of 30 adult red-backed salamanders completed the 4 versus 6 discrimination test by touching with their snouts one of the two tubes within the 10-minute trial. Sixteen salamanders touched the 6-fly tube and 14 touched the 4-fly tube, which does not yield a significant result. Three other replications of the same experiment, with fresh groups of different animals, yielded the same identical result, namely, that the salamanders were random at selecting one numerosity over the other. So explanations on the basis of amount of movement cannot hold, that is, that the same number of flies in both tubes, say, four in each, was moving at the time of choice, and therefore the salamanders' random selection was due to there being no difference in the amount of movement inside the tubes. This result seems to indicate that the limit on the highest numerosity chosen lies somewhere around 4.

In order to assess the exact limit for salamanders of the numerosity discriminated in forced-choice spontaneous conditions we ran the animals in a contrast of 3 versus 4. A group of 30 adult red-backed salamanders completed the 3 versus 4 forced-choice discrimination test by touching with their snouts one of the two tubes within the 10-minute trial. Fifteen salamanders touched the 3-fly tube, and 15 touched the 4-fly tube, which does not yield a significant result. The salamanders were random at selecting one numerosity over the other. This result indicates that the exact limit on the highest numerosity chosen lies at 3. Unlike nonhuman primates, and like human infants, salamanders will discriminate more from less in contrasts up to 3.

In order to draw a picture of the salamanders' capacity to choose the larger numerosity under conditions of spontaneous forced choice, it remains to be shown that salamanders indeed choose between two small numerosities, namely, those contained in sets of 1, 2, and 3. Now the question is—do the salamanders discriminate between 1 and 2? A group of 30 adult red-backed salamanders completed the 1 versus 2 discrimination test by touching with their snouts one of the two tubes within the 10-minute trial. Twenty-two salamanders touched the 2-fly tube, and 8 touched the 1-fly tube. This result goes along with the results from the 2 versus 3 discrimination task and replications. It shows that the salamanders have indeed a capacity to choose the larger numerosity in spontaneous forced-choice conditions of small sets containing 1, 2, or 3 fruit flies. These results parallel results with nonhuman primates and human infants showing comparable abilities.

Recently, my colleagues at the University of Bremen and I have developed computed generated stimuli to further these results. As proposed in

Uller et al. (2003), it is possible that amphibians use completely different mechanisms to assess prey and would therefore rely on information about mass (volume or surface area) or speed of movement to detect quantity. In order to rule out these possibilities, we devised computer-generated contrasts in which same mass versus different number stimuli were pitted against each other. The same was true for the "amount of movement" hypothesis: we pitted velocity against time, for example, in separate conditions such as two stimuli moving at 2 cm/sec. (regular amount of "walking" speed of a mid-sized real cricket) on the left side of the screen versus one stimulus moving at 4 cm/sec on the right side of the screen; or one stimulus at 2 cm/sec on the right side of the screen and another stimulus moving at 4 cm/sec. on the left side of the screen. These experiments are in the midst of being run and results should be forthcoming soon.

The Crab Experiments

This line of research has been carried out with collaborators in the Zoology Department, School of Biological Sciences, University of Aberdeen. The question is whether we can push the boundaries of evolutionary continuity even further. Suggestions made by Cummins and Allen (1998) take that the capacity for numerical competence is found not only in humans, but in many different warm-blooded vertebrate species. This therefore implies that a capacity for number may be a trait found only in warm-blooded vertebrate animals. Since then, research with red-backed salamanders (Uller et al. 2003) shows at least one kind of ability in a cold-blooded vertebrate species. Further research into animal cognition may be broadening our knowledge of the vertebrate kingdom, but has yet to break into the world of invertebrates, with no studies, to date, documenting investigations into number in invertebrate species. The aim here was to assess whether invertebrates show any kind of mechanism for the detection of quantity.

We are studying whether the common shore crab (*Carcinus maenas*) has a spontaneous ability to discriminate between various numbers of equally sized food items (Uller, Fraser, and Reeve, in preparation). For the first series of experiments, we used 200 common shore crabs, varying in age, size, and sex, obtained from the Ythan estuary. They were brought into the lab and were then left to adjust to their new environment for two months, being fed pellets and pieces of squid cut into small squares.

After two months, the crabs were removed from the original holding tanks and placed into a new holding tank of the same diameter, temperature, and environmental conditions to ensure as little disturbance as possible. They were then individually separated for testing. In order to disturb the crabs as possible, they were handled quickly and carefully.

Stimuli consisted of pieces of squid cut in cubes of approximately 2 cm × 2 cm which were placed into a 300 ml beaker filled with fresh water. Two pieces of the squid were placed into a test tube 15 cm in length and 1.5 cm in width. The squid pieces were greater than the width of the test tube so that they would not move. They were placed equidistant along the test tube with 3 cm empty on each side of the food and in between the pieces. The test tube was then filled with fresh water so that the image of the food pieces would not be distorted when placed into the filled aquarium. The test tube was then sealed with a cork bung to ensure that the scent of the food could not spread through the aquarium. Three pieces of squid were then placed equidistant into a second test tube with 2 cm empty on each side of the food and in between the food pieces. Again this test tube was then filled with fresh water and sealed with a cork bung. Held in place by small pieces of modeling putty, the two tubes were positioned across two adjacent corners of the aquarium.

The experimental conditions were identical to the salamander experiments, namely, 2 versus 3, 1 versus 3, and 4 versus 6. Like any other discrimination experiment, half the animals received (x) number on the left, the other half received (x) number on the right. Choice was coded as feeding behavior when the crab went to one selected tube and performed a behavior as if feeding.

Of the total 60 crabs tested in the 2 versus 3 condition, 46 (77%) exhibited feeding behavior on one of the food tubes. From these 46 crabs, 39 (85%) reliably selected the 3-food tube and 7 (15%) the 2-food tube, which is a significant result. In the 1 versus 2 condition, 43 (72%) of the 60 crabs that were tested exhibited a feeding behavior on either of the food tubes. Out of the 43 crabs, 31 (72%) reliably chose the 2-food tube while 12 crabs (28%) chose the tube containing the 1 piece of food, which is also a significant outcome. Finally, in the 4 versus 6 discrimination task, 27 (45%) of the total 60 crabs that were tested exhibited feeding behavior on either tube: 22 (81%) crabs chose the tube containing 6 pieces of food and 5 (19%) chose the tube containing 4 pieces of food. What is amazing about these crab experiments is that we obtained interpretable data, and the results seem to reveal a quite astonishing capacity in crabs to go for more.

Are these results comparable to those obtained with nonhuman primates and human infants? As we set out in the beginning of this chapter, and to emphasize again, we make no commitments as to whether these abilities have their origins by common descent and thus would be the reflection of similar (if not identical) mechanisms, or whether they happen to be expressed roughly the same way but have evolved independently. We choose to hypothesize that at least some of the numerical abilities described here may share evolutionary commonalities and are thus worth investigating. It is only with the aid of closely matched experiments that one will come closer to an understanding of the nature of such abilities, and trace any common evolutionary roots. Of course the research agenda sponsored here is still in its infancy, and many more experiments, including controls for amount of stuff, amount of surface area, amount of movement (in the case of the salamanders), density, mass, total area—are needed for a cleaner comparison among species.

A much more interesting and productive question regards the mechanism underlying these abilities. I have pointed out before (Uller et al. 2003) that there are five characteristics of the small system of number presumably in existence in a variety of species:

1. The system is limited. The limit on the spontaneous number representation in monkeys and human babies seems to lie between 3 and 4.
2. The system is precise. The system precisely tracks exact small numerosities that form the representations of small sets. It does not involve estimative capacities.
3. The system is spontaneously available. The representations revealed by visual attention and reach/touch tasks do not require training and thus are not learned.
4. The system is adaptively powerful. As such, these representations may be widespread in the animal kingdom.
5. This system is "entity based." The representations are constructed on the basis of one–one correspondences. For each entity encoded, one representation is formed and stored in short-term memory.

The nature of this ability, however, remains unclear. For human infants and nonhuman primates, researchers have proposed that an object-file model would be the best candidate to account for these and other results (see, e.g., Uller et al. 1999; Hauser, Carey, and Hauser 2000). This model is one originally taken from the literature on object-based attention (Kahneman, Treisman, and Gibbs 1992; Trick and Pylyshyn 1993, 1994) and later adapted to account for the young human infant's small number abilities (Uller et al. 1999). It assumes that objects are individuated according to principles of object individuation and identification and then encoded as object files maintaining one–one correspondence. For each object encountered in the world, one file is opened. A maximum of four object files can remain open simultaneously. Object files are discrete and precise. They do not rely on a capacity to estimate number. The counterpart of the object file model is an analog magnitude model that operates in concert with the former for number representation in humans and other animals and is used for larger numerosity encoding and estimation processes (see, e.g., Whalen, Gallistel, and Gelman 1999).

Further experiments that probe the animals' numerical capacities will help us decide if these abilities are or are not comparable to human and nonhuman primate numerical abilities. For example, experiments that address the ratio between two numerosities will clarify the nature of these abilities as far as a model is concerned. Contrasts between higher numbers in which the ratio is 1:2, namely, 6 versus 12, 8 versus 16, for example, are useful in this respect. Conversely, larger ratios that contrast a small numerosity with a rather large numerosity (e.g., 2 versus 20) will also determine the animals' capacity to go for more. That is, 2 may be an understandable/ tangible numerosity within their repertoire, but 20 may be "far too much," representing not a discernible quantity, but a rather confusing one, in which case the animal might spontaneously "go for less."

The evolution of cognition raises questions across domains of knowledge and provides us with material to speculate about how x sees the world and what it is like to be an x (Nagel 1974). I have always been in awe of the binary-ternary nature of our natural world and how we construct reality: the 3-D world, the basis of rhythms in music construction across cultures, the origins of graphic numerical notations (I, II, III, . . .), and so many more instances. I was once asked if I thought an octopus would have the same kind of "numerical capacity" as, say, a crab. "Interesting question"—I replied. Fortunately, an empirical one.

References

Antell, S. E., and D. P. Keating (1983). Perception of numerical invariance in neonates. *Child Development* 54: 695–701.

Bar-Yosef, O., and V. Valla (eds.) (1991). *The Natoufian Culture in the Levant*. Ann Arbor, Mich.: Ann Arbor Press.

Beran, M. (2001). Long term retention of the differential values of Arabic numerals by chimpanzees (Pan troglodytes). *Animal Cognition* 7: 86–92.

Bijeljac-Babic, R., J. Bertoncini, and J. Mehler (1991). How do four-day-old infants categorize multisyllabic utterances? *Developmental Psychology* 29: 711–721.

Bosma, A., N. de Haan, G. Arkesteijn, F. Yang, M. Yerle, and C. Zijlstra (2004). Comparative chromosome painting between the domestic pig (Sus scrofa) and two species of peccary, the collared peccary (Tayassu tajacu) and the white-lipped peccary (T. pecari): a phylogenetic perspective. *Cytogenetic and Genome Research* 105: 115–121.

Boysen, S. (1993). Counting in chimpanzees: Nonhuman principles and emergent properties of number. In *The Development of Numerical Competence*, ed. S. Boysen and E. J. Capaldi. Hillsdale, N.J.: Lawrence Erlbaum.

Boysen, S., and G. Berntson (1989). The development of numerical competence in the chimpanzee (Pan troglodytes). *Journal of Comparative Psychology* 103: 23–31.

Boysen, S., and G. Berntson (1995). Responses to quantity: Perceptual versus cognitive mechanisms in chimpanzees (Pan troglodytes). *Journal of Experimental Psychology*: Animal Behavior Processes 21: 82–86.

Bull, W., and C. Uller (under review). What do babies and piglets have in common? Numerical discrimination in a species of domestic pig (Sus scrofa).

Carey, S. (1995). Continuity and discontinuity in development. In: *An Introduction to Cognitive Science*, ed. D. Osherson. Cambridge, Mass.: MIT Press.

Clearfield, M. W., and K. S. Mix (1999). Number versus contour length in infants' discrimination of small visual sets. *Psychological Science* 10: 408–411.

Cullen, R., and C. Uller (under review). Origins of spontaneous numerical representations: Experiments with lemurs (Lemur catta, Varecia variegata variegata, Varecia variegata rubra).

Cummins, D., and C. Allen (eds.) (1998). *The Evolution of Mind*. Oxford: Oxford University Press.

Darwin, C. (1859). *On the Origin of Species by Means of Natural Selection, Or The Preservation of Favoured Races in the Struggle for Life. Facsimile of the First Edition*. Cambridge, Mass.: Harvard University Press.

Darwin, C. 1871/1981. *The Descent of Man and Selection in Relation to Sex*. Princeton: Princeton University Press.

Deaner, R. O., A. V. Khera, and M. L. Platt (2005). Monkeys pay per view: Adaptive value of social images by rhesus macaques. *Current Biology* 15: 543–548.

Duellman, W., and L. Trueb (1986). *Biology of Amphibians*. New York: McGraw Hill.

Feigenson, L., and S. Carey (2003). Tracking individuals via object files: Evidence from infants' manual search task. *Developmental Science* 6: 568–578.

Feigenson, L., S. Carey, and M. Hauser (2002). The representations underlying infants' choice of more: Object files versus analog magnitudes. *Psychological Science* 13: 150–156.

Flombaum, J., J. Junge, and M. D. Hauser (2005). Rhesus monkeys spontaneously compute addition operations over large numbers. *Cognition* 97: 315–325.

Gallistel, C. R. (1990). *The Organization of Learning*. Cambridge, Mass.: MIT Press.

Giuffra, E., J. Kijas, V. Amarger, Ö. Carlborg, J.-T. Jeon, and L. Andersson (2000). The origin of the domestic pig: Independent domestication and subsequent intro-gression. *Genetics* 154: 1785–1791.

Gould, S. J. (1977). *Ontogeny and Phylogeny*. Cambridge, Mass.: Harvard University Press.

Harper, D. G. (1982). Competitive foraging in mallards: "Ideal free" ducks. *Animal Behaviour* 30: 575–584.

Hauser, M., and S. Carey (1999). Building a cognitive creature from a set of primi-tives: Evolutionary and developmental insights. In *The Evolution of Mind*, ed. C. Allen and D. Cummins. Oxford: Oxford University Press.

Hauser, M., S. Carey, and L. Hauser (2000). Spontaneous number representation in semi-free-ranging rhesus monkeys. *Proceedings of the Royal Society of London* 267: 829–833.

Hauser, M., P. McNeilage, and M. Ware (1996). Numerical representations in pri-mates. *Proceedings of the National Academy of Sciences* 93: 1514–1517.

Hauser, M., S. Dehaene, G. Dehaene-Lambertz, and A. Patalano (2002). Spontaneous number discrimination of multi-format auditory stimuli in cotton-top tamarins (Saguinus oedipus). *Cognition* 86: B23–B32.

Held, S., J. Baumgartner, A. Kilbride, R. W. Byrne, and M. Mendl (2004). Foraging behaviour in domestic pigs (*Sus scrofa*): Remembering and prioritizing food sites of different value. *Animal Cognition* 8: 114–121.

Herrnstein, R. J. (1990). Levels of stimulus control: A functional approach. *Cognition* 37: 133–166.

Jaeger, R., D. Barnard, and R. Joseph (1982). Foraging tactics of a terrestrial salaman-der: Assessing prey density. *American Naturalist* 119: 885–890.

Jordan, K., and E. Brannon (2006). The multisensory representation of number in infancy. *Proceedings of the National Academy of Sciences* 103: 3486–3489.

Jordan, K., E. Brannon, N. K. Logothetis, and A. A. Ghazanfar (2005). Monkeys match the number of voices they hear to the number of faces they see. *Current Biology* 15: 1034–1038.

Kahneman, D., A. Treisman, and B. J. Gibbs (1992). The reviewing of object files: Object-specific integration of information. *Cognitive Psychology* 24: 175–219.

Kobayashi, T., K. Hiraki, and T. Hasegawa (2005). Auditory-visual intermodal matching of small numerosities in 6-month-old infants. *Developmental Science* 8: 409–421.

Koechlin, E., S. Dehaene, and J. Mehler (1997). Numerical transformations in five-month-old infants. *Mathematical Cognition* 3: 89–104.

Lankester, E. R. (1909). *Treatise on Zoology*. London: A & C Black.

Lipton, J. S., and E. Spelke (2003). Origins of number sense: Large number discrimination in human infants. *Psychological Science* 14: 396–401.

MacArthur, R., and E. Pianka (1966). On optimal use of a patchy environment. *American Naturalist* 100: 603–609.

Matsuzawa, T. (1985). Use of numbers by a chimpanzee. *Nature* 315: 57–59.

Matsuzawa, T., S. Itakura, and M. Tomonaga (1991). Use of numbers by a chimpanzee: A further study. In *Primatology Today*, ed. A. Ehara, T. Kimura, O. Takenaka, and M. Iwamoto. Amsterdam: Elsevier.

McComb, K., C. Packer, and A. Pusey (1994). Roaring and numerical assessment in contests between groups of female lions, *Panthera leo. Animal Behaviour* 47: 379–387.

McCrink, K., and K. Wynn (2004). Large number addition and subtraction by 9-month-old infants. *Psychological Science* 15: 776–780.

Moore, D., J. Benenson, J. S. Reznick, M. Peterson, and J. Kagan (1987). Effect of auditory numerical information on infants' looking behavior: Contradictory evidence. *Developmental Psychology* 23: 665–670.

Morgan, C. L. (1894). *An Introduction to Comparative Psychology*. London: Walter Scott.

Nagel, T. (1974). What is it like to be a bat? *Philosophical Review* 83: 435–451.

Nieder, A., D. Freedman, and E. Miller (2002). Representation of the quantity of visual items in the primate prefrontal cortex. *Science* 297: 1708–1711.

Pfungst, O. (1911). *Clever Hans. The Horse of Mr. von Osten. A Contribution to Experimental Animal and Human Psychology*. New York: Holt.

Pyke, G., H. Pulliam, and E. Charnov (1977). Optimal foraging: A selective review of theory and tests. *Quarterly Review of Biology* 52: 137–154.

Pylyshyn, Z. W. (1984). *Computation and Cognition*. Cambridge, Mass.: MIT Press.

Rumbaugh, D. M., S. Savage-Rumbaugh, and M. Hegel (1987). Summation in the chimpanzee (*Pan troglodytes*). *Journal of Experimental Psychology: Animal Behavior Processes* 13: 107–115.

Santos, L. R., J. Barnes, and N. Mahajan (2005). Expectations about numerical events in four lemur species (*Eulemur fulvus, Eulemur mongoz, Lemur catta* and *Varecia rubra*). *Animal Cognition* 8: 253–262.

Santos, L., G. Sulkowski, G. Spaepen, and M. Hauser (2002). Object individuation using property/kind information in rhesus macaques (*Macaca mulatta*). *Cognition* 83: 241–264.

Sayler, A. (1966). The reproductive ecology of the red-backed salamander, Plethodon cinereus, in Maryland. *Copeia* 1966: 183–193.

Sharon, T., and K. Wynn (1998). Infants' individuation of actions from continuous motion. *Psychological Science* 9: 357–362.

Simon, T., S. Hespos, and P. Rochat (1995). Do infants understand simple arithmetic? A replication of Wynn (1992). *Cognitive Development* 10: 253–269.

Starkey, P., and R. G. Cooper (1980). Perception of numbers by human infants. *Science* 210: 1033–1035.

Starkey, P., E. Spelke, and R. Gelman (1990). Detection of intermodal numerical correspondences by human infants. *Science* 222: 179–181.

Stephens, D., and J. Krebs (1986). *Foraging Theory*. Princeton, N.J.: Princeton University Press.

Strauss, M. S., and L. E. Curtis (1981). Infant perception of numerosity. *Child Development* 52: 1146–1152.

Sulkowski, G., and M. Hauser (2001). Can rhesus monkeys spontaneously subtract? *Cognition* 79: 239–262.

Treiber, F., and S. Wilcox (1984). Discrimination of number by infants. *Infant Behavior and Development* 7: 93–100.

Trick, L., and Z. W. Pylyshyn (1993). What enumeration studies can show us about spatial attention: Evidence for limited capacity preattentive processing. *Journal of Experimental Psychology* 19: 331–351.

Trick, L., and Z. W. Pylyshyn (1994). Why are small and large numbers enumerated differently? A limited capacity preattentive stage in vision. *Psychological Review* 101: 80–102.

Uller, C. (1996). Origins of numerical concepts. A comparative study of human infants and nonhuman primates. Unpublished doctoral dissertation, Massachusetts Institute of Technology, Cambridge, Massachusetts.

Uller, C. (2004). Disposition to recognize goals in infant chimpanzees (*Pan troglodytes*). *Animal Cognition* 7: 154–161.

Uller, C., S. Carey, M. Hauser, and F. Xu (1997). Is language needed for constructing sortal concepts? A study with nonhuman primates. *Proceedings of the 21st Annual Boston University Conference on Language Development* 21: 665–677.

Uller, C., S. Carey, G. Huntley-Fenner, and L. Klatt (1999). What representations might underlie infant numerical knowledge. *Cognitive Development* 14: 1–36.

Uller, C., P. Fraser, and E. Reeve (in preparation). Common shore crabs (*Carcinus macnas*) select the larger numerosity in small numerical choices.

Uller, C., M. Gaudin, and A. Fradella (in preparation). Infants search for one more in sets of 2 and 3, but not 4.

Uller, C., M. Hauser, and S. Carey (2001). Spontaneous representation of number in Cotton-top tamarins. *Journal of Comparative Psychology* 115: 1–10.

Uller, C., R. Jaeger, G. Guidry, and C. Martin (2003). Salamanders (*Plethodon cinereus*) go for more: Rudiments of number in an amphibian. *Animal Cognition* 6: 105–112.

van Loosbroek, E., and A. W. Smitsman (1990). Visual perception of numerosity in infancy. *Developmental Psychology* 26: 916–922.

Whalen, J., C. Gallistel, and R. Gelman (1999). Nonverbal counting in humans: The psychophysics of number representation. *Psychological Science* 10: 130–137.

Wood, J. N., and E. S. Spelke (2005). Chronometric studies of numerical cognition in five-month-old infants. *Cognition* 97: 23–29.

Wynn, K. (1992). Addition and subtraction by human infants. *Nature* 258: 749–750.

Wynn, K., P. Bloom, and W. Chiang (2002). Enumeration of collective entities by 5-month-old infants. *Cognition* 83: B55–B62.

Xu, F. (2003). Numerosity discrimination in infants: Evidence for two systems of representations. *Cognition* 89: B15–B25.

Xu, F., and S. Carey (1996). Infants' metaphysics: The case of numerical identity. *Cognitive Psychology* 30: 111–153.

Xu, F., and E. Spelke (2000). Large number discrimination in 6-month-old infants. *Cognition* 74: B1–B11.

Xu, F., E. Spelke, and S. Godard (2005). Number sense in human infants. *Developmental Science* 8: 88–101.

10 Cohabitation: Computation at Seventy, Cognition at Twenty

Stevan Harnad

On knowing how we generate our know-how   In the 1960s, my teacher, D. O. Hebb, had a standard example for his undergraduate Intro Psych course to show that (what we would now call) "cognition" cannot all amount to just inputs and outputs plus the reward/punishment histories that shaped them, as behaviorism (then already in decline) had taught. He simply presented two single-digit numbers, the first bigger than the second—say, 7 and 2. Next he would pause; and then he would remind us how different our "response" to those two "stimuli" would be if earlier— even much earlier—we had been told "subtract" (or "add" or "multiply").

Now Hebb's example was not decisive—as no refutation of push/pull behaviorism can be decisive. The input sequence "subtract," "7," "2," generating the output "5" can be interpreted as a rote I/O sequence no matter how long the delay between the input "subtract" and the inputs "7," "2." But Hebb's point was a double one, to the effect that, first, surely there is something intervening between the command, the digits, and the response; and, second, that surely whatever that intervening internal process is, *that* is the true object of study of (what would today be called) "cognitive science," and not just the inputs, outputs, and their reward histories.

Behaviorism begged the question   Or, to put it another way, the task of cognitive science is to explain what equipment and processes we need in our heads in order to be capable of being shaped by our reward histories into doing what we do. Skinner—whom Hebb had described (in part out of modesty, but surely also partly tongue-in-cheek) as the greatest contributor to psychology (*sic*) at the time—had always dismissed theorizing about how we are able to learn: Skinner regarded such theories of learning as either unnecessary or the province of another discipline (physiology), hence irrelevant (to psychology; Harnad 1985; Catania and Harnad 1988). Cognitive science has since rejigged the disciplinary boundaries somewhat, admitting neurosciences into the ecumenical circle, but it should be noted that Hebb's point was about internal processes in the head that underlie our capacity to do what we can do. There is still a Skinnerian uneasiness about counting the biomolecular details of brain implementation as part of cognitive science. We shall return to this when we discuss the hardware-independence of software.

In essence, Hebb's point was about question-begging: Behaviorism was begging the question of "how?" How do we have the behavioral capacity that we have? What makes us able to do what we can do? The answer to this question has to be cognitive; it has to look into the black box and explain how it works—but not necessarily in the physiological sense. Skinner was right about that. Only in the functional, cause–effect sense. And regarding the functional explanation of our behavioral capacities, behaviorism, in its brief against its predecessor, introspectionism had again been half-right. Behaviorists had rightly pointed out that sitting in an armchair and reflecting on it will not yield an explanation of how our mind works (except of course in the sense that explanation in all disciplines originates from human observation and reflection).

For this, Hebb had a companion example to his 7/2 task. He would ask the intro class: "What was the name of your third grade school-teacher?" When we triumphantly produced our respective answers, he would ask, "How did you do it? How did you find the name?" He might have asked the same of addition and subtraction: "How is it that your head, having been told 'subtract,' manages to come up with '5' in response to '7' . . . '2'?"

Beware of the easy answers: rote memorization and association. The fact that our brains keep unfailingly delivering our answers to us on a platter tends to make us blind (neurologists would call it "anosognosic") to the fact that there is something fundamental there that still needs to be accounted for. Memorizing single-digit sums, products, and differences by rote, case by case, covers the trivial cases, but it does not generalize to the nontrivial ones. Surely we have not pre-memorized every possible sum, product, and difference?

Is computation the answer? Computation already rears its head, but here too, beware of the easy answers: I may do long division in my head the same way I do long division on paper, by repeatedly applying a memorized set of symbol-manipulation rules—and that is already a big step past behaviorism—but what about the things I can do for which I do not know the computational rule? Don't know it consciously, that is. For introspection can only reveal how I do things when I know, explicitly, how I do them, as in mental long division. But can introspection tell me how I recognize a bird as a bird, or a chair as a chair? How I play chess (not what the rules of chess are, but how, knowing them, I am able to play, and win, as I do)? How I learn from experience? How I reason? How I use and understand words and sentences?

Skinner got another famous piece of come-uppance from Chomsky (1959), who pointed out how question-begging was the very idea that our linguistic capacity (in particular our syntactic capacity), which Chomsky called our competence, can be explained as having been "shaped" by our history of hearing, speaking, reward, and punishment. Grammar—at least the Chomskyan "universal grammar" portion of it—is a special case that I don't want to discuss here, because it seems to be complicated by a special condition called the "poverty of the stimulus" (Chomsky 1980), according to which the core grammatical rules are not learnable by trial and error and corrective feedback (i.e., reinforcement) based on the sounds the child hears and produces during the few years in which it learns language. That database is simply not rich enough for any inductive mechanism to learn the requisite rules on the basis of the data available and the time allotted to the child; hence the child must already have the rules built in, in advance.

But Chomsky's lesson to Skinner applies beyond syntax. Vocabulary learning—learning to call things by their names—already exceeds the scope of behaviorism, because naming is not mere rote association: Things are not stimuli, they are categories. Naming things is naming kinds (such as birds and chairs), not just associating responses to unique, identically recurring individual stimuli, as in paired associate learning. To learn to name kinds you first need to learn to identify them, to categorize them (Harnad 1996, 2005). And kinds cannot be identified by just rote-associating names to stimuli. The stimuli need to be processed; the invariant features of the kind must be somehow extracted from the irrelevant variation, and they must be learned, so that future stimuli originating from things of the same kind can be recognized and identified as such, and not confused with stimuli originating from things of a different kind. (Even "individuals" are not "stimuli," but likewise kinds, detected through their sensorimotor invariants; there are sensorimotor "constancies" to be detected even for a sphere, which almost never casts the identical shadow onto our sensory surfaces twice.)

So Chomsky already pointed out that it is not enough to say that learning words is just "verbal behavior," shaped by reward history. It is that too, but that much explains nothing. The question is: "How are we able to learn words, as shaped by our reward history? What is the underlying functional capacity?" (Chomsky called this the "competence" underlying our "performance.") The answer in the case of syntax had been that we don't really "learn" it at all; we are born with the rules of universal grammar already in our heads. In contrast, the answer in the case of vocabulary and categories is that we do learn the rules, but the problem is still to explain how we learn them: What has to be going on inside our heads that enables us to successfully learn, based on the experience or training we get, to identify categories, to which we can then attach a name?

Introspection won't tell us A misapplication of Wittgenstein (1953) (or perhaps a mistake of Wittgenstein's) is to conclude that if we cannot introspect the rules for categorizing things (today we would say "if their representation is not 'explicit'") then those rules do not exist. A more likely valid inference is that cognitive science cannot be done by introspection. If we are to explain our cognitive capacities, we must somehow come up with explicit hunches about how we are able to do what we can do, and then we have to test whether those hunches actually work: whether they can really delivery the behavioral goods. Our minds will have to come up with those hypotheses, as in every other scientific field, but it is unlikely that cognition will wear them on its sleeve, so that we can just sit in our armchairs, do the cognizing in question, and simply introspect how it is that we are doing it. In this respect, cognition is impenetrable to introspection (in a sense that is related to, but not quite the same as the sense that Zenon Pylyshyn [1980, 1999] had in mind with his "cognitive impenetrability" criterion—but I am getting ahead of myself).

One of the first candidate armchair theories of cognition was mental imagery theory: When we introspect, most of us are aware of images going on in our heads. (There are words too, but we will return to those later.) The imagery theorists stressed that, for example, the way I recall the name of my third-grade school-teacher is that I first picture her in my head, and then I name her, just as I would if I had seen her. Today, after three decades of having been enlightened on this score by Zenon Pylyshyn's celebrated "mind's eye" critique of mental imagery in 1973, it is hard even to imagine that anyone could ever have failed to see this answer—that the way I remember her name is by picturing her, and then identifying the picture— as having been anything but empty question-begging. How do I come up with her picture? How do I identify her picture? Those are the real functional answers we are missing; and it is no doubt because of the anosognosia—the "picture completion" effect that comes with all conscious cognition—that we don't notice what we are missing: We are unaware of our cognitive blind spots—and we are mostly cognitively blind.

It is now history how Zenon opened our eyes and minds to these cognitive blind spots and to how they help nonexplanations masquerade as explanations. First, he pointed out that the trouble with "picture in the mind" "just-so" stories is that they simply defer our explanatory debt: How did our brains find the right picture? And how did they identify whom it was a picture of? By reporting our introspection of what we are seeing and feeling while we are coming up with the right answer, we may (or may not) be correctly reporting the decorative accompaniments or correlates of our cognitive functions—but we are not explaining the functions themselves. Who found the picture? Who looked at it? Who recognized it? And how? I first asked how I do it, what is going on in my head; and the reply was just that a little man in my head (the homunculus) does it for me. But then what is going on in that little man's head?

Discharging the homunculus Imagery theory leaves a lot of explanatory debts to discharge, perhaps an infinite regress of them. Zenon suggested that the first thing we need to do is to discharge the homunculus. Stop answering the functional questions by just listing their decorative correlates, and explain the functions themselves. Originally, Zenon suggested that the genuine explanation has to be "propositional" (Pylyshyn 1973) but this soon evolved into "computational" (Pylyshyn 1984). If I ask you who your third-grade school-teacher was, your brain has to do a computation, a computation that is invisible and impenetrable to introspection. The computation is done by our heads implicitly, but successful cognitive theory must make it explicit, so it can be tested (computationally) to see whether it works. The decorative phenomenology that accompanies the real work that is being done implicitly is simply misleading us, lulling us in our anosognosic delusion, into thinking that we know what we are doing and how. In reality, we will only know how when the cognitive theorists figure it out and tell us.

There were other aspects to Zenon's insight too, not all of them quite correct, in my opinion. One was the implication that words and propositions were somehow more explanatory and free of homuncularity than images. But of course one could ask the same question about the origin and understanding of words in the head as of the origin and understanding of pictures in the head. Let's say that Zenon could have replied that words and propositions are nevertheless closer to computability than images, hence closer to an explicit, testable, computable functional explanation (Harnad 1982). Perhaps; but images can be "computed" too. Here Zenon would agree, but pointing out that a computation is a computation either way. He had famously argued that Shepard's mental rotation task (Shepard and Cooper 1986) could in principle be performed computationally using something like discrete Cartesian coordinates and formulas rather than anything like continuous analog rotation.

But at that point the debate became one about optimality (which of the two ways was the most general and economic way to do it?) and about actuality (which of the two ways does the brain in fact do it?) and not about possibility in principle, or necessity. It had to be admitted that the processes going on in the head that got the job done did not have to be computational after all; they could be dynamical. They simply had to do the job.

Zenon, in rightly resisting the functional question-begging of imagery theorists in favor of goods-delivering computational explanation, went a bit too far, first denying that noncomputational structures and processes could occur and explain at all, and then, when that proved untenable, denying that, if they did, they were "cognitive." Rightly impressed by the power of computation and of the Church–Turing thesis (Teuscher 2004)— that just about anything was computable, and hence computationally simulable to as close an approximation as one liked—Zenon relegated everything that was noncomputational to the "noncognitive." It occurred "below the level of the architecture of the virtual machine" that does the cognizing, implemented instead in "informationally encapsulated" sensorimotor modules that were "cognitively impenetrable"—that is, not modifiable by what we know and can state explicitly in propositions and computations (Pylyshyn 1984). The criterion for what was to count as cognitive was what could be modified by what we knew explicitly; what could not be modified in that way was "subcognitive," and the domain of another discipline.

(The similarity here to Skinner's dismissal of "how" questions as pertaining to physiology rather than psychology is ironic, but I don't think it is damning: Zenon, after all, was trying to make us face the problem of functional explanation, whereas Skinner was avoiding it. Moreover, both Pylyshyn and Skinner were right in insisting that the details of the physical [hardware] implementation of a function were independent of the functional level of explanation itself—except that Skinner had no functional explanation, whereas Pylyshyn had an all-powerful one: computation.)

Computational hegemony But Zenon was not the first to get a little too carried away with the power of computation. I think his attempt to formulate an impenetrable boundary between the cognitive and the noncognitive—in the form of his cognitive impenetrability criterion and the functional autonomy of the architecture of the virtual machine on which the computations were being performed—was not as successful as his rejection of imagery as nonexplanatory, his insistence on functional explanation itself, and his promotion of computation's pride of place in the explanatory armamentarium. Imagery was indeed nonexplanatory in and of itself. But an internal dynamical system that could actually generate some of our behavioral capacity (e.g., visual rotation judgments) certainly could not be denied out of hand; and if, as both brain imaging data (Kosslyn 1994) and considerations of functional optimality subsequently suggested, dynamical analog rotation-like processes in the brain really do occur, then there are certainly no grounds for denying them the status of being "cognitive."

A very similar point can be made about Zenon's celebrated paper with Jerry Fodor, which pointed out that neural nets were (1) uninteresting if they were just a hardware for implementing a symbol (computational) system, (2) irrelevant (like other dynamical systems) if they could be simulated computationally, and (3) subcognitive if they could be "trained" into becoming a symbol system (which then goes on to do the real work of cognition) (Fodor and Pylyshyn 1988).

So far, this sorting and labeling of functional "modules"—and I use the word "modules" here loosely, without any assumptions about the degree to which they are truly independent of one another functionally (Fodor and Pylyshyn 1988)—is more an exercise in semantics or taxonomy than anything substantive, functionally speaking. But there is one substantive issue on which I think Zenon has quietly conceded without ever quite announcing it, and that is symbol grounding—the ultimate question about the relation between the computational and the dynamical components of cognitive function (Harnad 1990).

Computation and consciousness But first, let us quickly get rid of another false start: Many, including Zenon, thought that the hardware–software distinction spelled hope not only for explaining cognition but for solving the mind–body problem: If the mind turns out to be computational, then not only do we explain how the minds works (once we figure out what computations it is doing and how) but we also explain that persistent problem we have always had (for which Descartes is not to blame) with understanding how mental states can be physical states: It turns out they are not physical states! They are computational states. And, as everyone knows, the computational "level"—the software that a machine is running—is independent of the dynamical physical level—the hardware of the machine on which the software is running. Not independent in the sense that the software does not need some hardware or other to run on, but in the sense that the physical details of the hardware are irrelevant for specifying what program is being computed. The same software can be run on countless, radically different kinds of hardware, yet the computational states are the same, when the same program is running. (I never thought there was much more to Zenon's "virtual machine" level than the hardware–software distinction, with a bit more interpretation.)

Well, this does not solve the mind–body problem, for many reasons, but here I will only point out that it does not solve the problem of the relation between computational and dynamical processes in cognition either: Computations need to be dynamically implemented in order to run and to do whatever they do, but that's not the only computational–dynamical relationship; and it's not the one we were looking for when we were asking about, for example, mental rotation.

Computation is rule-based symbol manipulation; the symbols are arbitrary in their shape (e.g., 0s and 1s) and the manipulation rules are syntactic, being based on the symbols' shapes, not their meanings. Yet a computation is only useful if it is semantically interpretable; indeed, as Fodor and Pylyshyn (1988) have been at pains to point out, systematic semantic interpretability ("systematicity"), indeed compositional semantics, in which most of the symbols themselves are individually interpretable and can be combined and recombined coherently and interpretably, like the words in a natural language—is the hallmark of a symbol system. But if symbols have meanings, yet their meanings are not in the symbol system itself, what is the connection between the symbols and what they mean?

Grounding the language of thought   Here it is useful to think of propositions again, Pylyshyn's original candidate, as the prototypes of Fodor's (1975) "language of thought": It is computation in both instances (propositions and the language of thought). The words in propositions are symbols.

What connects those symbols to their referents? What gives them meaning? In the case of a sentence in a book, such as "the cat is on the mat," there is no problem, because it is the mind of the writer or reader of the sentence that makes the connection between the word "cat" and the things in the world we happen to call "cats," and between the proposition "the cat is on the mat" and the circumstance in the world we happen to call "cats being on mats." Let us call that *mediated symbol-grounding*: The link between the symbol and its referent is made by the brain of the user. That's fine for logic, mathematics, and computer science, which merely use symbol systems. But it won't do for cognitive science, which must also explain what is going on in the head of the user; it doesn't work for the same reason that homuncular explanations do not work in cognitive explanation, leading instead to an endless homuncular regress. The buck must stop somewhere, and the homunculus must be discharged, replaced by a mindless, fully autonomous process.

Well, in Pylyshyn's computationalism, the only candidate autonomous internal function for discharging the homunculus is computation, and now we are asking whether that function is enough. Can cognition be just computation? The philosopher John Searle (1980) asked this question in his celebrated thought experiment. Let us agree (with Turing 1950) that "cognition is as cognition does"—or better, so we have a Chomskyan competence criterion rather than a mere behaviorist performance criterion—that "cognition is as cognition can do." The gist of the Turing test is that on the day we will have been able to put together a system that can do everything a human being can do, indistinguishably from the way a human being does it, we will have come up with at least one viable explanation of cognition.

Turing sets the agenda   Searle took Turing—as well as Pylyshyn—at their word. He said: Suppose we do come up with a computer program that can pass the Turing test (TT). Will we really have explained cognition? In particular, will the system that passes the TT really cognize, will it really have a mind? The classical TT is conducted by email (so you are not prejudiced by the way the candidate looks, which is irrelevant); it is basically a test—lifelong, if need be—of whether the system has the full performance capacity of a real pen pal, so much so that we would not be able to tell it apart from a real human pen pal. If it passes the test, then it really cognizes; in particular, it really understands all the emails you have been sending it across the years, and the ones it has been sending you in reply (Harnad 2007).

And of course, being implementation-independent, the winning software can be run on any hardware. If the TT-passing computational states are really the cognitive states, they will be the right cognitive states in every implementation. So Searle simply proposes to conduct the TT in Chinese (which he doesn't understand) and he proposes that he himself should become the implementing hardware, by memorizing all the symbol manipulation rules and executing them, on all the email inputs, generating all the email outputs. Searle's very simple point is that he could do this all without understanding a single word of Chinese. And since Searle himself is the entire computational system, there is no place else the understanding could be. So it's not there. The meanings are all just in the heads of the external users again—the real Chinese pen pals doing the Turing-testing. So the TT-passing program is no more cognitive than any other symbol system in logic, mathematics, or computer science. It is just a bunch of symbols that are systematically interpretable by us—by users with minds. It has again begged the question of how the mind actually does what it does—or rather, it has failed to answer it.

Newton still available   So what is still missing, then, if computation alone can always be shown to be noncognitive and hence insufficient, by arguments analogous to Searle's? Searle thought the culprit was not only the insufficiency of computation, but the insufficiency of the Turing test itself; he thought the only way out was to abandon both and turn instead to studying the dynamics of the brain. I think Searle, too, went too far (Harnad 2001). There is still scope for a fully functional explanation of cognition, just not a purely computational one. As we have seen, there are other candidate autonomous, nonhomuncular functions in addition to computation, namely, dynamical functions such as internal analogs of spatial or other sensorimotor dynamics—not propositions describing them nor computations simulating them, but the dynamic processes themselves, as in internal analog rotation; perhaps also real parallel distributed neural nets rather than just serial symbolic simulations of them.

The root of the problem is the symbol-grounding problem: How can the symbols in a symbol system be connected to the things in the world that they are ever-so-systematically interpretable as being about—connected directly and autonomously, without begging the question by having the connection mediated by that very human mind whose capacities and functioning we are trying to explain? For ungrounded symbol systems are just as open to homuncularity, infinite regress, and question-begging as subjective mental imagery is!

The only way to do this, in my view, is if cognitive science hunkers down and sets its mind and methods on scaling up to the Turing test, for all of our behavioral capacities. Not just the email version of the TT, based on computation alone, which has been shown to be insufficient by Searle, but the full robotic version of the TT, in which the symbolic capacities are grounded in sensorimotor capacities and the internal processes of the robot itself (Pylyshyn 1987) can mediate the connection, directly and autonomously, between its internal symbols and the external things its symbols are interpretable as being about, without the need for mediation by the minds of external interpreters.

We cannot prejudge what proportion of the TT-passing robot's internal structures and processes will be computational and what proportion dynamic. We can just be sure that they cannot all be computational, all the way down. As to which components of its internal structures and process we will choose to call "cognitive":

Does it really matter? And can't we wait till we get there to decide?[1]

Summary   Zenon Pylyshyn cast cognition's lot with computation, stretching the Church–Turing Thesis to its limit: We had no idea how the mind did anything, whereas we knew that computation could do just about everything. Doing it with images would be like doing it with mirrors, and little men in mirrors, so why not do it all with symbols and rules instead? Everything worthy of the name "cognition," anyway; not what was too thick for cognition to penetrate. It might even solve the mind–body problem if the soul, like software, were independent of its physical incarnation. It looked like we had the architecture of cognition virtually licked. Even neural nets could be either simulated or subsumed. But then came Searle, with his sino-spoiler thought experiment, showing that cognition cannot be all computation (though not, as Searle thought, that it cannot be computation at all). So if cognition has to be hybrid sensorimotor/symbolic, it turns out we've all just been haggling over the price, instead of delivering the goods, as Turing had originally proposed five decades earlier.

Note

1. One could ask whether grounded cognition ("sticky" cognition, in which symbols are connected to their referents, possibly along the lines of Pylyshyn's [1994] indexes or FINSTs) would still be computation at all: After all, the hallmark of classical computation (and of language itself) is that symbols are arbitrary, and that

computation is purely syntactic, with the symbols being related on the basis of their own arbitrary shapes, not the shapes of their referents.

References

Catania, A. C., and S. Harnad (eds.) (1988). *The Selection of Behavior. The Operant Behaviorism of BF Skinner: Comments and Consequences.* New York: Cambridge University Press.

Chomsky, N. (1959). A Review of B. F. Skinner's *Verbal Behavior. Language* 35: 26–58. http://cogprints.org/1148/.

Chomsky, N. (1980). Rules and representations. *Behavioral and Brain Sciences* 3: 1–61.

Fodor, J. A. (1975). *The Language of Thought.* New York: Thomas Y. Crowell.

Fodor, J. A. (1985). Precis of *The Modularity of Mind. Behavioral and Brain Sciences* 8: 1–42.

Fodor, J. A., and Z. W. Pylyshyn (1988). Connectionism and cognitive architecture: A critical appraisal. *Cognition* 28: 3–71.

Harnad, S. (1982). Neoconstructivism: A unifying constraint for the cognitive sciences. In *Language, Mind, and Brain*, ed. T. Simon and R. Scholes, 1–11. Hillsdale, N.J.: Lawrence Erlbaum. http://cogprints.org/0662.

Harnad, S. (1985). Hebb, D. O.—Father of Cognitive Psychobiology, 1904–1985. *Behavioral and Brain Sciences* 8: 765. (Obituary.) http://cogprints.org/1652/.

Harnad, S. (1990). The symbol grounding problem. *Physica* D 42: 335–346. http://cogprints.org/0615/.

Harnad, S. (1996). Experimental analysis of naming behavior cannot explain naming capacity. *Journal of the Experimental Analysis of Behavior* 65: 262–264. http://cogprints.org/1605/.

Harnad, S. (2001). What's wrong and right about Searle's Chinese Room argument? In *Essays on Searle's Chinese Room Argument*, ed. M. Bishop and J. Preston. Oxford: Oxford University Press. http://cogprints.org/1622/.

Harnad, S. (2005). To cognize is to categorize: Cognition is categorization. In *Handbook of Categorization*, ed. C. Lefebvre and H. Cohen. Amsterdam: Elsevier. http://eprints.ecs.soton.ac.uk/11725/.

Harnad, S. (2007). The annotation game: On Turing (1950) on computing, machinery, and intelligence. In *The Turing Test Sourcebook: Philosophical and Methodological Issues in the Quest for the Thinking Computer*, ed. R. Epstein and G. Peters. Dordrecht: Kluwer. http://cogprints.org/3322/.

Kosslyn, S. M. (1994). *Image and Brain: The Resolution of the Imagery Debate*. Cambridge, Mass.: MIT Press.

Kosslyn, S. M., S. Pinker, G. Smith, and S. P. Shwartz (1979). On the demystification of mental imagery. *Behavioral and Brain Sciences* 2: 535–548.

Pylyshyn, Z. W. (1973). What the mind's eye tells the mind's brain: A critique of mental imagery. *Psychological Bulletin* 80: 1–24.

Pylyshyn, Z. W. (1980). Computation and cognition: Issues in the foundations of cognitive science. *Behavioral and Brain Sciences* 3: 111–169.

Pylyshyn, Z. W. (1984). *Computation and Cognition*. Cambridge, Mass.: MIT Press.

Pylyshyn, Z. W. (ed.) (1987). *The Robot's Dilemma: The Frame Problem in Artificial Intelligence*. Norwood, N.J.: Ablex.

Pylyshyn, Z. W. (1994). Some primitive mechanisms of spatial attention. *Cognition* 50: 363–384.

Pylyshyn, Z. W. (1999). Is vision continuous with cognition? The case for cognitive impenetrability of visual perception. *Behavioral and Brain Sciences* 22: 341–364. http://www.bbsonline.org/documents/a/00/00/04/94/index.html.

Pylyshyn, Z. W. (2002). Mental imagery: In search of a theory. *Behavioral and Brain Sciences* 25: 157–182. http://www.bbsonline.org/documents/a/00/00/19/46/index .html.

Searle, John, R. (1980). Minds, brains, and programs. *Behavioral and Brain Sciences* 3: 417–457. http://www.cogsci.soton.ac.uk/bbs/Archive/bbs.searle2.html.

Shepard, R. N., and L. A. Cooper (1982). *Mental Images and Their Transformations*. Cambridge, Mass.: MIT Press.

Teuscher, C. (2004). *Alan Turing: Life and Legacy of a Great Thinker*. Dordrecht: Springer.

Turing, A. M. (1950). Computing machinery and intelligence. *Mind* 49: 433–460. http://cogprints.org/499/.

Wittgenstein, L. (1953) *Philosophical Investigations*. New York: Macmillan.

11 The Possibility of a Cognitive Architecture

Andrew Brook

What representation is cannot . . . be explained. It is one of the simple concepts that we necessarily must have.

—Kant, *Lectures on Logic* (trans. Michael Young, Cambridge University Press, 440)

Zenon Pylyshyn has made important contributions to a great many areas in cognitive science. One of them is cognitive architecture (hereafter CA). In fact, few people have paid more attention to CA than he has. Many researchers refer in passing to classical architectures, connectionist architectures, and so on, but Pylyshyn is one of few people who has come to grips with what CA is, what properties such a beast would have.

A number of arresting claims have been made about CA. Pylyshyn (1984, 1999) himself has claimed that it is that part of cognition which cannot be explained in terms of cognitive processes, and that CA is cognitively impenetrable, that is, that it cannot be influenced by beliefs or desires or other cognitive states and cannot be changed as a result of learning—it is the invariant framework of cognition, what persists through changes in cognitive contents. Fodor (1983, 2000) has claimed that we will never find a cognitive architecture for our central cognitive system. Against this claim, Anderson (2005) says that he has an existence-proof that Fodor is wrong, namely, the computational problem-solving system ACT-R. ACT-R does many things that central cognition does and clearly has a CA, so a CA for central cognition is not only possible, one actually exists. What are we to make of these claims?

A preliminary point. The term, "cognitive architecture," is used for two very different kinds of phenomena. (Multiple uses for a single term and other forms of terminological chaos abound in cognitive science.) By "cognitive architecture" some people mean the subsystems that make up a cognitive system: perception, reasoning, memory, and so on. Other people use the term to refer to the basic components out of which cognitive capacities and/or cognitive contents are built. Candidates include the compositional components of representations, "physical symbols," a system of weighted nodes, and so on. Yet others use the term to mean both things. A question that exemplifies the first use: "Must cognition have both procedural and declarative memory?" Some questions that exemplify the second: "Are representations the building blocks of cognition?" "Do representations all encode information in the same way or are there different encoding formats?"[1] As to the third option, people who talk about physical symbol systems and connectionist systems as two architectures have both notions in mind (perhaps only implicitly), since the two systems have both overall structure and basic building blocks.

In this chapter, I will focus on the building blocks notion of CA. The system structure issue has its interest, but the shape of the issue is fairly clear there, at least if described at the level of generality adopted above. Put another way, the conceptual situation with system structure is less vexed than is the building block issue.

What We'd Like a Story about a CA to Do

"Vexed? What's the problem? Obviously, representations are the building blocks of cognition. What more needs to be said?" Alas, a lot more. To begin with, there are some major problems about what representations are. For example, when I am conscious of the world and my own states in a single act of consciousness, how many representations do I have? I am representing many things, but that does not automatically determine how many representations I have. Indeed, there is a case for saying "just one," as James (1890) pointed out. If so, what about representations that are not part of current unified consciousness, something true of most memories, for example? Here "many" may be the better answer. And so on (Brook and Raymont, forthcoming, chapter 7). But we can set these problems aside. Even if we could develop a clear account of what representations are—how to individuate and count them and so on—representations still could not be the ultimate building blocks of cognition.

First, as Fodor put it in a pithy comment about propositions, "If they're real, they must really something else" (1990, 13). A single representation, no matter how "single" is cashed, is just too big a unit to be the basic unit of a science. In addition, the question, "What are representations made up of?," is just too interesting and too relevant to the question, "What is cognition like?," for representations to be treated as a rock-bottom primitive in cognitive science.

Second, we already know that radically different accounts can be given of the "architecture" of representations, for example, the account in terms of integrated bundles of information contained by a vehicle of some kind of classical cognitive science (and virtually all philosophy of mind until recently) and the account that holds that representations are distributed throughout the hidden nodes of a connectionist system. Since these differences of architecture are cognitively relevant—that is to say, would affect how the respective systems would perform cognitive tasks—they have to be viewed as part of cognition, and therefore on a plausible notion of CA (yet to come) as reflecting differences of CA.

Third and most important, representations have a number of kinds of component, each of which could have its own architecture. In particular, we can distinguish vehicles of representation—acts of seeing, acts of imagining, and so on—from the object or contents of representations—what is seen or imagined. We can also distinguish a third element, the apparatus for processing representations, the overall cognitive system. A fourth might be whatever it is in such a system that interprets representational content.

So if we could find the CA of the various elements of representation and representing just identified, we'd be done? Who knows?—we need to step back a step or two. What do we want a CA to do for us? And how could we recognize one if we saw it? Compare chemistry. In chemical reactions, the building blocks are atoms. Atoms are not the smallest units there are, not by orders of magnitude, but they are the smallest units that enter into chemical reactions. (Suppose that that is true, anyway.) Moreover, and more to the point, when they combine, chemical reactions ensue, in such a way that when we understand how atoms combine into molecules, interact with other atoms and molecules, and so forth, we see that the interactions in question must result in—or be—some larger-scale chemical phenomenon of interest.[2]

Atoms have other properties that we expect of building blocks. Their relationships to other atoms can be affected by chemical reactions, but they themselves are not thus affected. Moreover, atoms are universal across the domain of chemical reactions, so that all chemical reactions can be understood as combinations of and transformations in the relationships among atoms. In addition, the properties of atoms vis-à-vis other atoms are systematic and orderly (electron structure, atomic number, and so on). Finally, atoms have properties that allow us to systematically connect the properties that make them building blocks of chemistry to a science of a fine-grained structure the bits of which by themselves do not enter into chemical reactions, namely, particle physics.

We can now draw some morals about how we'd like a story about CA to go. We'd like it to be a story that:

1. identifies the units which when combined become but are not themselves cognitive processes;

2. in such a way that when we understand how these units interact, we see that the interactions must result in or be cognitive processes, and we understand what kind of cognitive processes they are;

3. where the units are universal across cognition or at least across one or more domains of cognition, so that all cognitive processes or all within a domain can be understand as combinations and transformations of them (plus, perhaps, other things);

4. if there are a number of kinds of such units, the various kinds systematically relate one to another; and,

5. all this helps us to explain how the physical system in which the units are located can "do" cognition—how these building blocks or more molar aspects of cognitive function are implemented in the brain. They should be points at which we can begin to bridge the gap between cognitive and noncognitive accounts.

A tall order!

Note that the CA that we seek is a functional architecture. We are interested in the functions that combine with one another and work together with other things to yield—or be—cognitive functions such as perceiving and thinking and imagining. I wish that I could now offer a nice, clean set of suggestions for what kind of functional unit might meet the five requirements just identified. Alas, I doubt that there is anything that will fill this five-part bill. But I may be able to clarify some of the context surrounding the issue. I will offer:

• a suggestion for a criterion by which we might recognize such functional units if we encountered them;

• some reasons why a couple of other suggestions along these lines won't work;

• some arguments claiming to show that Fodor's reasons for rejecting the possibility of a CA for central cognitive systems do not work; and,

• some reasons, I hope new reasons, for fearing that Fodor's conclusion is nonetheless right, that cognitive systems do not have a CA, not cognitive systems remotely like us at any rate.

Criteria for Being a CA

At least three suggestions, all due, more or less explicitly, to Pylyshyn (1984), have been made about what constitutes an item of CA, about the criteria for being a CA. They are related. Says Pylyshyn, we have got "down" to CA when:

1. The items in question remain unchanged, on the one hand through cognitive change (they are rearranged but are not changed), and on the other across implementations of cognition on different kinds of system.
2. The items in question are cognitively impenetrable. To be cognitively impenetrable is to be closed to influence by learning or by the semantic contents of beliefs or desires or other representations.
3. A cognitive account cannot be given of the items in question.

Start with (1). The first clause, unchanged through cognitive change, captures the idea that cognitive architecture is the framework or format of cognition, that within which cognitive change takes place. The second clause, across implementations, captures the idea of multiple realizability. On the standard functionalist picture of cognition, cognitive functioning can be "realized" in (can be done by) systems that otherwise have very different structures. Clocks can be realized in structures as different as sundials, hourglasses, weight-and-pulley systems, spring-and-escapement systems, the vibrations of a quartz crystal, and the wavelengths of certain elements. Like telling time, arithmetic, that is, encoding and manipulating numbers, can be done with pen-and-paper, a four-function calculator, an abacus, and so on. Similarly, thinking, remembering, perceiving, and so on could in principle be done by mammalian brains, silicon chips, quantum computers, and who knows what else. Or so many cognitive scientists claim.

There are problems with the first clause of (1). A great many things remain unchanged through cognitive change, the functioning of the glucose-distribution mechanism in the brain, for example. So the unchanged things must also be elements of cognition. Alas, the latter notion is just the notion "cognitive architecture" in different words, that is, the target of our search in the first place. But so restricted, would the remain-unchanged requirement at least be right? Not that I can see. So long as cognitive activity is built out of them, why should it matter whether the building blocks of cognition are (sometimes, under certain conditions) changed as a result of cognitive activity? (The cognitively impenetrable requirement, that is to say (2), would require—and explain—the remain-unchanged requirement; that may, indeed, be what gives (1) whatever attractiveness it has. See below.)

The second clause, about multiple realizability, does not work, either. Consider representations. For a host of reasons, it makes sense to say that serial computers, neural networks, and brains all have representations and can represent some of the same things using roughly the same forms of energy as inputs (i.e., in roughly the same sensible modalities). (I leave aside Searlean qualms about whether computers can really represent—it is enough for my purposes that all three function as though they represent in ways relevant to cognition.) Yet the three have very different architectures—binary encoding, encoding distributed across multiple nodes, and . . . who knows what, but neither of the above. Since how the representations are encoded is relevant to how they function cognitively in their respective systems, then even if kinds of representation of kinds of things can be constant across cognitive systems, their architectures need not be.

The attraction of (2), the idea of cognitive impenetrability, again lies in the distinction between the framework or structure of cognition, which it seems should remain unchanged as cognitive functioning develops and changes, and what learning and changes of belief and desire can change. (2) would be an explanation of why cognitive activity does not change CA. However, the explanation would be only as good as (2) itself. How is (2) supposed to work? Is cognitive impenetrability meant to be necessary for being a CA, sufficient for being a CA, or merely distinctive of CAs?

On a standard account of visual illusions, impenetrability could not be sufficient. The usual example illusion is the Müller-Lyer arrowhead illusion, in which two lines of the same length look to be of different lengths when one line has arrowheads pointing out, the other arrowheads pointing in. Such illusions are either completely impenetrable cognitively or can be penetrated only partially and incompletely. Even when subjects know that the lines are the same length, indeed even if they have measured them and have demonstrated to themselves that they are the same length, the lines will still appear to be of different lengths—and this will continue either forever or at least for many, many trials. Yet visual illusions are not part of CA on anybody's notion of CA. They may be a near result of something in visual CA but they are not themselves a part even of visual CA. (For an interesting exchange on visual illusions, see Churchland 1989, 255–279, and Fodor 1990, 253–263.)

If so, it would follow immediately that impenetrability is not distinctive of CA, either. A property P cannot be distinctive of a class of objects O unless the presence of P is sufficient for the presence of an O. The reverse is not true: P being sufficient for O is not sufficient for P being distinctive to Os. P may be sufficient for other things sometimes, too. But P must be sufficient for Os if P is distinctive to Os.

That leaves necessity. Is impenetrability at least necessary for something to be part of CA? The answer is less clear but it is far from clearly "yes." Consider an analogy. It is plausible to suggest that within the narrow domain of logic, the CA is symbols and rules, and within the narrow domain of arithmetic, the CA is numbers and rules. (We leave open the question of whether some more universal CA underlies these specific CAs.) Yet how rules in the two domains are applied can clearly be influenced cognitively. How they are applied can be influenced by false beliefs about how the rule applies and training in an incorrect version of the rules, that is, learning. Other influences such as tiredness, cognitive overload, distraction, and so on, can affect the application of such rules but they are not likely to count as cognitive. Clearly the two just mentioned do. The same is true of how symbols and numbers are perceived, and perception is cognitive if anything is. Why should the same not be true of other CAs?

One can think of at least two ways in which the impenetrability test might be patched up. (a) Rather than being utterly impenetrable, perhaps the test should merely be that a CA is something that is not usually penetrated. Or (b), rather than being utterly impenetrable, perhaps a CA is not cognitively penetrated when the system is functioning properly—as designed, in such a way as to be truth-preserving, or whatever. ((b) is just (a) with a normative spin.)

There is something to these revisions to the impenetrability criterion. Part of the CA of a computer (CA of the building-block kind that we are examining) is how inputs and instructions are formatted. Such a format would not usually be penetrated by the operation of the computer. It could be but it would not be. And any such penetration would probably not be compatible with the computer operating as designed, not for very long anyway.

Impenetrability faces another problem, however—it does not give us much guidance with respect to finding CA. What feature of a CA would impenetrability be reflecting? Why should mere penetrability rule out having such a feature? If something is cognitively impenetrable, perhaps that would be enough to show that it is noncognitive. But being noncognitive is not enough by itself to ensure that something is an element of CA.

So what about (3), Pylyshyn's suggestion that we have got "down" to CA when we cannot give a cognitive account of the units in question?

What do we, or should we, mean by "a cognitive account" here? Pylyshyn seems to have in mind an account in terms of beliefs and desires and similar things—in terms of a person's reasons for what they think and feel and do. I think that this proposal is too narrow; it would leave us with an implausibly large list of items for CA. Why? Because there are many influences plausibly thought to be cognitive that involve little or no influence of beliefs, desires, and the like. Our tendency to complete word-stems in line with a stimulus that we have just encountered even if we have no awareness of having encountered it is an example. This is a semantic, therefore on a plausible notion of "cognitive," a cognitive influence, but no belief or desire is involved. The process at work is semantic implication or semantic association or something along those lines.

In response, let us broaden Pylyshyn's suggestion to include accounts in terms of semantic relationships of all kinds. In addition to providing reasons for belief or action, Pylyshyn's preferred kind, semantic relationships include semantic implication and semantic association, and perhaps others. What ties the group together is that they are all either true–false or truth-preserving, or evaluable using these concepts.[3] His thought would then become something like this. Even though a cognitive system is what brings semantic relationships into being, we will not be able to give a semantic account of the behavior of the elements out of which that system is built. On this test, CA would be elements out of which semantic systems are built that are not themselves semantically evaluable (nor, therefore, cognitive, by the concept of the cognitive we have just developed). This strikes me as fine as far as it goes. Finally!

Central Cognitive Processes and CA

If cognitive accounts are accounts in terms of semantic (not just syntactic, synaptic, or whatever) relations, then CA would be the architecture of the semantic. Interestingly, this is exactly what Pylyshyn says (1999, 4). The content of representations is semantic content: It can be true or false, accurate or inaccurate, its lexical and propositional meaning has implications for what the content of other representations should be, and so on. One way to pin down what is meant by "semantic content" here is to say that the semantic content of a mental representation (MR) or other representation consists in the information in the MR (or . . .) about some object, event, or state of affairs that the MR would cart around with it even in the absence of the object, event, or state of affairs. If this content–vehicle distinction is sound, an account of the semantic must contain two parts: one about semantic content, what is seen, heard or imagined, and one about semantic vehicles. And our question now becomes, Is there an architecture for the semantic?

Let us start with a challenge to the very idea that cognition, or at least the brain, could deal with semantic relationships. Both Fodor (1994) and Dennett (1987) have urged that the brain, and therefore presumably the cognitive system, cannot process semantic content, only "syntactic" structures of various kinds. (By "syntactic," they don't mean syntax. They mean "physically salient," i.e., detectable by something that can detect only properties such as shape, spacing, order, and so forth.) As Dennett has put it (1987, 61), a semantic engine is impossible, so the most that a brain can do is to mimic this impossible thing. (Mimic? What is it to mimic something that is impossible? But we will let that pass.)

Even though some good philosophers have advanced this claim—Dretske also seems to accept or at least to presuppose it—not much by way of an argument is ever given for it. This is strange. On the face of it, the claim seems to be flatly false: we seem to process semantic information directly and effortlessly, indeed to do so every waking minute. If all that computational theories can account for is the processing of physically salient properties, so much the worse for such theories—but maybe what we really need is a less myopic picture of computation in the brain.

Even the philosophers who urge that the brain cannot *process* semantic content allow that the brain is *aware* of semantic content. But that requires that it be able to detect it. If so, skipping the details and cutting to the chase, it seems to me that what Fodor et al. should have said is that we can process semantic content only via the processing of "syntactic" information, that semantic content must be built out of the "syntactic" (i.e., the nonsemantic) in some way—in which case, syntactic structures would not mimic semantic structure, syntactic structures would make up semantic structure. And the job of finding a CA would be precisely the job of finding these semantic-composing but themselves nonsemantic elements. Put another way, it is silly to deny that we are aware of semantic content—but it would be far from silly to say that the content that we thus know is built out of the nonsemantic.

There are some well-known arguments, due to Fodor, that whatever brain processes can do, the processing activity involved could not have a CA.

As we saw earlier, more than one kind of function is involved in representing. The same is true of the semantic. There is semantic content—what sentences say, the content of representations, and so on. There are vehicles for semantic content—representations, sentences. There are mechanisms for processing semantic content—roughly, perceptual, memory, and reasoning abilities. As we said earlier in connection with representation in general, there may also be whatever it is in a cognitive system that interprets semantic content. In the terms of this scheme, the arguments of Fodor's that we will now consider focus on the mechanisms for processing semantic content.

He mounts two arguments that there could be no general, perhaps simply no, CA of processing semantic content. One is found in Fodor 1983, the other in Fodor 2000. (Whether Fodor mounted the arguments in the service of this implication or even meant them to have this implication does not matter.) The argument in Fodor 1983 starts from the idea that our central cognitive system is what he calls isotropic and Quinean. "Isotropic" means that information housed anywhere in the system could in principle increase or decrease the degree of confirmation of any belief in the system. "Quinean" means that if a belief comes into conflict with something that would tend, ceteris paribus, to disconfirm the belief, then ceteris need not remain paribus. To the contrary, there will always, or if not always certainly generally, be adjustments to be made elsewhere in the system, indeed all over the system, that would reduce or remove the tension.

Why do these observations entail the impossibility of a CA? The first move is to urge that we cannot even imagine a computational system able to handle the computational load that would be entailed by these two features of the central system. Even a heuristic account could not do the job; it could not explain how we know which heuristic to apply. (This latter point is perhaps made more clearly in Fodor 2000, 42, than in 1983.) If so, the central system could not be computational, not by any notion of the computational that we now have or can imagine. The second move is to urge that computational cognitive science is not only all the cognitive science we have; it is all the cognitive science that we have any notion of how to formulate. From this it would follow that we have no notion of how the central system is built or what a CA of such a cognitive system could be like.

But is the case so clear? The mathematics for determining probabilities is, or certainly seems to be, isotropic and Quinean. Indeed, probabilistic reasoning is a large part of central processing in general. Yet a strong case can be made for saying that, here as elsewhere, the mathematics has an architecture: numbers and variables, formation rules, transformation rules, and so on. The fact, if it is a fact, that central cognition is holistic in the two ways that Fodor identifies does not seem to be enough by itself to rule out the possibility of a CA.

So what about the argument in Fodor 2000? It is considerably more subtle (also harder to figure out). The first move is to argue that the syntax of an MR is among its essential properties (all claims by Fodor discussed in this paragraph are taken from chapter 2). If only syntactic properties of an MR can enter into causal cognitive processes, it would follow that only essential properties of an MR can enter into causal cognitive processes. And from this, says Fodor, it would follow that cognitive processes are insensitive to "context-dependent" properties of MRs.

The last statement, however, is not true, says Fodor. As conservatism about belief-change and appeals to simplicity show, we often use widely holistic properties of cognition in the process of determining which cognitive move to make. Which theory counts as simplest, which belief change counts as most conservative, will hinge on what other theories we accept. If so, the role of an MR can change from context to context. But the role of the syntax of an MR cannot change from context to context. If so, the cognitive role of an MR does not supervene on the syntax of that MR. Since the computational theory of the mind (CTM) models cognitive processes as causal relationships among syntactic elements, CTM cannot account for context-sensitive properties of syntax. So CTM is in trouble. Furthermore, since CTM is the only halfway-worked-out model of cognition that we have, we have no idea how to model context-dependent properties of cognition. If so, once again we have no notion of how the central system is built or what a CA of such a cognitive system could be like. QED.

Life is not so gloomy, thank goodness. Fodor twists his way through a couple of moves to try to save CTM, for example, building the whole of the surrounding theory into each MR (so that the essential syntax of the MR is the syntax of the whole theory) and concludes that such moves are hopeless. Who could disagree? What he didn't notice, however, is that there is a problem in his very first move. As we have seen, the term "syntax" as used in recent analytic philosophy is distressingly ambiguous. Sometimes it means, well, *syntax*—the kind of thing that Chomsky talks about. But sometimes it means all the physically salient properties of MRs and other representations. Unfortunately for Fodor, it is at most *syntax*, the real thing, that could be essential to MRs. There are lots of other physically salient properties—order, shape, and spacing, for example—that are not essential to MRs, that are context sensitive, and that computational processes could detect and make use of, including all the relationships of each MR to other MRs.

To get his gloomy result, Fodor would have to show that cognitive, context-sensitive properties could not supervene on any physically salient properties, not just syntactic ones, properly so-called. It is not clear how this could be done. Certainly he has not done it.

So where are we? At a bit of an impasse. On the one hand, Fodor's arguments, which are the leading arguments against the possibility of a CA, do not work. On the other hand, we do not have the foggiest notion how to think about CA.

Cognitive Architecture of What?

Let us return to the idea that a CA tells us what semantic content and processing is made of. If we want to know what a CA of the semantic might be like, the first thing we would need to know on this view is: What is the semantic like? Even Fodor would allow that our MRs have semantic properties, and that we often know them, even if CTM cannot account for them, indeed, even if, as Dennett (1987) maintains, the brain cannot process them, only syntactic analogues of them. So what has to be added to physically salient structures for them to take on semantic properties?

Fodor's own answer is: causal or nomological links to things in the world. Even without going into the details, it is clear that this approach faces major obstacles. It offers no account of semantic entailment. It has a huge problem accounting for nonreferring terms (e.g., "Santa Claus") (Scott 2002). It has next door to nothing to say about terms whose semantic role is other than to introduce things, events, or states of affair into representations (articles, etc.). And it has even less to say about semantic vehicles.

Semantic vehicles are the vehicles that provide our means of access to semantic content. Thus, an act of seeing is our means of access to the bluebird seen, an act of hearing is our means of access to the melody heard, an act of imagination is our means of access to a warm, sunny beach imagined.

Indeed, if our earlier suggestion about the structure of the semantic is right, a complete account would have to contain two more parts. The third would concern semantic processing—the cognitive equipment that processes the content in a given vehicle and relates it to other contents and vehicles, in memory, beliefs, and motives in particular.

A fourth would be this. The semantic content of an MR—what it represents with respect to what it is about—is not intrinsic to the representational vehicle in question. MRs not only represent something, they represent it to someone. And what they represent to the person who has them is a function, in part, of what else is going on in that cognitive system and in particular of how the cognitive system interprets the "raw materials" that have made their way into the vehicle in question. Perhaps the biggest failing in Fodor's account is that he has nothing to say about this aspect of the semantic.

There is a lot of hand-waving in the remarks I just made. Let me be a bit more specific. On the issue of what this interpretive activity might be like, Wittgenstein and Davidson are interesting guides. (My saying this may surprise some because Wittgenstein is often thought to be unremittingly hostile to cognitive science, and Davidson only slightly less so.) The kind of representing that most interested Wittgenstein (1953) is the kind that requires the mastery of a "rule"—a capacity to judge that something is something or to attribute properties to something. Wittgenstein argues that such mastery cannot consist, first of all, of behavior or being disposed to behave in any way, nor of a mental mechanism of any sort. The reasons are the same: Mastering a "rule" gives us the capacity to judge an infinite number of present, future, and merely possible cases, and a rule can be used correctly or incorrectly. No behavior, disposition, or mechanism could provide either by itself. However, mastery of a "rule" cannot consist in learning and then interpreting in accord with a proposition either. First, to apply the proposition, one would first have to interpret it, that is, figure out what it implies for the case in hand, which puts one on the edge of an infinite regress. Second, it will always be possible to think of yet-unencountered cases about which what to say is "intuitively" clear to us but about which the rule is silent.[4] So what does mastery of a rule, a judgmental capacity, consist in?

Just what was Wittgenstein's answer is, to say the least, controversial, but here is one reading that fits the texts fairly well. We have mastered a "rule," can apply it correctly, when the way we judge cases using it accords with how others do so. It is others' agreeing with us that makes our usage correct. This holds for all of present, future, and merely possible cases. (Once we have a base of agreement, we can then use the capacities thereby attained to generate new judgments. About these there need not be agreement.)

Davidson (2001) fills out this idea of intersubjective agreement in an interesting way. He introduces the idea of triangulation and urges that, not just to have grasped a rule but to have judgmental capacities at all, there must be triangulation. What he has in mind is that to have judgmental capacities, two or more organisms must be able find in others' and their own behavior a pattern of making roughly the same judgments about objects, events, and states of affairs in shared perceptual environments.

Implication? If there is anything to Wittgenstein's and Davidson's story, there is not going to be any mapping between semantic content and physically salient structures of any kind in the brain—vehicles, circuits, or anything else, of any kind short of insanely complicated, anyway. If so, the prospects of a CA of semantic content are bad. Worse is ahead.

Externalism

Many philosophers now believe that some essential element of representational content consists of something outside the representation, some relationship between the representational vehicle and something. This is *externalism* about representational content, the view that at least part of what is required for something to represent, to be about something, is external to the act of representing. The content of representations consists in part of some relationship between the representational vehicle and something else. (Fodor's view was already an example of this.) Hilary Putnam (1975, 227) famously said, "meanings just ain't in the head." Note that there are two issues here, whether meaning extends beyond the head (an issue to which we will return) and whether it extends beyond individual representations (but not necessarily the head). All forms of externalism maintain at least the latter.

There have been quite a few candidates for what the external element is. For Dretske (1995), it is the *function* of a symbol, symbol-stream, instrument, or the like, that is external to the representational vehicle in question. When I imagine something, while all the information about what I imagine is or can be carried by the representational vehicle (Dretske 1995, 103–104, 114), what determines what the representation has the function of representing is not carried by the representation. For Burge, the content of a concept and therefore of any representation using the concept is in part what is known or believed about the concept in one's society, not just by oneself (Burge 1979). For Putnam, the content of a concept and therefore of any representation using it is in part determined by the real nature of the kind of thing to which the concept refers, not just by what a given user of the concept happens to know (Putnam 1975). For Fodor, as we saw, the representational vehicle has to be in a causal or nomological (he goes both ways) relationship to something. And so on. Whatever the story, meanings extend beyond representational vehicles and many of them extend beyond the head. (We introduced this distinction just above.)

To be sure, not everything about representational content could be external to the representation. Here is one of the ways in which Dretske, for example, tries to strike the right balance. On the one hand, clearly there is something about representation that is internal to the cognitive system that has the representation. Dretske captures that this way:

When I close my eyes, I cease to see [the world around me]. The world does not vanish but something ceases to exist when I close my eyes. And this something has to be in me. (Dretske 1995, 36)

Furthermore, it is plausible to think, the contents of beliefs and desires have to be internal to me at least in part to explain the effects that they have on my behavior. I go to the fridge because I want something to drink and remember there being juice there. If the contents of the desire and memory were not internal to me, how could they—and why would they—affect my behavior as they do? Other phenomena are relevant, too. When I move to a new location, most of my external links change. Yet my mental contents move with me and most of them do not change: My thoughts, imaginings, emotions, memories, and so on continue to have the same contents. Even in a sensory deprivation chamber, I would still have thoughts and feelings, meanings and beliefs, even though I am cut off from all current causal contact with everything outside the tank. And so on. There is something important about representation and consciousness that is entirely "in the head."[5]

On the other hand, paying attention to how a representation is implemented in the brain is not likely to tell you either what you are representing or how you are representing it. Dretske uses analogies to argue for this. Staring at the face of a gauge is not the way to discover what information it provides or how it provides it (1995, 109). Peering at a meaningful symbol is not the way to discover its meaning or how it provides this meaning. Looking at neurons will not tell us what they do. (Neurons are all pretty much alike, or at any rate fall into a small handful of different kinds.) If so, why should we expect to learn what a psychological state represents by "peering" at it—even from the inside?[6]

What something represents, what it has the function of indicating, are not facts one can discover by looking in the cortex, at the representation. One could as well hope to figure out what, if anything, a voltage difference between two electrical contacts in a computer means (represents, signifies) by taking ever more exact measurements of the voltage. (Dretske 1995, 37)

Dennett (1978, 47) mounted a similar argument many years ago. Suppose that we devised a brain scanner with sufficient "resolution" to find the sentence, "America is the world's greatest country!" written in Hans's brain. We couldn't tell, merely from discovering the sentence, whether Hans is jingoistic, ridiculing the idea, using the sentence as an example of a proposition in a philosophy class, just liked the sounds of the words, or what.

To illustrate the point, consider an intention to refer to something. We are able to pick out and refer to particular examples of a kind of thing and distinguish them from even qualitatively identical other examples of this kind. No representation by itself contains anything that we could use to achieve such reference, or so it is plausible to claim. Rather, what accomplishes singular reference, as Perner (1991, 20) calls it, is that one intends to refer to that object and no other. We could not pick out referential intentions by gazing at the contents of representations.

Here is how, for Dretske, representing something requires something outside the representational vehicle. A thermometer has the function of representing, and so does represent, temperature, even though its indicating activities might correlate with other factors, level of charge in a battery or behavior of electrons in silicon, for example. Function is assigned by something outside a representation. The same is true for becoming a representation. Whether the indicator being at a certain place means so many degrees Fahrenheit or centigrade or . . . or . . . is determined by factors external to the thermometer, the factors that allow us to interpret marks and squiggles on it as indicators of particular temperatures.

On the externalist picture, then, representational content is fixed not just by the structure of the representational vehicle and the information entering it but also by complex relationships between that vehicle and the rest of the cognitive system. Beliefs about functional assignments, beliefs about causal and/or nomological connections, perhaps the functions and connections themselves, and who knows what else will enter into fixing representational content (not to mention good old-fashioned interpretation and intersubjective agreement, as we saw in the previous section).

If externalism of any form is true, the prospects for a CA of semantic content dim still further. Various instances of representational content consist of so many kinds of relationship between representational vehicle and what is around it, some vehicle–brain, some vehicle–world, that it is unlikely that the various contents will have many common constituents, will be built out of any single CA. Indeed, some of the relationships claimed to be relevant don't have any obvious simpler components at all—function assignment, for example. In addition, as Dretske puts it, even if representational content is globally supervenient on, for example, neural activity, so that for every change in representational content there will be some neuronal change, representational content will not be locally supervenient. That is to say, the contents of individual representings cannot be mapped onto specific, localized circuitry—which dims the prospects of them consisting of a single set of components even further.

Objection: "Didn't you say that not just representational vehicles but also representational content could persist through the closing of eyes and moves to new locations? If so, something crucial to representational content can be entirely inside the head—and the idea of CA for representational content would appear to get a new lease on life." Not a very big one, I am afraid. Distinguish:

1. Something being external to a representation

from

2. Something being external to the head (the person, subject, self, mind, consciousness)

The external element postulated by many externalists is something external only to representations, not to the head. Dretske is an example; function and even knowledge of function would be external to a representation, on his account, but they are not or certainly need not be external to the person who has the representation. If true, this would explain how we can cart representational content around with us even with our eyes closed.

What might be meant by saying that something external to a representation is still internal to a cognitive system? In my view, information is internal to a cognitive system when it is cognitively available to the system. Information is cognitively available to a system when the system can use it to structure cognitive activity: perceiving, thinking, remembering, emoting, and so on. (Just existing somewhere in the system in the way in which sentences exist in a book wouldn't be enough.) There are different ways in which a system can use information to structure cognitive activity, running from completely automatically without the system having any consciousness of what is going on, in the way that syntactic rules structure the parsing of sounds. to being known consciously and applied deliberately. On all of these alternatives, the information is internal to the system.

The external elements postulated by many forms of externalism are available to the system in this way. On Dretske's brand of externalism, for example, what is external to a representation is its function: what information it has the function of representing. The cognitive system not only has this information but has to have it to know (or even to have beliefs about) what information is being represented by this, that, or the other representational vehicle. And it does have it. Just as we grasp what the indicating function of a gauge is, we grasp what the indicating function of a perception is. We may not be conscious of what this function is, we may not be able to describe it, but we do grasp what it is—what a given perception represents and when it is misrepresenting.[7] If so, then the element external to representations in Dretske's picture can still be, indeed often must be, internal to the cognitive system that has the representations.

Notice that the same is true on Fodor's account. For Fodor, the external element is a causal or nomological link between symbol and referent-type. No matter. To use the symbol, the system must grasp (in the special sense we have been using, which does not imply being able to articulate) the kind of thing to which the symbol refers. Thus, on Fodor's account, it must grasp to what the symbol is linked. The same is true of Burge's social externalism. In general, many of the elements said to be external to representations will nonetheless be internal to the cognitive system that has the representations.

What about the externalists who deny that the element external to representations is graspable by the cognitive system that has them— Putnam (1975), for example? Here is how his story goes. Suppose that Adam and his twin on twin earth, Twadam, have beliefs about a certain clear liquid in front of them and both call it "water." One liquid is H_2O and one is XYZ. They have beliefs about different things and so, Putnam concludes, have different beliefs. Here the element of content external to the beliefs is not graspable by the cognitive system and so is external to the system in every way. Is this a problem for our story about how, even in the face of externalism, representational content can be internal to the cognitive system whose representation it is?

No. It is far from clear that the external element here does affect the content of representations. They both believe "This [the substance in front of them] is water." However, if they don't also believe, "This is water-rather-than-twin-water" (in their respective idiolects), then their concept of water in each case may well be broad enough to range over both substances. If so, their beliefs, representations, and conscious states would have the same content. (Brook and Stainton [1997] reach the same conclusion in a different way.)

Anyway, if some element of content is external to the cognitive system, that element could not be made up of anything in a system's CA, so it would be of no concern to us.

That on many accounts the element external to representational vehicles is nonetheless graspable by, and therefore in some way internal to, the cognitive system as a whole does not improve the prospects for there being a CA, unfortunately.[8] Why? For reasons we have already seen. To fix representational content, elements widely dispersed across a cognitive system would still be needed. This would make it unlikely that various instances of content have a common structure or are locally supervenient on anything that does.

If there is anything to the story that we have told in the previous section and this one, the prospects for representational content having a CA are bad. What about the prospects for a CA for representational vehicles, acts of representing? Here the prospects may be better—but the issue is also less important. How a given content is represented is much less important cognitively than what the content contains, is about. Why? Because whether something is seen or heard matters less for belief, memory, planning, action, and so on than the content of what is seen or heard. Moreover, since all acts of seeing, for example, are much alike, the nature of the representational vehicle won't usually have much influence on differences of representational or semantic content. Where representational modality does matter, as it does for example over whether something is seen or imagined, that is usually because the difference is an indicator of something else, in this case the status of the content as information about the world—and this status can be determined independently.

With these brief comments, I will leave the topic of a CA for representational vehicles. What about the fourth element, the thing that does the interpreting, enters into the intersubjective agreements, and so on—what used to be called the *subject of representation*? Given the centrality of interpretation in most current accounts of semantic or representational content and of triangulation or a related notion in not a few, it is remarkable how little attention this topic has received. Whether an interpreter/agent/subject has a CA has received even less. Could this entity have a CA? The whole issue has been so little explored that it is hard to know. There is a major piece of work to be done here.

Concluding Remarks: Prospects for Cognitive Science

Many theorists have suggested that if Dretske and Wittgenstein and Davidson are right, if content has elements that are external to representational vehicles, is a result of interpretation, has an ineliminable intersubjective element, and so on, then not just the prospects for a CA but the prospects for any science of cognition at all are dim. I do not think that this bigger conclusion is warranted.

There are more kinds of science than building models Lego-style. There is also correlational science. As recent brain-imaging work has shown, this kind of science can be exciting and revealing even without a story about building blocks and how they combine to yield target phenomena. A second kind of science flows from the point that Pylyshyn has repeatedly made (e.g., in Pylyshyn 1984 and 1999) that the only way to capture many of the similarities in cognitive function of interest to us across differences of implementation is to use the proprietary language of cognition, the language of representation, belief, desire, and so on. But phenomena thus described can be explored, models can be built, computational simulations can be created and tested, and so on with no story about CA. To return to where we began, even if Fodor is right that individual representations are too "big" a unit to be the ultimate constituents of a science of cognition, we can do a great deal of interesting and important work by treating them as though they are.

Notes

1. The latter question was one of the issues at stake in the mental imagery wars of the 1980s and 1990s in which Pylyshyn was a prominent participant.

2. I borrow this idea of mechanisms that must yield a target phenomenon from Levine's (1983) account of how we close explanatory gaps, gaps in our understanding of how and why something happens. For interesting suggestions about the kind of mechanisms that we should seek here, see Bechtel 2005.

3. Some would argue that there is more to semantic evaluation than being true–false or truth-preserving, including more kinds of satisfaction-conditions. The issue is not important here.

4. Drawn heavily from Mark Macleod, unpublished, with thanks.

5. It is not often noticed that the situations just described would not break the link to the external element on all forms of externalism. Causal links would be broken but nomological links would not. Social links would be broken but links to social practices would not. Functional links might be broken but function-assigning histories would not.

6. Compare this remark by Wittgenstein (1967, §612): "What I called jottings would not be a *rendering* of the text, not so to speak a translation with another symbolism. The text would not be *stored up* in the jottings. And why should it be stored up in our nervous system?"

7. Chomsky's (1980) way of putting the point that I am trying to make here is to say that one *cognizes* the function.

8. This external–internal mix does help in other places. It can be used to show that externalism is no threat to the view that consciousness is a kind of representation, for example (Brook and Raymont, forthcoming, ch. 4).

References

Anderson, J. (2005). The modular organization of the mind. Talk presented at Carleton University October 13, 2005.

Bechtel, Wm. (2005). Mental mechanisms: What are the operations? *Proceedings of the 27th Annual Conference of the Cognitive Science Society*, 208–201.

Brook, A. and P. Raymont (forthcoming). *A Unified Theory of Consciousness*. Cambridge, Mass.: MIT Press/A Bradford Book.

Brook, A., and R. Stainton (1997) Fodor's new theory of content and computation. *Mind and Language* 12: 459–474.

Burge, T. (1979). Individualism and the mental. *Midwest Studies in Philosophy* 4: 73–121

Chomsky, N. (1980). *Rules and Representations*. New York: Columbia University Press.

Churchland, P. M. (1989). *A Neurocomputational Perspective*. Cambridge, Mass.: MIT Press/A Bradford Book.

Davidson, D. (2001). *Subjective, Intersubjective, Objective*. Oxford: Clarendon Press.

Dennett, D. C. (1978). Brain writing and mind reading. In *Brainstorms*, 39–52. Montgomery, Vermont: Bradford Books.

Dennett, D. C. (1987). Three kinds of intentional psychology. In *The Intentional Stance*. Cambridge, Mass.: MIT Press/A Bradford Book.

Dretske, F. (1995). *Naturalizing the Mind*. Cambridge, Mass.: MIT Press/A Bradford Book.

Fodor, J. (1983). *Modularity*. Cambridge, Mass.: MIT Press/A Bradford Book.

Fodor, J. (1987). *Psychosemantics*. Cambridge, Mass.: MIT Press/A Bradford Book.

Fodor, J. (1990). *A Theory of Content and Other Essays*. Cambridge, Mass.: MIT Press/A Bradford Book.

Fodor, J. (1994). *The Elm and the Expert*. Cambridge, Mass.: MIT Press/A Bradford Book.

Fodor, J. (2000). *The Mind Doesn't Work That Way*. Cambridge, Mass.: MIT Press/A Bradford Book.

James, W. (1890). *Principles of Psychology*, vol. 1. London: Macmillan.

Levine, J. (1983). Materialism and qualia: The explanatory gap. *Pacific Philosophical Quarterly* 64: 354–361.

Macleod, M. (unpublished). Rules and norms: What can cognitive science tell us about meaning. Talk presented at Carleton University, November 24, 2005.

Perner, J. (1991). *Understanding the Representational Mind*. Cambridge, Mass.: MIT Press.

Putnam. H. (1975). The meaning of "meaning." In his *Mind, Language and Reality: Philosophical Papers*, vol. 2, 215–271. Cambridge: Cambridge University Press.

Pylyshyn, Z. W. (1984). *Computation and Cognition: Toward a Foundation for Cognitive Science*. Cambridge, Mass.: MIT Press/A Bradford Book.

Pylyshyn, Z. W. (1999). What's in your mind. In *What Is Cognitive Science?* ed. E. Lapore and Z. W. Pylyshyn. Oxford: Blackwell.

Scott, S. (2002). *Non-Referring Concepts*. PhD Dissertation, Institute of Cognitive Science, Carleton University, Ottawa, Canada

Wittgenstein, L. (1953). *Philosophical Investigations*. Oxford: Blackwell.

Wittgenstein, L. (1967). *Zettel*. Oxford: Blackwell.

12 Location, Location, Location

Austen Clark

1 Imagery, Round One

To understand Pylyshyn on perception, it is useful, and perhaps essential, first to understand his contributions on what might seem to be a distinct topic: mental imagery. The 1980s imagery debate was a portentous one for mental pictures, and Pylyshyn played a decisive role in it. Many of his recent (2001, 2003) arguments about the architecture of visual perception, and against "location-based" models, show a striking and admirable continuity with those earlier arguments about the forms of representation implicated in mental imagery. As he puts it near the beginning of his recent book:

we must dispense with the "picture in the head" . . . we must also revise our ideas concerning the nature of the mechanisms involved in vision and concerning the nature of the internal informational states corresponding to percepts or images. (Pylyshyn 2003, 3)

In the imagery debate we had bad inferences from experimental data to claims for a distinct, pictorial form of representation. Some of those same patterns of inference are found as well in the "objects versus locations" debate in visual perception.

What is the bad pattern of inference? The fundamental issue is: Do any available experimental results entitle us to believe that subjects in imagery tasks use a form of representation that is distinct in kind from the forms used in linguistic tasks? Do they provide *any reason at all* to think this? Pylyshyn says, forthrightly and firmly, "no." The question is whether results establish use of a distinct *form* of representation: of a "pictorial" or "depictive" form, as opposed to a "propositional" variety. To do this results must be traceable to a feature of the cognitive architecture, not simply to implicit knowledge, task demands, strategies, or some other labile cause.

What would it be to manifest a depictive form?

Let us try to be clear on what we take to be the central issue: Does visual mental imagery rely (in part) on a distinct type of representation, namely, one that depicts rather than describes? By "depict" we mean that each portion of the representation is a representation of a portion of the object such that the distances among portions of the representation correspond to the distances among the corresponding portions of the object (as seen from a specific point of view; see Kosslyn 1994). (Kosslyn, Thompson, and Ganis 2002, 198)

A depictive representation is a type of picture, which specifies the locations and values of configurations of points in a space. . . . In a depictive representation, each part of an object is represented by a pattern of points, and the spatial relation among these patterns in the functional space correspond to the spatial relations among the parts themselves. Depictive representations convey meaning via their resemblance to an object, with parts of the representation corresponding to parts of the object. (Kosslyn 1994, 5)

Pylyshyn's position:

what I shall argue is not true is that the information in the visual store is pictorial in any sense; i.e., the stored information does not act as though it is a stable and reconstructed extension of the retina. (Pylyshyn 2003, 15)

In the opinion of this spectator, the first round of the imagery debate ended roughly as follows. Two widespread, deep, and stubborn sets of reasons for holding to the pictorial form were by Pylyshyn isolated, illuminated, targeted, terminated, dissected, sliced, stained, and mounted. What was left was taken out back and buried. Unfortunately, those scraps seem to reanimate; they don't stay buried for long. The two, seemingly immortal, irrepressible reasons for mental pictures were (and are), first, that introspection reveals the pictorial form directly. The experience of having a mental image is like the experience of seeing something spread out in front of you. How can you deny that you seem to be looking at a picture? A good lawyer could make any witness who denies such a thing seem (at the very least) disingenuous; more likely a scoundrel and a liar, deserving to be convicted. Second, the intentionalist fallacy: When we talk about "the image" it can become almost impossible to tell whether we are talking about the thing imagined or the thing that does the imagining. Mental pictures suffer from the same queasy ambiguity. But in straightforward contexts, at least, it is straightforward: Places in the things one represents need not be represented by places in one's representings. If we carefully avoid these two mistakes, what is left of the argument for the claim that mental imagery must employ a distinct pictorial form? Not much. Pylyshyn also provided many arguments in detail about the inadequacies of "depictive" models. The most potent: that the content of the image depends on the subject's beliefs about the objects in the domain in question.

2   Imagery, Round Two

Round two of the imagery debate opened with the publication in 1994 of Stephen Kosslyn's *Image and Brain*, optimistically subtitled *The Resolution of the Imagery Debate*. (The analogy that springs to mind is a philosopher proposing a final resting place for *zombies*.) Accounts of depictive representation are amended, and the arguments acquire a neuroscience garnish. The key amendment is that the spatial properties and relations of the image are now construed as properties and relations in a "functional space." The basic idea: Talk of spatial properties and relations ascribed to the image should not be taken literally. Instead, all those attributions are a kind of "as if" talk, where what we're really talking about are the values returned by the procedures that read, write, and manipulate information in the image. Those procedures function in a way that is analogous to operations applied to a literal two-dimensional display. If the image is an array in a computer, we have procedures that access and manipulate distances between points. Those distances (the values returned by these procedures) would be true of a literal two-dimensional surface. But this doesn't require that values of adjacent cells in the array be physically next to one another. Basically this is a move to Roger Shepard's idea of second-order isomorphism: The image models spatial relations, but it need not itself employ spatial relations to do so.

Second, and more important for my purposes, neuroscience is claimed to provide evidence for some key features of depictions: first, that visual mental imagery uses some of the same brain mechanisms as does visual perception (in particular V1), and second, that neuroscience shows that those mechanisms use depictive representation. Kosslyn says:

Without question, topographically organized cortical areas support depictive representations that are used in visual perception. These areas are not simply physically topographically organized, they function to depict information. For example, scotomas—blind spots—arise following damage to topographically organized visual cortex; damage to nearby regions of cortex results in blind spots that are nearby in the visual field. Moreover, transcranial magnetic stimulation of nearby occipital cortical sites produces phosphenes or scotomas localized at nearby locations in the visual field. These facts testify that topographically organized areas do play a key role in vision, and that they functionally depict information. (Kosslyn, Thompson, and Ganis 2002, 200)

the actual physical wiring is designed to "read" the depictive aspects of the representation in early visual cortex. In so doing, the interpretive function is not arbitrary; it is tailor made for the representation, which is depictive. (Ibid., 199)

What defines round two as qualitatively distinct from round one is this appeal to neuroscience: the reference to topographically organized "feature maps," conjoined to the claim that some of the same mechanisms could support visual imagery.

Now the appeal to neuroscience adds yet another kind of image to the already confusing mix (fMRI images of the brain), and yet another kind of map ("feature maps"). If we can avoid being distracted by these pictures, however, the critical premise is easy to spot: that "topographically organized cortical areas support depictive representations." What are we to make of this premise? Pylyshyn gives a characteristically forthright response:

Even if we found real colored stereo pictures displayed on the visual cortex, the problems raised thus far in this and the previous chapter would remain and would continue to stand as evidence that these cortical pictures were not serving the function attributed to them. (Pylyshyn 2003, 388)

The scraps have reanimated and reorganized; the debate is up and running, once again. And with that I can state the point of this chapter. Theoretical objections to "depictive" representation, if they are cogent, would apply not just to imagery, but to everything, including visual perception. So, in particular, they would seem to rule out certain accounts of "location-based" effects in selective attention. If places in a mental picture are problematic, what are we to make (for example) of the notion of a "spotlight of attention" moving across the "master map," traversing intermediary locations as it moves, in its own inscrutable fashion, from *A* to *B*? For this to make sense we need places that the spotlight traverses, or across which the "window of attention" moves. Such places have alarming similarities to those found in mental images. How, if at all, can we make sense of the locations posited in location-based models? Perhaps the very notion of a "feature map" is at risk. Does any and every account of feature maps endorse some sort of "inner picture" model? In what sense, if any, are "feature maps" *maps*?

My goal in what remains is to sort some theoretical commitments on these topics into two bins: good and bad. The task is necessary and unpleasant. Theorists must sort out which aspects of an analogical model apply to the real system, and which do not. Here our analogical model for a visual state is a picture or a road map. When we talk of feature maps as "maps," which of the properties of maps must be taken literally? Which are meant only as metaphors?

The task can be unpleasant, but I hope here to render it less so by following the analytical lead of P. J. O'Rourke in his masterpiece of economic analysis, *Eat the Rich*. O'Rourke (1998, 1) says: "I had one fundamental question about economics: Why do some places prosper and thrive while others just suck?" Why indeed? The question applies to visual places too. O'Rourke follows this question with four chapters, entitled "Good Capitalism," "Bad Capitalism," "Good Socialism," "Bad Socialism." Here I shall try to distinguish Good Objects from Bad Objects, and Good Locations from Bad Locations. Because Pylyshyn's critique focuses on the badness of Bad Locations, I shall start there.

3   Bad Locations

Economically speaking, Bad Locations correspond to Bad Socialism: Cuba. O'Rourke visited Havana in 1997 and said it "looked like 1960 Cleveland after a thirty-seven year strike by painters and cleaning ladies" (1998, 80). A compelling candidate for a Bad Place! Visually speaking, Bad Locations are any of those found in models of visual perception that succumb to the same errors as models of pictorial or depictive representation. How can one succumb to the same errors? Let us count the ways.

3.1   The Ones in a Mental Image or in an Inner Picture

These are bad if they are stipulated to be not just places where the representation is located, or places that it represents, but places *in* it that represent places that the organism perceives. So these are stipulated to be places in the image or picture that "map" onto places in the world. The mapping is semantically significant. They are allegedly homomorphic to, and thereby depictive of, places in the world.

images are experienced as distributed in space. . . . Because they are experienced as distributed in space, we find it natural to believe that there are "places" on the image—indeed it seems nearly inconceivable that an image should fail to have distinct places on it. This leads naturally to the belief that there must be a medium where "places" have a real existence. (Pylyshyn 2003, 371)

But, as he argued mightily in the imagery debates, round one, this conclusion is not mandatory. No available evidence requires us to postulate representations of this form. Pylyshyn puts his conclusion these days even more firmly. "We will have to jettison the phenomenal image," he says (ibid., 47). What is tossed overboard is strictly the depictive form, not the phenomenology of imagery. That is, it is still true that to some people it seems as if they sometimes look at inner pictures. That's what they report. The claim is that this "phenomenon" (or appearance) of imagery is consistent with representations that are everywhere propositional.

3.2 Places in Your Percept not Within Your Current Field of View

A very similar point can be made about the phenomenology of visual perception. Though it might seem to common sense, and to some introspectors, that seeing things is a matter of apprehending an inner picture, Pylyshyn rightly insists that such appearances can be explained in ways other than by postulating internal pictorial representation.

We cannot escape the impression that what we have in our heads is a detailed, stable, extended, and veridical display that corresponds to the scene before us. . . . We find not only that we must dispense with the "picture in the head," but that we must also revise our ideas concerning the nature of the mechanisms involved in vision and concerning the nature of the internal informational states corresponding to percepts or images. (Ibid., 3)

One way to diagnose whether you suffer from an objectionable form of the "inner picture" model is to ask: Does that inner display extend, spatially or temporally, beyond the limits of what can, in a given moment, be seen? If the answer is "yes," your theoretical commitments clearly include some Bad Places. If the answer is "no," you might or might not be infected. As will be seen, further tests are necessary.

Pylyshyn does not deny the existence of retinotopically organized feature maps, as found in V1 to V4. But each of these is confined to registering information derived from the array of retinal receptors. They neither can nor need to register information about regions that currently cannot activate any receptors: all those regions in the ambient optic array whose light fails to intersect any part of the retinal array. Nevertheless, it might seem as if visual perception involves a comprehensive or panoramic inner picture, one that includes many of those momentarily unseen portions of the scene.

It has been suggested that what we "see" extends beyond the boundaries of both time and space provided by sensors in the fovea. So we assume that there is a place where the spatially extended information resides and where visual information is held for a period of time. (Ibid., 28)

This last assumption is one that Pylyshyn is most eager to deny. Although there might be retinotopic maps, there is, says Pylyshyn, no panoramic inner picture: no extension of the retinotopic maps so as to include, in the same map, portions of the distal scene that are currently unseen. So places in a retinotopic map are (tentatively) OK (more on this below); places represented by retinotopic maps are OK; but there the map talk stops. There is no further (much less final) comprehensive map, into which all the retinotopic versions—all the gleanings from each glimpse—can be arrayed. Gaze control and saccadic integration are not managed by larger and more comprehensive versions of the retinotopic maps found in V1 to V4.

3.3 The Ones Identified Using a Particular "Reference Frame" or Using Particular "Coordinates"

Talk of "reference frames" is often just a way of specifying a category of bodily motion invariance: which motions (of stimulus or of body parts relative to one another) will, and which will not, alter the proposed state (whether it be neural or representational). To say that a sensory state "employs a eye-centered reference frame" means that the state won't change as long as spatial relations between the stimulus and the eyeball are unchanged. To say that it employs a "head-centered reference frame" means that changes in that state are correlated instead with changes in the spatial relations between the stimulus and the head. Since the eyes can move in the head, these are distinct; a stimulus can have a fixed location in an eye-centered reference frame even while it moves in terms of the head, and vice versa. Such terminology is a useful and unobjectionable shorthand.

But talk of reference frames can have a more fulsome interpretation, where we assume there is an origin and some fixed points (axes) relative to which locations and other spatial properties and relations are determined. Often theorists can slide into this talk without even noticing that it says rather more than mere motion invariance. For example, Cohen and Anderson (2004, 104) say "A reference frame can be defined as a set of axes that describes the location of an object." Note that this description does not require the animal to use those axes! They then proceed to say

Sensory targets are often coded in different reference frames. For example, the location of a visual stimulus is initially coded based on the pattern of light that falls on the retinas, and is thus in retinal coordinates. . . . The location of a tactile stimulus is coded by the pattern of activation in the array of receptors that lie under the skin's surface and, consequently, it is coded in a body-centered reference frame. (Ibid.)

These inferences (the "thus, in retinal coordinates"; "consequently . . . in a body centered reference frame") simply do not follow, unless we read "coordinates" and "reference frame" very loosely. It might seem churlish to criticize what is here probably an innocent use of an analogy, and indeed, there is nothing to criticize as long as the theorist recognizes that this is *merely* an analogy. The danger with analogies, though, is that unintended portions of them creep unbidden and unnoticed into one's theories.

Similarly, talk of "coordinates" can just be a way of describing the data (as in spaces derived from multidimensional scaling); but it is dangerous if one presumes the animal actually employs them to identify anything. If we really mean "coordinates," then this presumes that we have an origin point, axes, and metrical-level measurements of distance along those axes (the real number plane, or perhaps polar coordinates). It also implies that mechanisms of spatial discrimination use those coordinates, as coordinates, to pick out the locations of things. This I think no one seriously believes, despite the occasionally fanciful diagrams.

3.4   The Place Lit Up by the Spotlight of Attention

If we assume that this is not a place in the world, but is rather one located on the master map of locations, then it *may* go onto the list of Bad Places. It depends on how one understands the "map" talk. If we presume that the master map is literally a map, or that differences in places in the map are used to represent differences in places in the world, then such places are heir to all the theoretical difficulties associated with places in a mental image, and are, indeed, Bad. If one endorses some semantically significant relation between places in the map and locations in the world, then it is prey to all the difficulties just noted for the fulsome sense of "coordinates."

One particularly clear diagnostic indicator: If one assumes that when attention shifts from stimulus A (in the world) to stimulus B, then the spotlight of attention must traverse locations on some "master map" intermediary between those used to represent the place of A and those used to represent the place of B, then one has endorsed some Bad Locations. Those "intermediary" locations are the Bad ones. The assumption that there are, and perhaps must be, such intermediary locations in the map indicates conclusively that one thinks of the spatial relations in the map as semantically significant. That satisfies the definition of "depictive." These implications are not evaded by the expedient of turning all the talk into talk of "functional" space.

3.5 Empty Ones

Sometimes Pylyshyn charges location-based models with the crime of representing empty space: places as such; unoccupied or unfilled places; places with nothing in them. These sound Bad indeed.

The theoretical question for us reduces to whether it is possible for visual indexes to point to locations *as such* (i.e., to unfilled places) and that question is not yet settled experimentally, although there is some evidence that the position of an object can persist after the object has disappeared . . . , and that at least unitary focal attention may move through the empty space between objects, though perhaps not continuously and not at a voluntarily controlled speed. (Pylyshyn 2003, 252)

The contrast is stark: The choice is between models that direct attention at empty places, and those that direct it at familiar, fulsome, objects:

there is reason to believe that at least some forms of attention can only be directed at certain kinds of visible objects and not to *unoccupied places* in a visual scene, and that it may also be directed at several distinct objects. (Ibid., 160)

the evidence . . . suggests that the focus of attention is in general on certain primitive *objects* in the visual field rather than on unfilled *places*. (Ibid., 181)

Is this a fair contrast? It is true that a location-based model worthy of the name should allow that differences in the direction of attention need not always be framed in terms of (or be resolvable into) differences in the objects to which attention is directed. Instead attention can be directed as finely as spatial discriminability allows. But do such models require or imply that attention can be directed to unfilled places?

Well, they might; but only if an animal sometimes encountered such locales. "Empty" can mean various things: (a) it contains nothing at all; (b) it contains nothing that would provide physical stimuli; (c) it contains nothing sufficient to stimulate any transducer of the organism in question; (d) it contains no perceptible physical objects. Case (a) is a literal vacuum. Case (b) is also extraterrestrial: It might include fields and particles that do not interact with any transducers. Not a vacuum, but filled with a soup of quarks, say. Strictly, (c) is more or less impossible to produce, unless it is the same as (b): Even a silent, pitch-black room contains stimuli for thermal sensation, as well as vestibular ones. In practice one must think of both (c) and (d) as confined to one modality. So a pitch-black room would give a *visual* example of (c). In contrast, (d) could include the ganzfeld, or for that matter a very foggy evening; the regions contain visual stimuli but no discriminable objects.

It would tax any animal to discriminate among places that are literally devoid of stimuli (as in (a) or (b)). An animal would have that capacity only if its forebears had routinely been challenged by the need to discriminate one empty location from another. The analogous burden to place on the other side would be to require the animal to be able to discriminate objects as such: objects that lack any properties at all. These are what philosophers call "bare" particulars: manifesting the pure objecthood of objects, isolated from all their distracting properties. I don't think it is fair to require object-based models to be able to tell two of these apart. Similarly, on this interpretation of "empty," a location-based model need not even try to satisfy the request to tell apart two empty places.

But if by "empty" one means simply that the animal has spatial discriminative capacity even if it is not confronted by any discriminably distinct objects, then I think the answer is yes, it does. The wafts of cloud in a white-out or a ganzfeld serve as examples. Different patches of cloud or portions of ganzfeld remain spatially discriminable from one another.

A better contrast might be between places that are filled with distinct *objects* and places that are not. An object-based model implies that where there fail to be distinct objects there cannot be differences in how selective attention is directed. A location-based model allows such differences as long as the organism still has the capacity to make spatial discriminations in that region. It asserts that when we write the operating principles for the directing of selective attention, the variables employed need not always refer to objects; they can range over any features that can be spatially discriminated from one another.

4 Good Objects

Visually speaking, good objects are all and only the ones fit to serve as values of variables in the true model of what the visual system represents. Economically, the analogue for Good Objects is Good Capitalism: Wall Street. O'Rourke says of this place: "The traders spend their day in that eerie, perfect state the rest of us achieve only sometimes when we're playing sports, having sex, gambling, or driving fast. Think of traders as doing all these things at once, minus perhaps the sex. . . . All free markets are mysterious in their behaviour, but the New York Stock Exchange contains a mystery I never expected—transcendent bliss" (O'Rourke 1998, 21).

The preceding problems with Bad Locations are used by Pylyshyn to argue for the thesis that visual indices are bound, not to locations, but to objects.

In what follows, discussion will be confined to . . . the view that focal attention is typically directed at *objects* rather than at *places,* and therefore that the earliest stages of vision are concerned with individuating objects and that when visual properties are encoded, they are encoded as *properties of individual objects.* (Pylyshyn 2003, 181)

Medium-sized package goods are good objects. Many visual proto-objects turn out to be identical to medium-sized package goods. So many visual proto-objects are perfectly OK.

5   Good Locations

Now the problem is just this: Are all locations posited in location-based models Bad Ones? Are any of them are good? Good Locations in O'Rourke's typology correspond to Good Socialism: Sweden. "Sweden was the only country I'd ever been to with no visible crazy people. Where were the mutterers, the twitchers, the loony importunate? Every Swede seemed reasonable, constrained, and self-possessed. I stared at the quaint narrow houses, the clean and boring shops, the well-behaved white people. They appeared to be Disney creations" (O'Rourke 1998, 56).

My question is whether there are any Good Locations in the intentional domain. How can we construe the talk of locations in location-based models, or the talk of maps in feature maps, so as to avoid the very real dangers of which Pylyshyn has warned us? Specifically, is any theorist who wants to pitch a tent somewhere in the location-based domain (or on a feature map) necessarily camping in a Bad Location?

To start, it helps to note that Pylyshyn does endorse some Good Locations—some unproblematic spatial domains. They include:

5.1 Locations of objects and of their parts.
5.2 The location of the brain.
5.3 Location of mental representations within the brain.
5.4 Locations in topographically organized areas in V1 to V4.
5.5 Locations as represented in retinotopic maps.
5.6 Locations of "feature clusters."

But what then of feature maps? Must these contain, or be maps of, Bad Locations? V1 is one of many alleged "feature maps" in the cortex. What's going on in those? And is Kosslyn right to say that "without question" they support depictive representations?

The core notion of a "feature map" in neuroscience is, I think, a region of cortex organized topographically. But everything hangs on how one understands the term "topographical." The simplest interpretation is anatomical. The fibers coursing into the cortical area come from some source region or regions, also within the nervous system. In a "topographical" organization, there are local regions in the source within which neighboring cells project, more or less, to neighboring cells in the destination. There might be several such local regions, between which there can be abrupt discontinuities in the projections. A prominent example is found in the retina: The left side of each retina projects to the left side of the brain, and the right to the right. So we find a topological "tear" right down the middle of the retina. But within each region, neighborhood relations are (pretty much) retained.

In sensory areas, cells in a feature map can often be associated with receptive fields: regions in circumambient space within which stimuli of a specified kind can affect the activation level of the given cell. This yields a second way to understand the topographic organization. Cells that are neighbors in the cortical region in question often have receptive fields that are neighbors in circumambient space. When they do, one can see a very strong reason to call the thing a "map": It is a topographically organized array within the organism that seems to represent places outside and around the organism. But as will be seen shortly, the notion that cells in feature maps preserve neighborhood relations among points in space is never strictly speaking true, and it is often very misleading.

It should be obvious that mere topographic organization is not by itself sufficient to show that the cortical region in question employs pictorial or depictive representation. That way of organizing the fiber bundles can be better ascribed to physiological economy (fewer crossovers and shorter bundles) or neural development (easy ways to grow the things) than to features of our cognitive architecture. Furthermore, the cortical region may be representing something other than location altogether. For example, an auditory feature map can be topographically organized, respecting neighborhood relations on the basilar membrane, but this makes it a *tonotopic* map, of different frequencies, not different places. Mustached bats have auditory maps across which we get systematic variation in Doppler shift (see Suga 1990). It is not mapping space, but rather relative velocities.

What, then, is needed for these regions of cortex to be, also, maps of space? This conclusion is not automatic! A second obvious necessary condition can be put as follows: The region must enable some spatial discriminations. It carries information about spatial properties and relations of its targets in such a way as to allow the organism to navigate. Without this it wouldn't contribute to what I think of as "feature-placing."

Is that enough? Are these regions of cortex "without question" depictive? If we consider V1, for example, the best possible case for calling it a "feature map" gives us three premises. First, we have an orderly projection of fiber bundles from its source (mostly LGN) to V1. So, second, neighbors in V1 typically have receptive fields that are neighbors. (And it functions in accord with this principle, as Kosslyn points out. Damage to V1 causes scotomata whose perimetry can help the neuropsychologist identify where the damage took place.) Third, thanks to V1, the creature can make certain spatial discriminations that it otherwise cannot make. If you doubt this, just consider what it loses in those scotomata.

These three premises, so far, do not imply that the map is a "map of space," that is, that points and distances within V1 map homomorphically onto points and distances within the ambient optic array. For it to be a literal map of space, it would have to sustain those spatial discriminations in just one way, via a homomorphism with spatial properties. As Kosslyn puts it, it must be such that "distances among portions of the representation correspond to the distances among the corresponding portions of the object" (Kosslyn, Thompson, and Ganis 2002, 198). The pattern of inference here seems eerily familiar. In fact, thanks to Pylyshyn, we can recognize it. It is exactly the pattern used to sustain the idea that mental imagery must involve inner pictures.

That V1 is required for certain sorts of spatial discriminative capacities shows that information in V1 is used by the organism to improve its steerage. It does not show that the information in V1 is organized just like a map or a picture. The structure might enable spatial discriminations (of some particular sort) without itself modeling space. If you look at its finer structure, I think it's pretty clear it does *not* model space. In fact, perhaps no feature maps are maps of space in the "depictive" sense. V1 is certainly a big array of measurements, but values in adjacent cells are not invariably measurements of adjacent places.

Details of the structure of V1 make this clear. The details in question are not subtle or contentious; most of them have been known since the work of Hubel and Wiesel. In particular, the ocular dominance pattern, and the arrangement of "orientation slabs," royally messes up the neighborhood relations. In a given orientation "slab" within (layer III of) a cortical column, all the cells will fire maximally to an edge, bar, or slit of a given orientation. Cells in the neighboring slab do not register the same orientation in neighboring receptive fields, but instead a different orientation (in different receptive fields). And we have a block of orientation slabs for the left eye immediately adjacent to a block for the right eye. These are the left-eye view and the right-eye view of the same location in external space.

The critical point: If you move half a millimeter in one direction, you might not change the receptive field at all, but instead move to a region receiving input from that same receptive field, but from the other eye. Move in another direction and the receptive field will shift, but so will orientation. Move in a third direction and only the optimal orientation shifts. These distances do not map uniformly onto distances in the ambient array. Ergo, homomorphism fails. V1 is not depictive.

6 Hypothesis: How a Feature Map Represents

How then does a feature map represent? One minimal but plausible description of the content of a feature map is: It indicates the spatial incidence of features. It might do more than this, but it does at least this. That is, it registers information about discriminable features, in such a way as to sustain varieties of spatial discrimination that can serve to guide the organism. The latter two conditions focus on downstream consumers of the information, not what causes it. Registration of information in a feature map endows the creature with some spatial discriminative capacity. If that map is used, the steerage can improve. To carry on its other business, the animal relies on the constellation of features being as therein represented.

One way to get at the spatial content of a feature map, guaranteed to work for every feature map, is to ask: What sorts of spatial discrimination does this particular feature map enable? That is, which spatial discriminations are possible using this map that were not or would not be possible without it? For some cortical regions dubbed "feature maps" by neuroscientists, the answer could well be "none"—in which case the map is not a representation of the spatial incidence of features at all. (Such a map will not employ the representation form I identify below as "feature placing.") The idea: If feature map M is representing the spatial incidence of features, then it is being used as a representation of the spatial incidence. The information in it about spatial properties and relations is exploited. One way to show that it is exploited is to show that certain kinds of spatial discriminations could not be made without it; without map *M* working normally, the guidance and steerage system—the navigational and spatial competence of the organism—suffers some decrements.

The focus on downstream consumers is a way of showing that the registration of information is used as a representation; that it has a content that is used. To tie representations to the world, show that they improve the capacity to get around. But feature maps can do this without necessarily being pictorial or depictive; they can satisfy the condition without being, literally, maps or inner pictures.

Psychological theory right now lacks any deductive proofs, or even compelling arguments, that establish how information *must* be organized to endow creatures with some new spatial discriminative capacity. It's too early to invoke a priori principles in this domain. (It follows that there's never a good time to be a priori—but that's another question.) So, in particular, there is no compelling reason to think that information *must* be organized depictively in a feature map if that feature map enables a creature to make spatial discriminations that it otherwise could not. Here again we should thank Pylyshyn: His work on mental imagery showed how, in principle, a set of propositions could do the job.

What then does V1 represent? To answer this question, analyze what use downstream consumers make of the information registered in it. A first stab: These cells in layer III of V1 represent "(edginess of orientation theta) (thereabouts)." Edginess is the feature; "thereabouts" indicates its incidence. Those cells in layer III of V1 have the job of registering differences in orientations, in such a way as to allow spatial discrimination of them. If they do that job, the animal can rely upon those indicators, and thereby steer a bit more successfully than if it lacked them.

More generally, I have proposed that we call this form of representation "feature-placing." It "indicates the incidence of features" in the space surrounding the organism. The name is partly in honor of Sir Peter Strawson's (1954, 1974) work on "feature-placing languages," which contain just a few demonstratives ("here" and "there") and nonsortal universals (feature terms, like "muddy" or "slippery.") A paradigm feature-placing sentence is "Here it is muddy, there it is slippery." Such sentences indicate regions and attribute features to them. Strawson argued that these languages could proceed without the individuation of objects. The same seems true of the representations employed in feature maps. It seems a bit much to claim that V1 "refers" to places, "identifies" regions, or "demonstrates" locales. All the latter locutions arguably invoke some portion of the apparatus of individuation. Feature-placing is prior to, and can provide the basis for, the introduction of that rather heavy machinery.

Another way to put it is that feature maps in V1 to V4 transact their business in a location-based way. A particular feature map can endow a creature with new spatial discriminative capacities without also endowing it with an ontology of objects. It can get the spatial discriminative job done without investing in that sort of machinery. A skimpy basis can suffice; the business can be run on an ontological shoestring. It is also important to insist that the regions visually discriminated are not inner, or mental, ones. They are not inside the organism or inside the mind. If the job is to guide spatial discriminations, then representing those places will not help. Visual "thereabouts" are always, resolutely, in the ambient array, not in the retina. The cortical feature map might be retinocentric (it uses an "eye-centered" reference frame) but it is *not* retinotopic. It is not about the states of the retina, but instead about features in the world.

If V1 were representing places on the retina, then it should represent the blind spot as empty. But patterns are completed "across" the blind spot, as shown by Gatass and Ramachandran's experiments on scotoma and "filling in" (see Churchland and Ramachandran 1994). The filling in across the optic disk can give a veridical "perception" of the distal place, even though it would be a nonveridical representation of what is going on at the retina. V1 cells in the "Gatass condition" fire just as they would if there were a stimulus stimulating the nonexistent receptors in the optic disk. If we were representing places on the retina, this would be a nonveridical representation (Churchland and Ramachandran 1994, 82).

So I think there is good reason to say that what these parts (of layer III) in V1 are representing is something akin to "(edginess of orientation theta) (thereabouts)." "Thereabouts" indicates a region of circumambient space— a region of visual perimetry, in the ambient optic array. "Edginess of orientation theta" indicates a feature discriminable in some portion of that space. The orientation is of an edge in external space, not across the eyeball. It is feature-placing, and both the features and the places are distal.

7 Bad Objects

That concludes my plea for the possibility that not all Locations in the intentional domain are Bad. Symmetry demands that we also consider the possibility that not all Objects are Good. This is our last quadrant: Bad

Objects. In O'Rourke's typology, it corresponds to Bad Capitalism: Albania. Albania, he says, "has the distinction of being the only country ever destroyed by a chain letter—a nation devastated by a Ponzi racket" (O'Rourke 1998, 36). Chain letters and Ponzi rackets in completely unregulated markets can be tough on widows and orphans. Likewise, visually speaking, Bad Objects are the kinds of objects to which a purely object-based model is at least somewhat vulnerable.

7.1 Merely Virtual Ones

By "merely virtual" I mean an object that seems to exist, or appears to exist, but does not. The ogres, wizards, and dragons displayed on computer screens in some computer games are paradigm examples. The experience of looking at such a screen can be very much like seeing a dragon, but there is no dragon there to be seen.

It is a bad idea ever to allow merely virtual objects to serve as the referents of visual indices. Such an index is supposed to be entirely nondescriptive, gaining all its representational capacities from direct access to the referent itself. So if in fact there is no referent, there is nothing to which the index can be attached. An index attached to such a thing is attached to no thing.

Now in many of the experiments in multiple-object tracking, subjects are not in fact tracking objects, in any ordinary sense of the word. Instead they are looking at a computer display and tracking figures on the screen. What exactly is the object to which a visual index is attached? Pylyshyn says "the observer may be indexing clusters on the screen or, more likely, a virtual distal object, where only the part of the chain from the scene to the observer is real" (2003, 217). I think the latter alternative invites indoors some Bad Objects. Suppose one can index a merely virtual object. Then in one episode of a computer game an index might be attached to a dragon, and in another, to an ogre. But indices are supposed to be nondescriptive, and neither dragons nor ogres exist. So what is the difference between indexing a dragon and indexing an ogre?

An index gets it content entirely from what it points at. It does not encode any properties, contains no description, and so on. So if it is pointing at nothing, it should have no content. So if it is pointing at an object that is a merely virtual object, there should be nothing that differentiates one such pointer from another. So there can't be a difference between pointing at inexistent object *A* versus inexistent object *B*.

For this reason it seems preferable to keep the door shut, and adopt the other alternative: What is indexed must be something literally seen, on the screen. Similarly, for the same reason, it is hard to see how an index could ever get attached to a nonvisible object. Pylyshyn wonders "What exactly the index points to when the object is temporarily out of view" (2003, 268 n20). Nothing comes to mind!

The problem in both cases is that reference failure is catastrophic for an index. In such a case there is nothing to which it points, and reference does not succeed by description. So in what sense is it "referring" or pointing at all? This should be a case of an indexical without a referent. How could it have any content at all? If we style these pointers on those found in programs, this one should give an "out of bounds" memory error, cause the blue screen of death to appear before the mind's eye, and make the mind itself lock up. Abort, retry, fail?

7.2 Nonindividual Ones

If vision is to be object based through and through, from the get-go, then the values of variables in all of its representations, everywhere, are always, and only, objects. Even at the earliest stages, the representanda are objects. The worry here is simply that some of those earliest stages do not have the wherewithal to represent their objects *as* objects. In particular, they lack the wherewithal to represent that which makes one of them one, and not two.

To use the technical terminology: These "objects" lack criteria of individuation. And if they lack individuation, it will seem feckless, at least to some philosophers, to call them "objects" at all. If "this" and "that" are bound to objects, then one can distinguish the possibility of encountering first this one and then that one from the possibility of encountering this same one twice. Otherwise the application of the apparatus of individuation—count nouns, identity, sortals, indefinite pronouns, articles, and the like—is not required.

Consider the early stages of visual representation, in V1 to V4. You, the neuroscientist, laboriously describe how one of them works. Someone in the audience rises to ask, "but does this particular state, at this stage, represent exactly one x, or does it represent both one x and a y such that y is not identical to x?" Even though the question is probably from a philosopher, and I am a philosopher, I would sympathize with your plight. Such a question seems somehow maladroit, ill-informed, out of place. In these stages there is nothing available yet that would be, or could be, sufficient to answer the question of what makes one thing one, or distinct things distinct. These stages operate in a regime that is prior to, and free from, such worries.

If this sympathy is not entirely misplaced—if the notion of such regimes is at all plausible—then these stages are representing the "things" they represent without representing them as falling under criteria of individuation. If we insist that even these stages are representing objects, these will be "nonindividual" objects. They lack individuation. Nothing is such as to make one of them one, and not two. Common sense would cavil at calling such things (such values of variables) "objects." If we cannot count them, what justifies the distinction between singular and plural? Quine (1974), Geach (1980), and Wiggins (2001) have argued, at length, that the acquisition of the apparatus of individuation is no mean feat. Unless we think that V1 (for example) can acquire such a thing, the variables therein range over features or regions, but not objects.

7.3 The Ones Numbered More Than Six

This is the most variegated kind of Bad Object, because it is not a kind at all. Like vulnerability to Ponzi schemes, the problem here seems to be a structural limitation of visual indices. Indices are limited to five or six. What happens when we run out?

In particular: Can we account for the spatial discriminative capacities that become possible when a creature acquires a feature map by supposing instead that all the reference of its representations proceeds through five or six visual indices? To be object based all the way down, all such information can be registered in a system of object-files (or, more broadly, a system in which all the variables are bound to objects). Consider, for example, V1. In order to explain how this map (V1) endows the creature with (say) the ability to discriminate horizontal lines from slightly off-horizontal lines, we have to think of feature detection and registration across a vast swath of space, sensitive throughout to minute differences in orientation. It has somehow to register that there is an edge or bar or stripe extending from x to y; and then register orientation of that edge from point to point.

How would a FINST system represent a pattern of (say) nine parallel lines, tilted slightly? We have more lines than we have FINSTs, yet even registration of the features of one line seems to require lots and lots of terms and relations (edginess, connectedness, continuity, straightness, orientation, parallelism, etc.).

Location-based theorists surmise that at least some of the information must be registered in data structures that contain variables that range over something other than objects. The books can be organized differently; the business might be transacted in an ontologically skimpy, location-based way.

8 Conclusion

To sum up. Some clearly Bad Locations are: the ones in a mental image or in the inner picture; places in your percept that are not within your current field of view; and, finally, the ones identified using coordinate systems or reference frames.

In contrast, the presumption is that almost any Object is Good, particularly if it is one that can be bought or sold in a capitalist economy—things you can track, and, when the funds become available, purchase. Medium-sized package goods are, therefore, the paradigm Good Objects.

There are also some Bad Objects, however. Merely virtual ones qualify: the ones that do not exist, even though they have an index attached to them. Sadly, these too are sometimes bought and sold in capitalist economies. Other Bad ones include objects that lack individuation. If you buy one of these you don't know what you bought. Finally, those numbered more than six. These are bad because they can't be indexed.

Close examination of Pylyshyn's theory shows that it allows for the existence of at least some Good Locations. These include: locations of objects and of their parts; the location of the brain; locations of mental representations within the brain; locations in topographically organized areas in V1 to V4; locations as represented in topographic maps; and locations of "feature clusters."

In terms of this typology, are "feature maps" Good or Bad? I have argued that they can be Good, though to stay that way they must eschew any claim to be depictive.

The upshot? Let us leave the last word to P. J. O'Rourke: "Money turns out to be strange, insubstantial, and practically impossible to define... economic theory was really about value. But value is something that's personal and relative, and changes all the time. Money can't be valued. And value can't be priced.... I should never have worried that I didn't know what I was talking about. Economics is an entire scientific discipline of not knowing what you're talking about" (O'Rourke 1998, 122–123).

References

Churchland, Patricia S., and Vilayanur S. Ramachandran (1994). Filling in: Why Dennett is wrong. In *Consciousness in Philosophy and Cognitive Neuroscience*, ed. Antti Revonsuo and Matti Kamppinen, 65–91. Hillsdale, N.J.: Lawrence Erlbaum.

Cohen, Yale E., and Richard A. Andersen (2004). Multimodal spatial representations in the primate parietal lobe. In *Crossmodal Space and Crossmodal Attention*, ed. Charles Spence and Jon Driver, 99–121. Oxford: Oxford University Press.

Geach, P. T. (1980). *Reference and Generality*, 3rd ed. Ithaca: Cornell University Press.

Graziana, Michael S. A., Charles G. Gross, Charlotte S. R. Taylor, and Tirin Moore (2004). A system of multimodal areas in the primate brain. In *Crossmodal Space and Crossmodal Attention*, ed. Charles Spence and Jon Driver, 51–67. Oxford: Oxford University Press.

Konishi, Masakazu (1992). The neural algorithm for sound localization in the owl. *The Harvey Lectures*, Series 86: 47–64.

Kosslyn, S. M. (1994). *Image and Brain: The Resolution of the Imagery Debate*. Cambridge, Mass.: MIT Press.

Kosslyn, Stephen M., William L. Thompson, and Giorgio Ganis (2002). Mental imagery doesn't work like that. (Reply to Pylyshyn 2002.) *Behavioral and Brain Sciences* 25(2): 198–200.

O'Rourke, P. J. (1998). *Eat the Rich*. New York: Atlantic Monthly Press.

Pylyshyn, Z. W. (2001). Visual indexes, preconceptual objects, and situated vision. *Cognition* 80: 127–158.

Pylyshyn, Z. W. (2002). Mental imagery? In search of a theory. *Behavioral and Brain Sciences* 25(2): 157–237.

Pylyshyn, Z. W. (2003). *Seeing and Visualizing: It's Not What You Think*. Cambridge, Mass.: MIT Press.

Quine, W. V. O. (1974). *The Roots of Reference*. La Salle, Ill.: Open Court.

Stein, Barry E., Terrence R. Stanford, Mark T. Wallace, J. William Vaughan, and Wan Jiang (2004). Crossmodal spatial interactions in subcortical and cortical circuits. In *Crossmodal Space and Crossmodal Attention*, ed. Charles Spence and Jon Driver, 25–50. Oxford: Oxford University Press.

Strawson, P. F. (1954). Particular and general. *Proceedings of the Aristotelian Society* 54: 233–260.

Strawson, P. F. (1974). *Subject and Predicate in Logic and Grammar*. London: Methuen.

Suga N. (1990). Cortical computation maps for auditory imaging. *Neural Networks* 3: 3–21.

Suga, N., J. F. Olsen, and J. A. Butman (1990). Specialized subsystems for processing biologically important complex sounds: Cross correlation analysis for ranging in the bat's brain. *The Brain: Cold Spring Harbor Symposia on Quantitative Biology* 55: 585–597.

Wiggins, David (2001). *Sameness and Substance Renewed*. Cambridge: Cambridge University Press.

13 Visual Objects as the Referents of Early Vision: A Response to *A Theory of Sentience*

Brian P. Keane

Our sensations are for us only symbols of the objects of the external world, and correspond to them only in some such way as written characters or articulate words to the things they denote.

—Hermann von Helmholtz (1853/1995)

1 Introduction

In his book *A Theory of Sentience* (2000), Austen Clark offers an ambitious account of the representations that are required for sentience. He delves into contemporary cognitive psychology and neuroscience to develop a view of sensory representation that he takes to be both empirically and metaphysically adequate. An aim of this chapter is to show that Clark's theory, which applies to all sense modalities of sentient creatures, fails to provide an adequate account of early vision in human beings.

In the first part of this chapter, I defuse a series of arguments that Clark offers for the view that the early visual system identifies and attributes features to physical space-time regions. In the second part of the chapter, I provide positive arguments for the view that the visual system identifies and attributes features (including location) to visual objects rather than space-time regions. Both sets of arguments are constructed to deal specifically with Clark's account, but are expected to bear on any theory of early visual representation. I ultimately conclude that something similar to a theory of sentience could characterize visual sensory representation if visual objects, rather than space-time regions, serve as the referents of those representations.

1.1 Preliminaries on Feature-Placing and Spatial Sensing

According to Clark's "feature-placing" hypothesis, sentient beings come equipped with the capacity to represent a range of regions and a range of features. Characterizing a sensory representation is done by filling the two "place-holders" in the expression: appearance of qualities Q at region R (Clark 2000, 60).[1] Q specifies what appears; R specifies where in space-time the appearance obtains. Elements that can fill a place-holder vary over some number of dimensions. In the case of space-time elements, they presumably vary over four dimensions—three dimensions of space, and one of time. The qualitative character of a sensation can vary over a far greater number of dimensions, limited only by the representing capacities of the neurological equipment and the variety of features physically available. Specifying a value for every dimension is tantamount to specifying which element fills a place-holder. If m dimensions of space-time can be represented $(R_1 \ldots R_m)$, then representing a feature's location will involve specifying a determinate value for each one of those dimensions $(r_1 \ldots r_m)$. The resulting content can be expressed as a vector, $[r_m]$. On this view, two features are represented as being at the same place-time[2] if and only if they are specified by the same place-time vector. The same story is given for the quality of a feature: Given n dimensions along which a feature can qualitatively vary $(Q_1 \ldots Q_n)$, and given a determinate value for each one of those dimensions $(q_1 \ldots q_n)$, there will be exactly one qualitative character, vector $[q_n]$. Two sensory representations are qualitatively the same, according to Clark, if and only if they have the same qualitative vector. It follows that two sensory representations are identical in content if and only if[3] they have the same qualitative and space-time vectors.

The second place-holder, which will be of interest in the present chapter, plays an important explanatory role. It explains, for instance, why we never sense anything without sensing it as having or as coming from some location (p. 61). If a red dot is sensed, it is also sensed as having a location; a perceived pinprick can be identified as coming from a particular part of the body, and so on. The second place-holder also allows us to understand how scenes, differing only by the layout of features, can be differentiated.[4] Without it, a scene half matte red on the left and half glossy green on the right will not be represented any differently than one that is half matte green on the left and half glossy red on the right. In each case, the same conjunction of qualities will be represented: red, matte, green, and glossy. The second place-holder also allows an explanation of how two qualitatively identical sensations can be had at the same time. Two colored dots of the same hue, saturation, and brightness flashed on a screen produce not just a representation of that hue, saturation, and brightness, but a hue, saturation, and brightness here and hue, saturation, and brightness there.[5] Distinguishing qualitatively identical dots would be impossible if they were not represented as being at different locations.

Justifying a partition between spatial and qualitative variation also requires showing that one kind of variation cannot be reduced to the other. Most accept that qualitative variation is not a kind of spatial variation. What might be contested is whether spatial variation is a kind of qualitative variation, whether perceiving spatially is a matter of having spatial qualities. Though qualities are invariably coupled with locations, Clark claims that it makes sense to consider the character of a sensation independently of its location (e.g., pp. 54–61). An indicator that spatial variation is its own kind of variation is that token representations of locatedness cannot be multiply instantiated at a time. Whereas there can be various simultaneous tokens of, say, red, there can only be one instantiation of a located at x-ness at a time. This of course does not preclude located-at-x-ness from being a quality—only certain kinds of qualities might get to be multiply and simultaneously instantiated. But it does give reason to treat spatial variation differently from qualitative variation.

1.2 The Controversy and the Claims

I now turn to the issue of what elements fill the second place-holder in the feature-placing schema "appearance of quality Q at region R" (p. 164). For the sake of argument, I shall agree with Clark that place-time discrimination is necessary for normal human vision; that place-time variation is irreducible to qualitative variation; and therefore that there must be a place-holder that allows for place-time variation. Clark draws the conclusion that accounting for the spatial character of sensing requires identifying something outside the head. I will agree with Clark on this point too: What sensory systems pick out are neither retinal images, nor brain states, nor phenomenal episodes. I further agree that sensory representation is nonconceptual representation. I shall not argue for this view; I assume it for the sake of argument. For all of the agreement, we disagree with respect to *what kind* of spatially extended entity is represented. Sensory representations could get their spatial character in virtue of picking out physical space-time regions; or they could get it in virtue of picking out occupants of space-time regions. Clark opts for the former. He holds that feature-placing representations "name space-time regions of finite but definite extent. These regions are physical regions: the very same ones about which physics may have something useful to say" (p. 81, cf. p. 155). When a feature of quality Q is placed, it is placed onto the space-time region named by the representation. My *primary* claim goes against this view and asserts that feature-placing representations identify and attribute features to visual objects, construed as *occupants* of space-time regions (hereafter, the "broad construal"). I shall not specify what exactly these occupants must be or how they are individuated; all that will be said is that they take up space-time, they are physical, and hence describable via physical laws, but they are not space-time regions themselves. An occupant could be a baseball, a surface of a picture, a gaseous cloud, or a photon. A *secondary* claim of this chapter is that the early visual system identifies and attributes features to visual objects, as construed in cognitive psychology and cognitive neuroscience (hereafter, the "*narrow* construal").[6] In the scientific literature, visual objects are taken to be the units into which the visual system parses the distal world. They are occupants of space-time and roughly correspond to what we would conceptually consider to be objects (Feldman 2003; Scholl 2001; Scholl and Pylyshyn 1999). Though I think there is good reason to think that the secondary claim is true, I acknowledge the possibility that the final science may not bear this out. It is possible that visual objects as understood by scientists may not be what fills the second placeholder of Clark's feature-placing schema. But in no case, I shall argue, will space-time regions be suitable to serve as the entities identified at the lowest levels of visual processing. Any adequate theory of visual sensory representation must have visual objects serve as the referents of those representations. Ultimately, I recommend that—if a theory of sentience is to be salvaged—the feature-placing schema should be changed to: "appearance of quality Q on visual object O" or "(quality Q and location L) on visual object O," where O is construed broadly, if not narrowly.

2 Region Arguments and Defusing Arguments

2.1 Argument: Object Identification Requires Conceptual Resources Unavailable in Sentience (aka the Concept Argument)

Clark offers a number of arguments why physical space-time regions occupy the second slot of the feature-placing schema.[7] He acknowledges the import of objects in cognition, but he thinks that sentience must first identify space-time regions, before objects can be identified. Identifying objects, he says, requires conceptual resources, and concepts cannot be accessed by sentience (pp. 144–145).

A problem with Clark's simple concept argument, as I shall call it, is that it relies on a conflation of terms. There are at least two kinds of "identification." There is identifying *tout court* or individuating or picking out, on the one hand, and identifying *as such* or recognizing, on the other.[8] The second sort of act is more sophisticated than the first insofar as the second presupposes the first. Clark himself should agree: Space-time regions, he claims, are identified (individuated) at the level of sentience, but they are not identified *as* space-time regions (p. 145). The latter operation would most likely involve knowledge of the physical properties of space, which is unavailable at the sensory level.

Now Clark may be right in claiming that identifying an object as an object requires concepts. It might require, inter alia, understanding that objects take up space, have momentum, bear or collect properties, persist through time, or produce certain sorts of experiences when we interact with them. For the concept argument to be correct, however, object individuation (unlike region individuation) must additionally be shown to be conceptually driven. If individuation does not require concepts, then the concept argument will not suffice to show that objects cannot be identified in sentience.

2.1.1 Defusing argument: Visual objects can be nonconceptually individuated

Prima facie, object individuation certainly *appears* to be possible without concepts. If a computer vision system uses only edge-detection algorithms to determine the presence of an object, it will be successful in picking out just one object when just one object lies within an edge and when just that edge is registered. Although only a conceptually stocked agent could subsequently identify the object as an object, that agent could not do any better than the computer in individuating. It is true that in more complex cases the edge detector would not function as well as a full-blown concept-stocked mind for individuation. But the question at present is whether nonconceptual systems can individuate objects, not how good they can be. That nonconceptual systems are built to automatically individuate indicates that individuation need not involve concept possession.

Some philosophers and psychologists argue for the plausibility of preconceptual object individuation. Pylyshyn (2003) spends considerable time developing a visual indexing theory, according to which visual objects are picked out by way of automatic "data-driven" mechanisms in the earliest stages of vision. He says:

Primitive visual processes of early vision segment the visual field into something like feature clusters automatically and in parallel. The ensuing clusters are ones that tend to be reliably associated with distinct token individuals in the distal scene. I refer to the distal counterparts of these clusters as *primitive visual objects* (or sometimes just as *visual objects*), indicating my provisional assumption that the clusters are, in general, proximal projections of physical objects in the world. (Pylyshyn 2003, p. 211, italics in original)

A visual object, on Pylyshyn's account, is individuated in virtue of its visible features being automatically segmented into a cluster. The segmentation process "operates without regard to knowledge and expectation" and "without involving the conceptual system" (2003, p. 214). Pylyshyn is careful to point out that visual objects individuated by the early visual system are not always identical to what we would ordinarily consider to be objects. Whereas an object might include "more usual notions of object, such as tables and chairs and people," visual objects "are defined in terms of the special sort of primitive nonconceptual category of objecthood induced by the early visual system" (ibid.). A horse and rider galloping past a sensory system, for instance, might be counted as one visual object, when conceptually there are at least two objects moving along the trajectory. A large or small object (e.g., the Earth or a molecule, respectively) might not be individuated as any object, since the boundaries cannot be visually determined for an ordinary observer. Still, the objects that the sensory system segments a visual scene into roughly correspond to the objects that we might conceptually consider the scene to have.[9]

An objection might be that the visual system, while picking out visual objects, does not individuate objects. If an individuator is defined by the class of entities that it picks out, then because the class of objects differs from the class of visual objects, there are two individuators: a visual object individuator, which functions at preconceptual levels, and an object individuator, which functions at the conceptual level. Thus, the claim that no object individuation obtains at the level of sentience can be preserved.

This objection ignores, first, the close relation that visual objects bear to objects; second, the fact that imperfect *Y*-individuators are individuators of *Y* nevertheless; and third, that Clark's account requires also precluding preconceptual regional occupant individuation. Concerning the first point, a visual object certainly *appears* to be a breed of object: It has causal powers (to affect the visual system in certain ways) and it fills space-time. Moreover, the prevailing opinion among vision scientists, as indicated in the passages above, is that there exists considerable overlap between the entities that a visual system detects, and the objects that we might conceptually consider a scene to have. It is no surprise that in many papers cognitive psychologists use the term "object" and "visual object" interchangeably (Feldman 2003, p. 252; Kahneman, Treisman, and Gibbs 1992, p. 178; Carey and Xu 2001, p. 207).[10] Concerning the second point, we must be sure not to impose an overly strict criterion for what counts as a Y-individuator. It makes perfect sense to say that X possesses a mechanism for individuating Y, even though X's mechanism occasionally generates misses and/or false positives in doing so. The situation is analogous to testing for a disease. A given test might have a certain specificity and sensitivity such that when it is administered to a group of people, it usually, but not always, turns out positive for those who have the disease, and usually, but not always, turns out negative for those who do not. But just because the hit rate is lower than desired, or the false positives higher than desired, does not mean that the test is testing for something other than the disease in question. The test is a reliable, albeit imperfect, indicator of the presence of that disease. In the same way, low-level visual mechanisms are reliable, albeit imperfect, indicators of the presence of objects. They may occasionally pick out nonobjects and they may occasionally fail to pick out objects, but they nevertheless reliably indicate when an object is in view. And this, I take it, is enough for an individuator to be considered an object individuator.

An objection to this second point might be that the visual system's object individuators are seriously deficient in their hit rate—there are many more invisibly small objects (e.g., particles of dust, molecules of air) in a typical field of view than visible objects (e.g., chairs), and since visual objects are defined as having a disposition to interact with the visual system, there will be many more objects than visual objects in such a scene. A visual object detector will be, at best, a very poor object detector. This line of reasoning, although correct, is in no way problematic. The important claim is that most external entities individuated by the visual system are objects, not that the visual system individuates most objects in a field of view. Only the former needs to be shown to argue that object individuation needs to be accounted for in a theory of sentience.

More important, even if visual object individuators could not be considered rough-and-ready object individuators, the *most* that Clark could conclude is that objects play no role in sentience. He could not thereby infer that space-time regions fill the second place-holder of the feature-placing schema. He must also rule out the possibility that visual objects play a role in sentience by showing that there can be no individuation of occupants of space-time at that level. And this is something that he does not do.

2.1.2 Defusing argument: Individuation of regions and objects could involve same resources In this section it will be argued that to adequately explain our behavioral capacities, there must be an apparatus for individuation *and* individuative constraints within sentience itself. The visual system needs a mechanism that picks out external entities in a non-random way such that we can succeed in discriminating presented scenes. Because the constraints that need to be present on a region-based account will need to carry out the same functions as the constraints present within an object-based account, it is reasonable to think that individuating objects does not require more resources than individuating space-time regions.

An apparatus of individuation is necessary for visual sentience. On Clark's view, selected regions have "definite extent"; placing a feature presupposes the determination of *which* region the feature is being placed onto (pp. 81, 155). Consider, for example, the task of differentiating the following scenes presented sequentially to an observer (p. 46):

Scene 1: Red square next to green triangle
Scene 2: Red triangle next to green square

To discriminate the foregoing scenes, observers must be able to pick out a region containing a square, and another region containing a triangle. But this requires individuating two different regions.

In criticizing adverbialist theories (pp. 61–65), Clark explicitly acknowledges the need for an individuative apparatus at the level of sentience. According to those theories, there are no features of sensation, no mental objects, and certainly no sense data; there are only ways of sensing. Rather than sensing that something is red, one senses redly. Rather than sensing that something is curvilinear, one senses curvilinearly. Problems arise for the adverbialist in accounting for how we discriminate scenes like those above. Because adverbialists do not posit mental objects to explain scene discrimination, and because sensing at multiple locations seems to require mental objects, locations, like features, must be treated adverbially. On this account, scene 1 would be characterized as sensing redly, squarely, leftly and greenly, triangularly, rightly; and scene 2 as sensing redly, triangularly, leftly, and greenly, squarely, rightly. The problem, of course, is that the two characterizations are identical.[11] The modifiers of sensing are conjoined, and the principle of commutativity guarantees equivalence of the two combinations.[12] Discriminating scenes 1 and 2 requires sensing in one fashion at a place, sensing in another fashion at a place, and determining that those two fashions of sensing are of two places. But the adverbialist's account lacks the resources to make the last determination. Numeric differences cannot be described purely qualitatively. To avert the "many properties" problem,[13] Clark holds (pp. 64–65):

It seems that we need an apparatus of identification within the content of sensation itself. We cannot construct such an apparatus using only adverbs. . . . The placing component [of feature-placing] serves an individuative role, which adverbs cannot provide.

Clark is right to say that individuation is necessary for success in many ordinary scene-discrimination tasks.[14] But he is wrong to say that the adverbialist account involves no individuation. The region that characterizes a sensing is the region that is picked out. Its boundaries, which are determined by the position of the observer's eyes relative to her environment, enclose a specific portion of space-time. The adverbialist's failure to account for discrimination of scenes 1 and 2 arises because the individuated region is too coarse grained. The same problem can arise in feature-placing. Suppose that the only region (region R) on which features were placed encompassed both the square and the triangle. In such a case, all the features (e.g., square-ness, triangularity, etc.) would be placed onto the same region, and a viewer would be unable to discriminate them. To avoid these unsavory consequences, and to ensure that scene discriminations can happen as they do, Clark and the adverbialists must posit individuative principles or constraints at the level of sentience that ensure proper individuation of a field of view. An individuative apparatus, by itself, does not suffice for having adequate individuative constraints, since there could be apparatuses that randomly individuate. Sensory systems that randomly pick out regions will doubtfully afford success in even the simplest discrimination tasks. The problem is that Clark does not indicate what the individuative constraints are. He does not even argue that they exist. In looking at some additional simple scene-discrimination tasks, it will become evident that on either a space-based or visual-object-based account of sentience, the constraints necessary for success in simple scene discrimination tasks will need to be functionally the same.

It was shown in the adverbialist example that there ought to be individuative constraints to prevent picking out regions that are bigger than the relevant objects available; in this example, it is shown that there must be a constraint that prevents picking out regions that are smaller than the objects available. Consider the task of discriminating the following scenes:

Scene 3: Big red square to right of little red square
Scene 4: Little red square to right of big red square

For each scene the visual system might view two spatial regions: exactly
the region of the little square, and a square-sized portion of the big square
that is exactly the size of the little red square. Rather than detecting a big
red square, the visual system detects a small red square, since the edges of
the spatial region require such a perception. In such a case, the two scenes
will be indistinguishable. There are two regions, and they both contain
redness and squareness of a certain magnitude. In each case what is viewed
is: little red square to right of little red square.

The possibility of undersized regions can also allow alternative explanations that we would ordinarily want to rule out. Consider again scenes 1
and 2. In order to discriminate the scenes, Clark claims that "the creature
must divvy up the features appropriately, and perceive that Scene 1 contains something that is both red and square, while Scene 2 does not"
(p. 46, my italics). Without individuative constraints, the last two conditions are unnecessary. Suppose that for each scene, the observer picks out
exactly two spatial regions, and the same spatial regions are picked out for
both scenes. (Again it is assumed that the scenes are presented in succession, and that the observer places features properly onto individu ated
regions.) The observer picks out a square region inside the triangle, and a
triangle region inside the square. In such a case, the observer will see a red
triangle next to a green square in the first scene, and a green triangle next
to a red square in the second scene. Discrimination succeeds even though
the observer did not perceive that Scene 1 contained something that is
both red and square or that Scene 2 did not.

Clark claims rightly that features must be divvied up "appropriately" for
the many properties problem to be solved and for discrimination tasks to
be passed (p. 46). But he never specifies what appropriate feature assignment is. One possibility is that a feature is assigned appropriately iff a
region is represented as containing it and that region contains it. In light
of the foregoing considerations, the former definition, which is consistent
with undersizing a region and oversizing a region, is wrong. It fails to
ensure that scenes like those discussed above will be correctly discriminated.[15] Clark must intend that appropriate feature assignment requires
not assigning it to some regions that contain the feature—in which case
he must concede that there ought to be constraints on how to individuate
regions.

Individuating constraints must do more than ensure that a picked-out region is properly sized; they must ensure that it is properly located. Consider again scenes 1 and 2. Suppose that the background behind the triangle and square is purple. The visual system might form a spatial region just the size of the square and just the size of the triangle, but those regions might be placed to the right or left of the colored shapes. In both cases the visual system represents a purple square next to a purple triangle. The regions are adequately sized for the task in question, but they are not properly placed.

Even if the visual system properly sizes and locates a spatial region so that it encompasses a relevant feature or feature cluster, the visual system's task is finished only if it repeats the individuation for all the feature or feature clusters relevant to the task. Consider:

Scene 5: A red circle next to a red circle
Scene 6: A red circle

The visual system might pick out the region of exactly one red circle in each case, and it might pick out the same region in each case. The region picked out is properly placed and sized, but not enough regions are considered to succeed in the discriminatory task. I leave it as an exercise for the reader to construct an alternative task that the subject will fail because too many regions are individuated.

Any adequate theory of sentience cannot make it improbable that we perform as well as we do on discrimination tasks. Having individuative constraints is necessary to ensure that we discriminate normally. A theoretical reason to believe that object individuation and region individuation could occur on the same level is that, in both cases, the individuative constraints must perform identical functions:

1. *locating*, which might involve specifying the geometrical center of an object/region;
2. *circumscribing*, which involves determining how far from the geometrical center the object/region extends; and
3. *iterating*, which involves repeating steps 1 and/or 2 until all relevant objects/regions are accounted for.[16]

After steps 1–3 are executed, feature-placing proceeds in a more or less identical fashion. In the one case, features are placed onto objects; in the other case, features are placed onto regions. From the point of view of a sensory system, there seems to be no added difficulty in placing features onto objects. If there is any difference at all between an object-based and a region-based account, then it must occur in one of the individuation stages. But where is the difference? Because region and object individuation involve functionally the same constraints, one process need not be considered more complex than the other. At whatever level or stage in visual processing one process occurs, the other could also.

An obvious objection is that functionally equivalent processes can still be carried out at different levels. The argument could be stated thus: (a) only preconceptually individuated entities can fill the second placeholder in the feature-placing hypothesis; (b) objects are only conceptually individuated; and therefore (c) features of quality Q must not be placed onto objects. Of course, for this argument to work, Clark needs evidence for (b). My argument in this section simply shows that (b) could reasonably be false, since individuating regions in the right sort of way is going to involve performing the same kinds of functions as when we individuate objects. In section 3, I provide theoretical evidence to think that (b) is false.[17] If objects can be individuated according to sophisticated, high-level conceptual processes, they can also be individuated (in rougher fashion) by the early visual system.

2.2 Argument: If There Are Preconceptual Object Representations, Then They Are Postsensory

Clark holds that conceptual representation cannot be sensory. But at points he also holds that there can be nonsensory, nonconceptual representations. By positing such intermediate representations, he can conveniently claim without contradiction both that objects are individuated preconceptually and that sentience does not represent objects. He can also accommodate the considerable empirical evidence for preconceptual object representation (e.g., Pylyshyn 2003; Keane 2006).

To spell out the difference in the levels of visual processing, Clark adverts to Treisman's feature hierarchy theory (Treisman 1988). At the bottom, the visual system recognizes features that can be ascribed to single points (e.g., hue and saturation). Slightly more complicated are features that can be attributed to two-dimensional surfaces, such as texture (smooth, rough), orientation (vertical, horizontal), and segment features (curved, straight). A third level of complexity involves both two- and three-dimensional shape. Reaching higher levels of the hierarchy requires first going through the lower levels. Representing a square, for instance, requires first being able to represent line segments as having particular orientations. Object representations, therefore, are thought to be built up out of feature-placing processes; they arise as a result of sufficient "overlap" of features.[18]

The difference between features and objects, on Clark's view, is not sharp. Whereas features found at the first two levels of the hierarchy "fully qualify as 'sensory features'—as features extracted by 'early vision'" (p. 186), that is not the case with the third level:

Perhaps shape perception lies in the interesting contested transition zone between "early vision" and "visual perception" proper. To use the older terminology, shapes lie in the no-man's land between sensation and perception. Features in the more complicated layers of the hierarchy certainly begin to take on some of the characteristics of object-based perceptual categories. Feature-placing does not deny the existence or importance of the latter processes, but since the goal is simply to give an account of sensation, it can stay safely on the sensory side of that no-man's land. (p. 187)

The "more complicated" level of the hierarchy may represent objects, according to Clark, because shapes in a visual field can be counted, and items that can be counted might be particulars (ibid.). A related reason, not mentioned by Clark, to think that shape perception involves representing objects is that shape individuation appears to be object individuation. Determining that there is a shape of definite extent in the field of view could be tantamount to determining the boundaries of an object. Thus the desire to eliminate shape representation from feature-placing is understandable.

2.2.1 Defusing argument: Distinction between sensory and nonsensory stages not well placed

As noted, the advantage to having nonsensory, nonconceptual stages in visual processing is that Clark can allow for the possibility of preconceptual object representation without thereby relinquishing the space-based understanding of sensation. From what I can tell this is Clark's only reason to draw the line where he does between sentience and the "no-man's land." For the distinction to be legitimate, he must offer an independent reason for excising shape determination from sensation. No such reason is given.

Moreover, Clark has reason to draw the line *higher* in the hierarchy than where he does. The added complexity of going from two-dimensional segments to two-dimensional shape pales in comparison to that required to go from two- to three-dimensional shape. In the former case, the difference can be accounted for without going far beyond what appears on the retina; in the latter case, the difference involves drawing complicated inferences from what appears on the two-dimensional proximal image. An infinite number of three-dimensional scenes could correspond to a single retinotopic image, leaving the visual system with the daunting task of having to infer the most probable scene that it confronts. The same magnitude of informational poverty is absent for immediately represented two-dimensional shape.[19] Wherever postsensory visual processing stages begin, it is most plausibly at some point after the determination and placement of immediate two-dimensional shape.

2.2.2 Defusing argument: Regional occupants can be individuated, even if objects cannot

Suppose that Clark could present a convincing argument that the feature-placing hypothesis does not need to account for two-dimensional shape. He could say: Certain features need to be placed at or prior to a level in visual processing for an object to be individuated at that level. Shape is arguably a feature that must be represented to individuate a visual object. But shape is not represented until postsensory stages, so there can be no object individuation within sentience. It should first be noted that some researchers argue that neither shape features nor any other feature need to be encoded for objects to be individuated. According to Pylyshyn (2003, p. 180), individual objects are detected first, before any of their properties (including their shapes) are encoded. This view may not be pervasive in the field, but it cannot be cavalierly dismissed. More important, even if a shape feature needs to be represented for an object to be individuated, there is no threat posed to the primary claim of this chapter, namely, that space-time regional occupants are individuated at sensory levels. There will still be within sentience more rudimentary features that need to be individuated, and because these features are of actual bits of matter or energy in space, individuating these features will be tantamount to individuating space-time regional occupants. Determining a token feature of curvilinearity, for instance, involves picking out just that token, and the bits of matter or energy causally giving rise to that token. The same can be said of orientation or closure or numerous other two-dimensional features. Thus even if sensory systems could not individuate by carving along the edges of shapes, they might carve along smaller joints—that is, those of features or feature bunches. More will be said on this point in section 3.1 below.

2.3 Argument: We Can See Even When a Field of View Does Not Contain Objects; Ergo Our Visual System Does Not Pick Out Objects

In some passages Clark argues against an object-centered view of sentience on the ground that some viewable regions contain no objects:

Not all things seen can be classified as physical objects; we also see the sky, the ground, lightning flashes, shadows, reflections, glares, and mists. But all such sights can be classed as physical phenomena located in regions around the sentient organism. For each of them physics can contribute something to the story of what one sees. (p. 88; see also p. 135)

Clark's argument appears to be as follows. If the early visual system operated over only objects, that is, if feature-placing representations were formed only when objects were present, then we could not represent any object-less region. But we regularly view objectless, feature-full regions like the sky and ground without any problem. It must be that features are placed onto nonobjects, that is, regions of space-time.

2.3.1 Defusing argument: Regional occupants can still be picked out One of the problems with the foregoing argument is that it turns on a questionable ontology. It is not obvious why mists, skies, reflections, and the like cannot be counted as objects. Nor is it obvious why objects must have clear borders. Indeed, an object-based proponent might construe Clark's argument as a reductio: If a theory of sentience implies (absurdly) that no objects are viewed in the aforementioned scenes (of the sky, etc.) by assuming that objects have sharp borders, then the assumption is false— objects *can* lack sharp borders. Without giving a story of what a proper ontology might look like, Clark is not in a position to rule out the reductio.

Suppose, however, that Clark is right to say that certain viewable regions lack physical objects. Those regions still might contain visual objects, construed broadly. Although it is not clear whether objects will have sharp borders, it *is* clear that regional occupants need not have sharp borders. If I see a mist, a sky, or a glare, I see what I do in virtue of picking out occupiers of space-time, whether they are bits of matter or photons of energy. Interestingly, Clark unwittingly endorses this view in the previously quoted passage. It is worth reciting: "But all such sights can be classed as physical phenomena located in regions around the sentient organism. For each of them physics can contribute something to the story of what one sees" (p. 88). Clark is right: All such sights can be classed as physical phenomena located in space-time regions. And it is for exactly this reason that I recommend a visual object view of sentience. Physical phenomena occur at a region in virtue of there being occupants at those regions. Physics can tell us something about why we see as we do, in virtue of telling us about the occupants to which our sensory systems causally connect.

3   Arguments for a Visual Object View

So far, I have attempted to defuse Clark's arguments that space-time regions are the referents of visual sensory representation.[20] I hope to have shown that the reasons motivating a space-based understanding of visual sentience are not so convincing. I now provide positive arguments to adopt a visual object view of sentience. These arguments, as before, are directed at Clark's theory, but are expected to apply generally to any theory of visual sensory representation.

3.1   Argument: Only a Visual Object View Can Make Sense of Individuative Constraints

As noted in section 2.1.2, Clark wants to give an account of sensory representation that explains how we engage in simple discrimination tasks. I have argued that because a region view and an object view require functionally equivalent individuative constraints to explain success in such tasks, no additional resources need to be available for object individuation. Therefore, object individuation cannot be barred from sentience on the ground that it is more complex than sentience can handle or requires greater resources than sentience can afford. I now argue that to make sense of the capacities that we have, visual object individuation *has* to occur at the same level as region individuation, and thus if any individuation occurs in sentience, visual object individuation does. I first begin with the two steps that Clark claims are involved in the sensory identification of objects:

(a)   the sensory identification of a place-time,

and

(b)   the individuation of the object in question on the basis of its occupation of that place-time (p. 141).

Object individuation, which is not a "strictly sensory" (ibid.) process, proceeds only after sensory machinery individuates the region containing it. Objects are identified in virtue of the identification of their containing region. The problem is that Clark's account is unable to explain how (or in virtue of what) the right region is selected. He cannot explain, without adverting to occupants, how identified regions end up approximately "coextensive" with objects, or even why they happen to contain a "portion" of an object (ibid.). Because individuation is *necessarily* arbitrary with respect to occupants on Clark's account, it will be inexplicable as to how occupants are consistently identified successfully, and how simple discrimination tasks, like those described above, can be performed. Consider again the discriminatory task involving scenes 1 and 2.

Scene 1: Red square next to green triangle
Scene 2: Red triangle next to green square

The explanation that Clark proposes for how subjects distinguish the two is that they pick out a region containing just the triangle, and another region containing just the square. But Clark cannot explain why the sensory system identified those regions. He cannot explain why a different number of regions were not picked out or why differently sized or differently located regions were not picked out. Prima facie, there seem to exist many more incorrect region selections than correct region selections (where "correct" region selections are [roughly] those that allow for proper scene discrimination; incorrect selections are [roughly] those that do not). It seems highly improbable that the visual system will successfully discriminate by chance on a given occasion, and even more unlikely that there will be successful discrimination over many occasions. To get to stage (b), there must be a stage (a), but it is a mystery as to how (a) can be satisfied in a way that allows for (b) to be reliably satisfied.

On the basis of these considerations, it seems that individuative constraints are decidedly visual object centered. A region is properly individuated if (a) the contours of an individuated region correspond to those of a visual object; (b) the location of a region corresponds to that of a visual object; and (c) there is iteration of the first two steps, so that there is exactly one individuated region for each visual object in the field of view. In a word, a region is properly individuated if the visual object in that region is. There might be additional constraints needed to produce proper region individuation (i.e., as when scenes become more complicated), but any set of constraints will ultimately have to take into account what occupies space-time. I suggest, therefore, changing Clark's account so that sensory identification proceeds in the following stages:

(a*)   the sensory identification (i.e., individuation) of a visual object O;
(b*)   the individuation of a space-time region on the basis of its being occupied by O.

Whereas (a) causally preceded (b) (since (b) was carried out "on the basis of" (a)) at postsensory processing stages, (b*) logically accompanies and hence is carried out simultaneously with (a*). Explanatorily, (b*) is posterior to (a*). Regions are individuated as they are because their occupants are individuated as they are. In contrast to Clark's account, both (a*) and (b*) are "strictly sensory." Regions are the only external entities that can serve as referents for visual systems, according to feature-placing, but on my account if there is any individuation at the level of sentience, then visual objects are individuated at that level. Stage (a*) and a fortiori (b*) may be necessary for visually identifying three-dimensional objects *as* objects, but they are not jointly sufficient. Identifying an object as such very well may require access to concepts or some other higher-level processing within or beyond the "no-man's land" that follow sensory stages.

There are advantages for Clark's own agenda to revise the stages of sensory identification in the way I've suggested. Clark perceives a "kinship" between sensory reference and linguistic varieties of reference (p. 134). He believes that causal theories of direct reference "require that sensory processes be endowed with specified capacities of identification" (p. 131). One of Clark's goals is to describe these capacities, and in particular "the ancient engines of spatial discrimination" requisite for successful deployment of (what Kaplan dubbed) perceptual demonstratives (Kaplan 1989, p. 582; cf. Clark 2000, p. 133). Perceptual demonstratives require, in addition to "immediate sensory contact" (p. 131) with the demonstratum (visually, aurally, etc.), a discrimination of the demonstratum from other items in the scene. Perceptual demonstratives refer when "one sees or hears or feels or (in general) senses *which* of the many currently sensible space-time regions is *the* region containing the target" (p. 162, italics in original). To understand "That is a critter," on this view, "that" must denote a critter-containing region, which in turn requires a presentation of the critter-containing region, and a discrimination of that region from other regions in the scene. But Clark's account cannot explain how this happens. When sensory systems pick out regions by picking out visual objects, by contrast, it *can* be explained how or why "that" refers to a critter, rather than to a patch of space containing only a critter nose or no critter at all. In general, it can be explained how or why perceptual demonstratives are successfully deployed in ordinary referring expressions. The same point can be made by looking at another passage:

The strategy is to exploit location in space as the fundamental ground of difference of an object. Strictly all that is required [to locate an object in space] is a location that serves to individuate. If one manages to identify a place-time such that the object in question is the occupant of that place-time, then the demonstrative identification is secured. What I will argue can be readily anticipated: *that sensory processes are perfectly fitted on their own to identify the requisite place-times.* (p. 138, my italics)

A "ground of difference" is a term borrowed from Evans (1982, p. 107), and distinguishes an object from all other objects of its kind at a particular time. A fundamental identification is an identification that correctly attributes a ground of difference to an object (p. 137). If spatial dimensions of objects or visual objects do not constrain what locations are chosen, it will be a wonder that a person can manage to identify the object in question. It will be a wonder that there can be a fundamental identification because it will be a wonder that there can be a ground of difference. Distinguishing an object from all other objects of the same kind in a scene requires picking out the place-time of the object in question, but again, there is no way to reliably do that on Clark's account. Sensory processes are "fitted" to identify the place-times requisite to secure a perceptual demonstrative only if those processes are sensitive to what fills those regions, that is, only if regional occupants in some sense recommend region identification. If Clark wants to posit a kinship between sensory reference and linguistic reference such that sensory processes can afford fundamental identifica tions, then individuating constraints will need to take into account visual objects.

3.2 Argument: Only Occupiers of Space-Time Enter into Causal Relations

One of the simplest, though perhaps strongest, arguments in favor of my primary claim—that feature-placing representations name and attribute features to space-time regional occupants—concerns the causal relation between external entities and sensory systems. I agree with Clark that such a relation exists (it would be impossible for a distal feature to be detected otherwise), but we disagree with respect to *what kind* of external entity is eligible to enter into that relation. Clark claims that regions tout court enter into that relation (p. 116; see also p. 165). But space-time regions, all by themselves, do not cause anything, much less invoke sensory representations. Without occupants of space-time filling our field of view, no sensory representation could ever be externally caused, and no feature could ever be successfully attributed. So while space-time regions may *host* the causes of sensory representations, they do not thereby *constitute* those causes. At points, Clark appears to unwittingly admit that this is the case:

The region where the brown of the brown table appears is occupied by a cloud of elementary particles, which cause the visual sensation as of a brown surface. Such common sense "causes" are typically inconstant distal conditions, proximate to the sense organs, and at least partially determinative of the variations in the qualities sensed. (p. 112)

The cloud of elementary particles making up a table, rather than the region that encloses that cloud, produces within us a feature-placing representation of brownness. The inconstant distal conditions are conditions of the table, not the region containing it, and these produce what we represent visually. Thus Clark seems to be conceding exactly what his account is supposed to deny, namely, that sensory systems enter into causal/informational relations with occupants of space-time regions. If sensory representations name what they are causally connected to, then they name visual objects, broadly, if not narrowly, construed.

3.3 Argument: Features Are of Regional Occupiers, Not Regions

Yet another simple and, I submit, powerful argument for my primary claim concerns the relation between features and regions. A feature can belong to an entity only if the entity, by itself, can sustain that feature. Regions, by themselves, cannot sustain features.[21] But if features do not belong to regions, then the propositions expressed by sensory representations on a region framework will almost always be false.[22] They will say that a feature X belongs to such-and-such region when in fact the feature belongs to the occupant of the region. To use the example from above (section 3.1), if exactly a critter-containing region is identified by a sensory system and the critter's color were attributed to the referent, the resulting proposition would absurdly entail that the creature's region possesses the critter's color.

Clark may very well be right to hold that a feature must be located and placed *at* a particular region during the process of feature placing. What I dispute is that features must be located *on* a particular region; that they belong to (in some strong sense) or are sustained by a region. Strictly speaking, our visual system does not say "Here it is red, there it is green." It says "This is red, that is green," where the demonstratives pick out (but do not identify as such) visual objects, broadly, if not narrowly construed. Only on this visual-object-centered framework can we ensure that sensory representations by and large express true propositions regarding the external world.

3.4 Argument: Collecting Principle Can Hold Only If Features Are Placed Onto Visual Objects

In arguing that there need to be two kinds of terms or two different placeholders to characterize sensory representation, Clark utilizes Strawson's observations about subjects and predicates. A symmetry is observed with these terms—the same subject term can be matched with different predicates and the same predicate term can be matched with different subject terms (Strawson 1959, pp. 168–175). The predicate "is laughing," for example, can be tied to each of my friends when they hear me try to sing; and, when trying to sing, I can be predicated of "is making some unharmonious noises," "won't quit his day job," and the like at the very same time. But there is no such symmetry when dealing with what Strawson calls "attributive ties," which obtain inter alia between *instances* of predicates and subjects (ibid.). Whereas a subject can be tied to or collect many different instances of predicates, an instance of a predicate can be tied to or collected by at most one subject. The asymmetry, in addition to allowing an intuitive distinction between subjects and predicates, also can be used to make sense of the different roles of the two place-holders in the feature-placing schema. On Clark's view, binding a token feature to a location or binding multiple token features to the same place-time can all be considered examples of attributive tying:

The same patch can be both red and glossy, smooth and warm. . . . But to this particular instance of red we can tie exactly one place-time. And if that one place-time is red, it cannot also be green. . . . Without the many–one character of this collecting principle (many features, one place time), it would not be possible to sense that same place-time as red, glossy, smooth, and warm: to sense it as characterized by multiple features. Our two kinds of term—our two kinds of place-holder, or two kinds of dimensions of variation in sensory appearance—can thus be differentiated from one another by the asymmetry in the collecting principles that govern their association. (p. 73)

It is understandable that Clark wants to import the collecting principle into feature-placing, since there appears to be a genuine distinction between subject and predicate within sensation. An instance of redness belongs to just one external entity, not several. When an instance of curvilinearity and an instance of redness belong to one external entity, they do not simultaneously belong to other distinct entities. It is partly in virtue of this many–one relation between features and feature-bearing entities that we sense just one entity to bear a number of features. The problem is that if regions collect features, the collecting principle's asymmetry no longer holds. For any given instance of a feature, an indefinite number of regions contain it. Given, say, an oriented line, there is an indefinite number of concentric spherical regions that can be drawn around just that feature. A feature does not belong uniquely to any one of those individuated regions, but it does belong to each of them in virtue of being contained by each.

The feature-containing regions are all different in virtue of having different expanses, different parameters relative to the observer.

If objects bear features, the foregoing problem can be eliminated by stipulation, if not by metaphysical principles. When dealing with subjects qua objects, it makes sense to say that an external entity falls into one of two categories: object or nonobject. The instance of the predicate "having a seedy center" might be tied to an apple, but "having a seedy center" would not *have* to simultaneously belong to other objects within or outside the apple, if any story could be given as to why those other entities are nonobjects or why they are not different objects. Presumably some story could be given. That is not the case when dealing with subjects qua space-time regions. A spatially larger feature-containing region is not the same as the smaller, since the larger one will be characterized by a different vector $[r_m]$. A temporally longer feature-containing region is not the same as a shorter one, for the same reason. So there is a large, if not infinite, number of regions that all collect the same feature. Moreover, when dealing with the external entities that the visual system picks out, there is no dichotomy between regions and nonregions; there are only regions. Clark himself utilizes this fact to tout his theory as one that is not plagued by missing referents (p. 195). Thus a story cannot possibly be given that shows that exactly one region contains an instance of a feature, because that requires showing impossibly that the other instance containing regions are in fact nonregions. It follows by Clark's own admission (above) that the same region cannot be sensed to be glossy, red, and warm. Whereas the problem of a token feature simultaneously belonging to several could-be objects can be eliminated by finding some story as to why all but one could-be objects do not count as objects or different objects, the same move is impossible for space-time regions. If Clark wants to preserve the intuitions that the external entities picked out by early vision play the role of Strawson's subjects and that instances of predicates can be of at most one subject, and thereby further bolster the view that the second place-holder serves a function distinct from the first, then Clark is better off invoking visual objects in his theory of sentience.[23]

3.5 Argument: Visual Object View Makes Better Sense of Binding

In the passage quoted at length in the foregoing section, Clark holds that if the collecting principle were false, then "it would not be possible" to sense one and the same place as being, for instance, warm and glossy (p. 73). Call this conditional statement the *binding principle*. As already shown, the entities that collect features cannot be space-time regions, since such regions cannot uniquely collect features. But might Clark save the binding principle by couching it in terms of *represented* regions? It might read as follows:

If we can sense a space-time region R as having features X, Y, and Z, then no other represented region collects X, Y, and Z.

It would not matter, on this view, whether there were other distinct space-time regions bearing or collecting the same features; all that matters is whether exactly one represented space-time region bears or collects features. Because a sensory system can, according to Clark, represent only one external entity at a time, binding—and the sensation it gives rise to—can obtain. Unfortunately, this move will not help. There might be one represented feature-containing entity at one point in time, another spatially identical feature-containing entity at another point in time, and therefore two distinct represented space-time regions that contain or collect the same features. Neither represented entity uniquely collects the features, but both entities are sensed as having the features in question. The collecting principle fails and so too does a binding principle that requires it. Regional occupants must be what collect features since only they can uniquely collect them. Here is another attempt to appropriately modify the binding principle:

If we can sense an object or regional occupant O as having features X, Y, and Z, then exactly that object or occupant collects X, Y, and Z.

It is one thing to say that a given feature must be collected by or belong to exactly one external entity for that feature to be sensed. It is quite another to say that exactly one external entity must collect two or more features for those features to be sensed together. This latter claim is false. Bound features can belong to uniquely different entities. Illusory conjunction studies show that when presented a green "X" next to a red "O," subjects might occasionally sense (and thus represent) a red "O" (Treisman and Schmidt 1982; Prinzmetal, Diedrichsen, and Ivry 2001). It seems uncontroversial that the "X" and "O" are different objects, but a member of the set of features that I sense with the "O" object can be borrowed from the "X." Clearly two features do not have to be collected by exactly one entity for those features to be bound and sensed together.

Clark is right to employ the collecting principle to understand how the binding problem is solved, but he does not use it in the right way. The collecting principle is important not so much because it imposes a necessary condition on sensing bound features; it is important because it explains binding teleologically. If "having such-and-such shape" and "having such-and-such color" belonged to no single external entity, there would be no good reason for the visual system to fuse together those two features. The visual system could just as well bind each of those features to other features present in the visual scene (e.g., features of background, say, C and D). When two or more features are uniquely collected by external entities, by contrast, the binding routines can be understood. Sensory systems tend to bind together A and B because A and B uniquely belong to the same external entity. Sensory systems do not tend to bind A and C because A and C belong to two distinct external entities. Because our systems do solve the binding problem, and because that solution renders a by-and-large veridical picture of the world, there must be some reason grounded in reality as to why we bind together some features but not others. That reason cannot be that the features belong to the same space-time region, nor can it be because they belong to the same represented region; the reason is that they belong to the same *space-time regional occupant*. I therefore reformulate the binding principle thus:

If standard observers in standard conditions can sense an object or occupant O as having features X, Y, and Z, then exactly O collects X, Y, and Z.

In our kind of world when we sense features as being together, it is because we are sensing an object that has just those features. This does not mean that all of us sense this way. Nor does it mean that we might usually sense this way in strange conditions (such as those under which illusory conjunctions arise). But it does mean that in our kind of world, for the average observer, sensing two or more features together indicates that those features are indeed of the same thing.

What about location? Does not an individual's whereabouts play some role in determining how features are bound? If so, how does a visual object version of feature-placing accommodate for the representation of location? Much can be said on this topic, but for now it should be stated without hesitation that location does indeed strongly guide what features the visual system binds together. One feature might be represented as having a certain location, another feature might be represented as having a certain location, and the relation between the locations that those features are represented as having will help decide whether binding occurs. In particular, a Gestalt proximity constraint might apply (Wertheimer 1923/1958): If the location of one feature is encoded as being similar to the location that another feature is encoded as having, then those two features will more likely be represented together (i.e., bound) than features that are represented as having less similar locations (see figure 13.1; see also Keane and Pylyshyn 2006).

Though location is a major factor that guides binding, it is not the only factor.[24] An element $X1$ might more likely be bound to a more distant element $X2$ rather than a closer element $X3$, if only $X1$ and $X2$ have edges that allow for contour interpolation or if only X1 and X2 share the same polarity. Similarly, if $X1$ has one shape or expanse, it might be bound to an element $X2$ of the same shape or expanse rather than a differently sized, more closely located element $X3$. These facts and others indicate that binding is a complex process that cannot be understood solely in terms of location.

If location does indeed play the role that I just spelled out, it is one that differs radically from the one Clark envisions. *Location is important not because it is a subject to which features are attributed, but because it itself is a feature that is attributed to visual objects.* A more appropriate feature-placing schema is "quality Q on visual object O," where locations, like colors and orientations, can specify values of each dimension of the Q vector. Alternatively, if one wishes to preserve the special status of location in sensory representation (reasons for which are found in the introduction) and if location features and nonlocation features are always attributed to an individual together at once (as feature integration theory appears to suggest, Treisman and Gelade 1980), an alternative schema might be "(quality Q and location L) on visual object O," where L can be expressed as a m-element vector $[l_m]$. Either of these alternatives will turn out to be superior to schemas invoking space-time regions as referents.

4 Concluding Remark

In this chapter, I offered two sets of arguments. In the first set, I attempted to undermine major arguments for the view that regions are the fundamental referents of visual sensation. I argued that individuating regions does not require more resources or more complex processing than individuating objects. Features that might be associated with object representation, such as shape, probably can be represented at the level of sentience. And even if sensation represents only more basic features, that does not mean that those features are attributed to space-time regions. Finally, the claim that we sense regions because we occasionally sense "object-less" scenes turns on a questionable ontology, which itself does not imply that sensed scenes can lack space-time regional occupants.

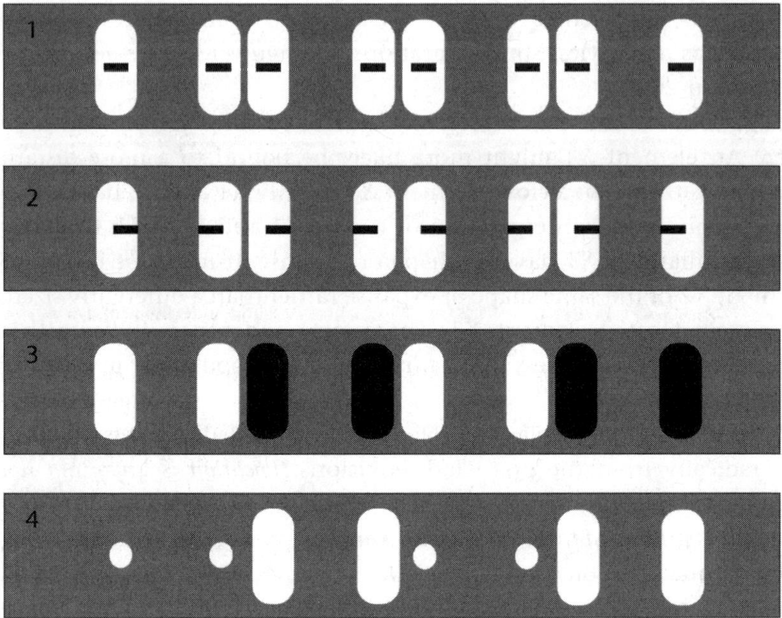

Figure 13.1

In panel 1, when black segments are enclosed by white rectangular ovals, the ovals and segments closest to one another tend to group into three central pairs. In panel 2, when each segment is shifted slightly toward the edge of its respective oval, segments no longer group with their nearest neighbors, and instead form partly occluded contours with more distant segments. The perception is of four black horizontal bars partly occluded by a gray holed surface. In panel 3, more distant elements sharing the same polarity are more likely to be grouped than more proximal elements with opposite polarity. Finally, in panel 4, more distant elements with the same shape or expanse are more likely to be grouped than closer elements with different shapes/expanses. These kind of phenomena, which have been well established at least since the Gestalt movement in the early part of the twentieth century, indicate that location is just one feature among many that helps determine how we bind elements of a visual scene.

In the second part of the chapter, I provided positive arguments for the claim that visual objects are the referents of early vision. I argued that visual objects are needed to explain how the visual system parses simple scenes so that we discriminate normally, and deploy perceptual demonstratives successfully. I argued that visual objects are also necessary for explaining how sensory systems causally connect with, attribute properties to, and form veridical representations about the external world. Finally, visual objects are required to explain our binding practices—why it is that some features are bound together, and not others. Arguments throughout the entire chapter, though constructed specifically to address the theory of sentience, should apply to any theory of visual sensory representation.

In targeting Clark's theory, I did not mean to imply that there was nothing valuable to be salvaged from his rich and insightful exposition of sensation. The good ideas in his book were many: that sensory representations can be characterized propositionally by multiple place-holders; that those place-holders are filled by one of a number of elements each of which can have one of a number of values; that location plays a special role in visual sensation; and so on. Yet for all of its merits, the theory cannot survive as it stands. Any theory of sentience—to be adequate—must have visual objects serve as the referents of visual sensory representation.

Acknowledgments

Special thanks to Brian McLaughlin and Jerry Fodor for helpful comments on earlier drafts. Most important, I am indebted to Zenon Pylyshyn for inspiring me to examine questions of the mind experimentally and philosophically.

This chapter is an abridged version of chapter 1 of the author's unpublished dissertation, "Visual Objects: Philosophical and Cognitive Science Perspectives" (2006).

Notes

1. All page references are to Clark 2000, unless noted otherwise.

2. Clark interchanges the terms "place-time" with "space-time." I follow suit and use the terms synonymously.

3. Hereafter, the term "if and only if" will be abbreviated as "iff."

4. One counterexample, which Clark acknowledges, is olfaction. It is doubtful that we can have two qualitatively distinct smells at the same time.

5. Clark calls this the "partition argument" (p. 58). It was also noted by Ernst Weber in 1846, as noted by Boring (1942).

6. The terms "narrow" and "broad" were chosen according to entity class size. All visual objects, as typically construed in the scientific literature, are space-time regional occupants, but not the other way around. Thus the first construal is narrow; the second, broad.

7. Clark (2004a) and (2004b) suggests minor modifications and additions to his original theory. For discussion on why those arguments fail to make a region view superior to a visual object view of sensation, see chapter 2 of Keane 2006.

8. From what I can tell, Strawson (1959) does not acknowledge this distinction either.

9. Driver et al. (2001, p. 62) also explicitly express this view.

10. From time to time in the chapter, I follow suit and simply speak of objects rather than visual objects. When I do, I employ the scientific (narrow) conception of visual object, viz. an entity that roughly corresponds to what we conceptually consider to be an object, and that interacts with visual systems in particular sorts of ways.

11. Jackson (1977) points this out.

12. Another problem with adverbialism is the incoherence of how one manner of sensing can be of (or shaped by) two logically incompatible features. How can there be, say, a triangularly, squarely sensing?

13. Successfully carrying out the discrimination task involves solving what Jackson (1977, p. 65) called the "many properties problem."

14. At the same time, when discussing sensory identification and divided reference, Clark claims that sensory systems do not have any access to an apparatus of individuation: "For [divided reference] one needs sortals, the identity predicate, counting and count nouns, singular terms: a substantial portion of what Quine calls the 'apparatus of individuation.'. . . As already noted, a feature-placing language . . . lacks sortals, count nouns, identity, plurals" (pp. 158–159). On pains of inconsistency, there must be two apparatuses of individuation: Clark's, which operates at the level of sentience, and Quine's, which functions only for higher-level cognitive processes.

15. Indeed, placing features into the correct region is not necessary for scene discrimination; features might be systematically swapped and placed onto neighboring regions. In scene 1, the greenness of the triangle might be placed on the square, and the redness of the square onto the triangle. If the same swapping of colors ensues for the second scene, then scene discrimination does not depend on the proper divvying up of features. This possibility, which is analogous to the inverted spectrum problem, will be bracketed for the time being.

16. I remain noncommittal on how the stages are ordered in time. I also remain noncommittal on whether iterating occurs serially or in parallel, although I strongly suspect that the latter is true.

17. Psychophysical and/or neurobiological evidence for the existence of object representations in early visual processing can be found in chapter 4 of Pylyshyn 2003, in section 3.2 of chapter 1 of Keane 2006, and in a variety of other places (e.g., Keane 2008).

18. A hierarchical view of visual processing is not obsolete. Malach et al. (2002) write in an online document: "How is the information transformed from the retinal image to recognition-related representations?—a central principle in visual cortical research is that such transition occurs in a hierarchical manner—by gradually building more and more complex organizations."

19. There are at least two ways to consider represented shape. First, there is the immediately represented shape that derives directly from the retinal image. This shape varies with almost any slight movement of the object or observer. Next there is the mediately represented shape, which is extracted from the retinal image and allows for shape constancy (see Palmer 1999, pp. 327–332). A square, for instance, when viewed from different angles will continue to look like a square (the mediate representation), though it does not always cast a square shape on the retina. The section above considers immediate shape representation.

20. Space limits prevent discussion of another major argument that Clark offers, namely, that sensation lacks the representational vocabulary to refer to or attribute properties to objects. Clark's argument, and my rebuttal, are provided in Keane 2008.

21. Regions can have features in an abstract sense. For example, a region may have the feature of having such-and-such volume, or such-and-such relation to an object. Nevertheless, these are not the features that our sensory systems are sensitive to. We are visually sensitive to only features of occupants (location, color, luminance). When we are surrounded with nothing but an empty region, we visually represent no external region, much less a feature of an external region.

22. I say "almost always" rather than "always" since regions and visual objects can share features (e.g., size).

23. In this section, I focus on objects, but a parallel story can be given for the more general notion of space-time regional occupant. External entities fall into one of two categories: occupant or nonoccupant (empty space). Given an occupant O bearing a feature Q, O can uniquely collect Q iff a story can be given why other occupants bearing Q either are not different occupants or are nonoccupants. I submit that some story can be given stipulatively, if not by way of metaphysical principles.

24. This point was made by classic Gestalt psychologists (Wertheimer 1923/1958) and also by contemporary philosophers and psychologists (e.g., Campbell 2000; Prinzmetal 1995).

References

Boring, E. G. (1942). *Sensation and Perception in the History of Experimental Psychology*. New York: Appleton Century Crofts.

Campbell, J. (2000). *Reference and Consciousness*. New York: Oxford University Press.

Carey, S., and F. Xu (2001). Infants' knowledge of objects: Beyond object-files and object tracking. *Cognition* 80: 179–213.

Clark, A. (2000). *A Theory of Sentience*. New York: Oxford University Press.

Clark, A. (2004a). Feature-placing and proto-objects. *Philosophical Psychology* 17(4): 443–468.

Clark, A. (2004b). Sensing, objects, and awareness: Reply to commentators. *Philosophical Psychology* 17(4): 553–579.

Driver, J., G. Davis, C. Russell, M. Turatto, and E. Freeman (2001). Segmentation, attention, and phenomenal visual objects. *Cognition* 80: 61–95.

Evans, G. (1982). *Varieties of Reference*. Oxford: Clarendon Press.

Feldman, J. (2003). What is a visual object? *Trends in Cognitive Sciences* 7(6): 252–256.

Helmholtz, H. V. (1995). On Goethe's scientific researches (E. Atkinson, Trans.). In *Science and Culture: Popular and Philosophical Essays*, ed. D. Cahan, pp. 1–17. Chicago: University of Chicago Press. (Lecture delivered before the German Society of Königsberg, 1853).

Jackson, F. (1977). *Perception: A Representative Theory*. Cambridge: Cambridge University Press.

Kahneman, D., A. Treisman, and B. Gibbs (1992). The reviewing of object files: Object specific integration of information. *Cognitive Psychology* 24: 175–219.

Kaplan, D. (1989). Afterthoughts. In *Themes from Kaplan*, ed. J. Almog, J. Perry, and H. Wettstein, 565–614. New York: Oxford University Press.

Keane, B. P. (2006). Visual objects: Philosophical and cognitive science perspectives. Ph.D. dissertation, Rutgers, The State University of New Jersey, New Brunswick, New Jersey.

Keane, B. P. (2008). On representing objects with a language of sentience. *Philosophical Psychology* 21: 113–127.

Keane, B. P., and Z. W. Pylyshyn (2006). Can multiple objects be tracked predictively? Tracking as a low-level, non-predictive function. *Cognitive Psychology* 52: 346–368.

Malach, R., G. Avidan, I. Goldberg, U. Hasson, M. Harel, Y. Lerner, I. Levy, and R. Mukamel (2002). Topography of human visual object areas revealed by functional magnetic resonance imaging. *Life Sciences Open Day Book*, 258–259. http://www .weizmann.ac.il/Biology/open_day/ book/rafael_malach.pdf.

Palmer, S. (1999). *Vision Science: Photons to Phenomenology*. Cambridge, Mass.: MIT Press.

Prinzmetal, W. (1995). Visual feature integration in a world of objects. *Current Directions in Psychological Science* 4: 90–94.

Prinzmetal, W., J. Diedrichsen, and R. B. Ivry (2001). Illusory conjunctions are alive and well: A reply to Donk (1999). *Journal of Experimental Psychology: Human Perception and Performance* 27: 538–541.

Pylyshyn, Z. W. (2003). *Seeing and Visualizing: It's Not What You Think*. Cambridge, Mass.: MIT Press/A Bradford Book.

Scholl, B. J. 2001: Objects and attention: The state of the art. *Cognition* 80(1/2): 1–46.

Scholl, B. J., and Z. W. Pylyshyn (1999). Tracking multiple items through occlusion: Clues to visual objecthood. *Cognitive Psychology* 38: 259–290.

Strawson, P. F. (1959). *Individuals*. London: Methuen.

Treisman, A. (1988). Features and objects: The Fourteenth Bartlett Memorial Lecture. *Quarterly Journal of Experimental Psychology* 40A: 201–237.

Treisman, A., and G. Gelade (1980). A feature integration theory of attention. *Cognitive Psychology* 12: 97–136.

Treisman, A., and H. Schmidt (1982). Illusory conjunctions in the perception of objects. *Cognitive Psychology* 14: 107–141.

Wertheimer, M. (1923/1958). Principles of perceptual organization. In *Readings in Perception*, ed. D. C. Beardslee, and M. Wertheimer, pp. 115–135. Princeton, N.J.: Van Nostrand.

Contributors

John Bickle Department of Philosophy and Neuroscience Graduate Program, University of Cincinnati

Darlene A. Brodeur Psychology, Acadia University, Canada

Andrew Brook Professor of Philosophy, Director of the Institute of Interdisciplinary Studies, and Chair of the Cognitive Science Management Committee, Carleton University, Ottawa, Ontario, Canada

Austen Clark Philosophy, University of Connecticut

Michael R. W. Dawson University of Alberta, Edmonton, Alberta, Canada

Jerry Fodor Rutgers University

Mel Goodale CIHR Group on Action and Perception, University of Western Ontario, Canada

Stevan Harnad Canada Research Chair in Cognitive Sciences, Université du Quebec au Montréal, Canada, and Department of Electronics and Computer Science, University of Southampton, U.K.

Heather Hollinsworth University of Guelph, Canada

Lisa N. Jefferies Psychology, Simon Fraser University, British Columbia, Canada

Brian P. Keane Rutgers Center for Cognitive Science, Rutgers University

Zenon W. Pylyshyn Rutgers Center for Cognitive Science, Rutgers University

Charles Reiss Linguistics, Concordia University, Montreal, Quebec, Canada

Brian J. Scholl Psychology, Yale University

Lana Trick University of Guelph, Canada

Claudia Uller Psychology, University of Essex, U.K.

Marla Wolf   CIHR Group on Action and Perception, University of Western Ontario, Canada

Richard D. Wright   Psychology, Simon Fraser University, British Columbia, Canada

Index